EARTH-SHELTERED HABITAT

History, Architecture and Urban Design

EARTH-SHELTERED HABITAT

History, Architecture and Urban Design

Gideon S. Golany

VNR VAN NOSTRAND REINHOLD COMPANY
NEW YORK CINCINNATI TORONTO LONDON MELBOURNE

Library of Congress Catalog Card Number: 82-1936
ISBN: 0-442-22992-5
ISBN: 0-442-22993-3 pbk.

Manufactured in the United States of America

Published by Van Nostrand Reinhold Company Inc.
135 West 50th Street, New York, N.Y. 10020

Van Nostrand Reinhold Publishing
1410 Birchmount Road
Scarborough, Ontario M1P 2E7, Canada

Van Nostrand Reinhold
17 Queen Street
Mitcham, Victoria 3132, Australia

Van Nostrand Reinhold Company Limited
Molly Millars Lane
Wokingham, Berkshire, England

15 14 13 12 11 10 9 8 7 6 5 4 3 2 1

Library of Congress Cataloging in Publication Data
Golany, Gideon.
Earth-sheltered habitat.

Bibliography: p. 217
Includes index.
1. Underground architecture. I. Title.
NA2542.7.G64 721 82-1936
ISBN 0-442-22992-5 AACR2
ISBN 0-442-22993-3 (pbk.)

For my son Amir, whose
systematic approach and
reliability give promise.

Introduction

Among the many types of climates there are three which concern us most, for there the use of subterranean construction can be suitable.

1. Very warm and dry climate, such as in the arid zones of most of the southwest USA and most of north Africa
2. Very cold and dry climate, such as the continental arid zone of central Asia or regions of central Canada
3. Temperate zone climate, cold and snowy in winter and rainy and relatively warm in summer, as in Pennsylvania, Minnesota, and Maine in the USA

It is our belief that subterranean structures do not effectively suit the humid, warm climate of the tropics or of Florida because of the high humidity, the necessity for extensive ventilation, and the potential for a high water table in the soil.

This book discusses an alternative housing and settlement design solution, earth-covered or subterranean living, especially for regions of stressed climate in the belief that those regions can make the most effective use of this system. At the same time, many of the solutions offered here are applicable for housing in nonstressed climates, too, if necessary adjustments are made.

Stressed climate is defined here as that extreme type which has a very low or very high temperature in the daytime, a great differentiation between daytime and nighttime temperatures, and little or no precipitation; it would require special treatment in housing design and urban configuration. Areas of stressed climates are those which are hot and dry (such as the arid zones), very cold and dry (such as in the continental plateau), or cold and humid (such as snowy, northern regions).

People living in regions of stressed climates have serious problems to cope with. In most cases, the conventional solutions have been high energy consumption for cooling or for heating. This book offers unconventional solutions for the design and construction of houses for the stressed climate, using concepts of subterranean or earth-covered housing, passive energy, ventilation, and evaporative cooling.

The book treats these subjects comprehensively and suggests details for the design of such housing. The mass of soil around the house acts as an efficient diurnal insulator between the enveloped house and the harsh climate; and, more importantly, it has a heat storage capacity. Moreover, the long time lag between soil surface temperature and house temperature offers a remarkable system in which the house generally receives the desirable natural temperature at the right season. The design also addresses the issues which arise from the new positioning of subterranean houses, such as those of natural light and sunshine penetration needs, direct eye contact with the outdoor environment, environmental impact, and code requirements while offering a variety of applicable and practical solutions.

More than any other geographical areas, climatically stressed zones, especially arid ones, are characterized by delicately balanced ecosystems. The scarcity of water, the aridity of the climate, and the extremes of temperature have linked all creatures and natural forms in a state of ever-adapting equilibrium. All these constituents of the ecological system are constantly poised at the threshold of survival; therefore, their sensitive ecosystem has evolved in this state of both balance and dependency. Because of the delicate nature of this equilibrium, it takes a long time to correct any imbalance that develops. Furthermore, serious ecological disturbances can often result in losses that are irreparable. Precipitation is the key to both maintaining and restoring ecosystem balance, but the limited rainfall in the hot, dry arid and semiarid zones we are concerned with here is unpredictable and unregulated.

Sections of this book discuss the feasibility of emulating desert flora and fauna and developing a unique type of settlement and a special kind of house, specifically designed to cope with the stressful hot, dry climate of the arid zone. The proposed settlement and house combine three different ancient systems: (1) underground housing used in ancient northern China and North Africa; (2) a system of passive ventilation used in ancient Mesopotamia and Persia; and (3) humidification or cooling by evaporation used throughout the Middle East and many other places in the world. The ultimate goal is to create a settlement pattern with housing using energy-free home cooling. Such a house can create a moderate microclimate inside which can ease the impact of the hot, dry macroclimate outside the house, use natural environmental processes, and be affordable to large numbers of arid-zone, low income residents. Most of the very recent innovative designs for subterranean construction have been in response to the needs of the middle and upper income classes. However, the majority of the world population lives in climatically stressed zones and has low incomes and societies characterized by low technology. Furthermore, most researchers have ignored the vast and impressive achievements of ancient civilizations in this field.

Until recently, we used the subsurface of the earth primarily for extracting natural resources, such as water, oil, gas, ores, and other minerals. Recently, we have renewed an interest in subsurface usage which has been sporadically applied throughout history—for living. Certainly, we have just begun to explore and evaluate the nature and the potentialities of this resource for living and other multiple land uses. However, this evaluation must be comprehensive and multifaceted for the diversified land uses which have just started. The contemporary movement toward living in and using subsurface space is a sign of a culture which is characterized by advanced technology.

In today's cities there are already many underground activities. Underground structures include oil or liquid storages, shelters for civil defense, utility tunnels (such as telephone, sewage and stormwater lines, electricity and water systems, heating pipes, gas pipes, TV cable), power plants (especially hydraulic ones), subways, bus and railroad stations, parking areas, shopping centers, refrigeration facilities, theaters and other public gathering places, libraries, offices, and, frequently, military installations. The latter require great financial resources and expertise for intensive and extensive experimentation in the laboratory and in the field in order to reach optimal results under various soil types and conditions. The army reports on this subject are diverse and practical and could be important sources for planners/designers.

The work included here is prepared for planners, designers, architects, engineers, developers and real estate people and for social scientists and economists as well. Moreover, the large number of illustrations and their introduction of alternatives makes the book useful for laymen who would like to build such a house or become knowledgeable in the subject. However, there are still guidance and input needed from construction engineers relating to the construction and materials to be used. On the other hand, the suggestions made here, especially those on design, should guide the engineer when he begins his work. Although the approach is comprehensive, attempting to include much information and analysis, it cannot be all inclusive because the subject embraces nearly all aspects of life—technical, economic, social, and political. The information in this book is considered to be basic for professionals who may be just starting to design for the underground. It would be necessary to refer to more specialized literature when more detail is required for such aspects as soil, thermal conduction, construction, or ventilation. Also, every site has its own unique character which must be studied thoroughly. Thus, although there is no universal solution which can apply for all sites, this book treats these complex subjects with a much more comprehensive approach.

Subterranean houses have many recognizable advantages which are discussed in this book. They are virtually weatherproof and fire resistant and are protected from surface hazards such as earthquakes, cyclones and solar radiation. On the other hand, they have almost no heat loss or gain, thus saving on energy consumption for heating and cooling and providing a moderate microclimate compared with the extreme macroclimate of their environs. For example, in cases of interruption in electric service in very cold climates, the temperature inside a subsurface home will usually stay tolerable. When such houses are properly designed, owners can enjoy a pleasing and adaptable environment: there is no noise, no vibration; floors bear heavy loads; expansion is almost limitless; and construction and maintenance costs are, in many cases, lower than for conventional housing. In addition, with subterranean housing, the land can be saved for other uses; the external environment is minimally affected. The Underground Construction Research Council of the American Society of Civil Engineers has concluded in its research report: "Conservative analyses indicate that new technology for locating civil works facilities in sub-surface space could lead to the release of approximately $60 billion per year to other social and environmental needs." (Robert F. Baker, et al. *The Use of Underground Space to Achieve National Goals.* New York: American Society of Civil Engineers, 1972, p. iii.)

One of my prime conclusions throughout this study is that the use of subterranean housing yields more favorable results in extreme climates, especially arid hot and cold or very cold climates, than in temperate or moderate climates.

This book considers the ancient lessons as well as today's economic realities and, in light of them, introduces the idea of developing a compact, mixed under- and aboveground neighborhood. Such a neighborhood forms part of a city of clustered, continuous cells. In addition to residences, however, each urban cell will feature facilities for most daily social and economic activities as well as some places of employment—all within walking distance. This new urban cell is thus designed to reduce the use of motor vehicles within its confines, thereby diminishing or eliminating air pollution and increasing safety.

The two concepts of the compact urban cell and the subterranean house are inseparable. We will treat them discretely, but this rhetorical necessity should not distract the reader from seeing this innovative approach to stressed climate living as a necessarily integrated one. Every year cities expand, in some countries an equivalent of hundreds of square kilometers. Also, the city aggregate increases land value to a thousand times higher than that of the farmland. These two processes will continue in the future. The development of the subterranean settlement responds to those two issues and tries to ease them.

The preparation of this book has taken me more than one decade. The seeds of this project were planted during my youth in a desert region in a semisubterranean house and also during the late 1940s when I was a founder and member of the kibbutz settlement of Beari in the arid zone of southern Israel. The design of this volume inevitably began there when I dug into the loess soil to build shelters and trenches for defense; but only recently have I been able to transfer the vision to reality.

The mother soil gives us food and ore materials to support our existence and finally becomes our cradle when we die. All our life cycles around the soil, and it can certainly offer us shelter which is protective, quiet and warm.

I would like to thank the many people who have helped me in the preparation of this book: graduate students Chung-Do Pang and Dan Harrigan and senior student in architecture Celeste Klasic, who have drawn the final drafts of most of the illustrations; graduate student Peter Rieck who has helped in the compilation of statistical tables; students in the Department of Architecture, Patricia C. Kucker, Juan Reyes, and Lee Washesky who have helped in the final drafting of other drawings; Mrs. Linda Gummo, who carefully typed the manuscript; and Ms. Beverly Jones, who has been particularly helpful throughout the preparation process.

Further thanks go to Dean Walter H. Walters who has given his support for the completion of the book. My colleague, Architect Raniero Corbelletti, reviewed all the illustrations; his valuable comments and his support of my efforts were always timely and encouraging.

G. Golany

Department of Architecture
The Pennsylvania State University
University Park, PA

Contents

EARTH-SHELTERED HABITAT

History, Architecture and Urban Design

Section I—Historical Lessons

Ancient Experiences

It is highly probable that man's first shelters were underground caves. And subterranean structures have been built and used throughout human history to meet a variety of needs:

1. To achieve a moderate ambient indoor climate in regions with stressed climates such as dry and cold or dry and hot. Examples are the Eskimo houses and the Matmata subterranean houses in Tunisia.
2. To meet religious ceremonial needs, such as the kivas of the Indians of the Southwest United States. The subterranean space offered a secluded atmosphere for meditation and reflection.
3. To meet defense needs, such as the cliff dwellings in Arizona or the Maginot defense line in France.
4. To save agricultural land, such as the villages in northern and eastern China.
5. As storage space for agricultural products (especially wheat) as exemplified by the seminomads in southern Israel.
6. As civilian shelters against air attacks.
7. For educational purposes, such as Abo Elementary School in Artesia, New Mexico; or Reston, Virginia.

Thus, man has used underground structures throughout history for a wide variety of practical purposes, most of which have involved defense, protection from extremely stressful climates and conservation of land for agricultural needs.[1] Existing natural land forms or ideal soil conditions often have made subterranean structures the most practical alternative for human shelter. Such was the case for the ancient Nabatean city of Avdat in Israel which was partially carved out of easily cut limestone.

Of all the reasons for selecting an underground shelter, the first one mentioned—stressful climates—is the one with which we are most concerned. By climatic stress we mean the results of an extremely high or low temperature, high daily amplitude between day- and nighttime temperatures, extreme seasonal temperature differences as in plateau areas, dust-laden winds and intense radiation and ground reflection which are detrimental to human comfort and health. As it is described in this chapter, most of the historical underground dwellings developed by man have been in areas characterized by stressful climates, primarily arid zones. In the following pages, we will discuss some of the important underground shelter designs once used or still in use throughout the world.

Fig. I-1. Generalized cross section of the degrees of underground placement in semisubterranean and subterranean construction.

During the discussion, it is important to remember that many terms are used for structures constructed under the ground: semisubterranean, subterranean, geotectural, earth covered (generally used to mean "subterranean"; sometimes specifically used to mean the "enclosing of above-surface structure by earth"), petratectural (i.e., covered with rock), psammotectural (i.e., covered with sand), argillatectural (i.e., covered with clay), and terratectural.[2] There are also different degrees of building underground and different relationships to landforms (Fig. I-1).

Troglodyte people are cave dwellers, and the term is usually associated with technologically primitive cultures. Stone Age people lived in natural caves because they were the most easily available spaces for use as protection against weather, wild animals and enemies. Today there are still some tribes and communities which prefer the caves they dig for their climatic advantages, for climatic or physical protection or for saving agricultural land. In any case, troglodyte settlements have not been confined to a special geographical region or time. They are found throughout the Middle East, Europe, Asia, Africa and America, mostly in arid regions.

Catacombs are subterranean cemeteries, generally used in the past and consisting of galleries with recesses for tombs (cata—down; comb—a hollow, a cavity). Like troglodyte caves, catacombs are also found in many places; but the term is primarily used to describe the large number of sepulchers located about three miles from Rome in which were buried the bodies of Jews and early Christians. The catacombs were also used for a variety of other purposes such as worship and refuge. Catacombs are also to be found at Naples, Syracuse, and Chiusi. In addition, there are extensive

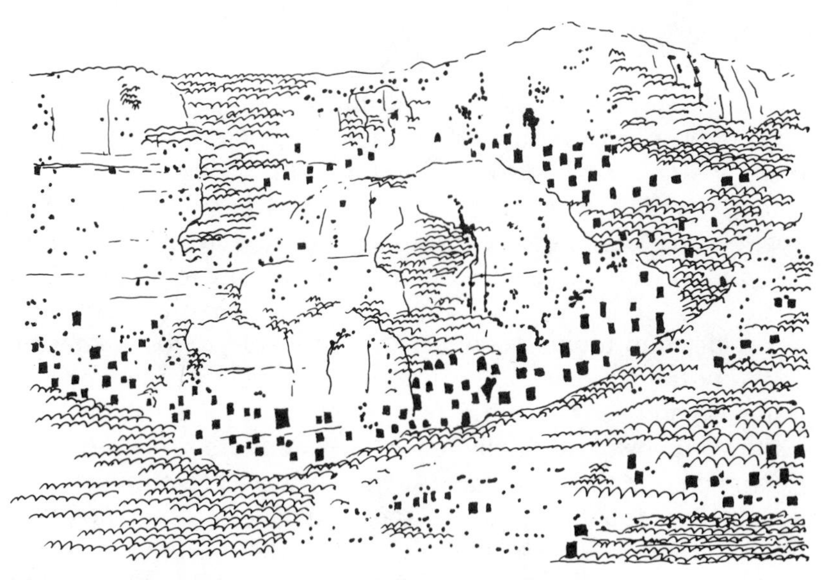

Fig. I-2. Ancient trogolodyte settlement of Necropoli di Pantalica (*City of the Dead*) in the Anapo Valley of Sicily, Italy. There are many of these settlements in Sicily, used as early as one millennium B.C. They were also used as burial grounds and later during the Middle Ages, for living. The connections between the houses are underground.

catacombs near the coastal city of Sousse in Tunisia and in Egypt at Alexandria; in Iran at Persepolis; and also on the Malta Islands, in Spain and in Hungary.

Cave dwellers were also found in the medieval town of Pantalica in the Anapo Valley near Syracuse, Sicily (Fig. I-2). The town was composed of caves used by the island tribes as early as 1000 B.C. Similar caves were found in Sicily near Siculiano, Bronte, Maletto, Calablotta and in the valley of Ispia near Modica.

Some underground settlements still exist in our time; their continuity does not necessarily stem from technological primitiveness or a lack of knowledge about other ways of living, but rather from practicality and a full understanding of the strengths and weaknesses of the environment and effective uses of it. The most commonly known are those of North Africa and China, Cappadocia in Turkey, and others in the southwestern parts of the United States.

TUNISIA: ARID ZONE CASE

In the lowlands and lower altitudes of the mountain regions of southern Tunisia, a few thousand people are still living in artificial caves. Accessible through an open courtyard, the housing units are 20 to 40 feet deep and range in area from 30 to 40 square feet.[3] In Matmata, a mountain village of southern Tunisia where most shelters are underground, the people live in the underground structures, while their animals are sometimes sheltered aboveground. Their grain is also kept in a special structure aboveground since caves are not always suitable for keeping the grain dry or adequately ventilated.[4] In fact, as Hallet reports in "Mountain Villages of Southern Tunisia," ". . . in one village only the recently built schoolhouse and the granary are visible [aboveground]."[5] There are 10 such villages 20 to 40 kilometers southwest of Gabes, the best known of which is Matmata (Fig. I-3).

These villages have been built deep within the earth. The general house design is a deep patio surrounded by rooms on different levels for living and storage areas. Entrance to these shelters is through stairways or graded tunnels. In the walls of the tunnels, spaces have been dug out for housing the animals. There have been some shelters located in the hill and built in a half-circle form with entrances from the side leading to tunnels. Socially, such quarters are arranged by kinship relations among the residents. Subterranean pedestrian tunnels connect the different groups of the tribe to maintain the system of social order. The rooms, which have curved ceilings, are rectangular and curved at the corners. Room size is usually 2 by 2½ meters with columns occasionally left in the middle to support the ceiling. Rooms connected to the patio have slightly lowered thresholds.

The mountains in southern Tunisia are composed of alternating layers of hard and soft limestone, each layer 6 to 8 feet thick. Thus, digging out the soft layer results in a well-constructed subterranean space. The ceilings can be flat by virtue of their natural horizontal stratification. The walls usually have been dug straight with a door-sized opening to the room. Some spaces have two levels. Although most of the rooms were not painted, the formal

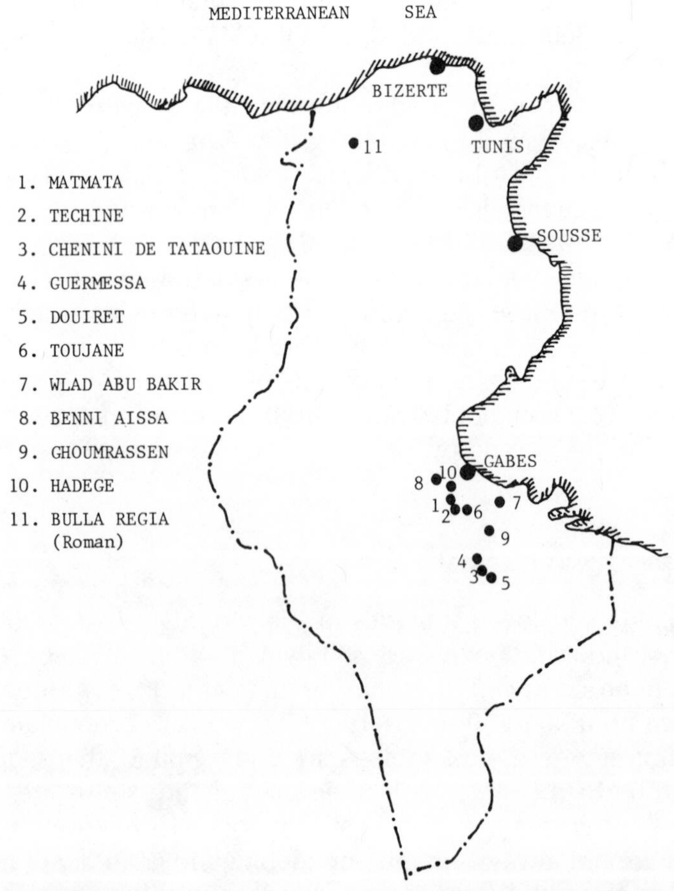

Fig. I-3. Locations of subterranean settlements in southern Tunisia, developed by the barbarian tribes after the pre-Islamic invasion in the seventh century. All these settlements are still populated. Bulla Regia is a subterranean settlement developed by the Romans following the Berber tradition.

area where the men and guests meet together is usually whitewashed and decorated. Food is stored in a special smaller room dug deep inside one of the rooms.

In some subterranean units (still in use), the Berbers who populated those villages also used the aboveground space by constructing some units as stables or storage rooms. The earth mound structures built aboveground for the storage of grain were called ghorfas (Fig. I-4). The wall of the ghorfa formed a row of niches and smaller arches. Originally, such structures (sometimes five to six stories high) were developed in the southern mountains before the Arab invasion in the seventh century and were placed higher up the mountain side than the living quarters. A series of ghorfas could become special fortified structures by attaching one to another to form an extended wall integrated with the mountain landscape. These ghorfas could house the inhabitants of an entire village. Each family unit dug a cistern inside its granary to collect water and store it in the event of a siege. For this same reason, tunnels connected one family with another. Most of these mountain villages were used seasonally by the specific tribe which had sovereignty in the region (Fig. I-5). However, when the region became stable with the strengthening of ties between Arabs and Berbers, the dwellers began to move to the lowland where they again developed subterranean units and ghorfas above the ground. The people of Matmata, Hadege, Techine,

Fig. I-4. Ghorfas for grain storage aboveground. The ghorfa was designed and built by the Tunisians.

Fig. I-5. Subterranean settlement of Guermessa, Tunisia. The troglodyte dwellers are located 20 km southwest of the city of Gabes. Guermessa is one of 10 subterranean villages in this region.

Bled Kebira and other tribes built such villages (Fig. I-6). A group of ghorfas is called Ksar.

Historically the most commonly known house form has been that of Matmata (Fig. I-7). At first, settlers dug a cave as a family shelter, and they later expanded it to a full house unit. The courtyard, in a circle, square or rectangular form, was surrounded by rooms which had been dug out slightly lower than the courtyard level. Some rooms were on two levels with stairways. The upper rooms were used to store grain with tunnels leading to the surface. Entrance to the living unit was made through a tunnel leading from the surface to the courtyard. The unit could be expanded by digging more rooms or by digging a tunnel leading to a new courtyard. The settlement could not be observed, even at a close distance; and only smoke or fire traces from the courtyard signaled its existence.

It is estimated that today 5–6000 people live in Matmata, while Guermessa has 700 people and Douirat, fewer than 500. Today's underground houses have basically the same design as those of the past. Usually now, water is stored in an elaborate system of waterpits sited in the middle of the patio where a cistern has been built and plastered. The patio, which is generally on the hillside, is at a depth of 10 meters (Figure I-8). For ventila-

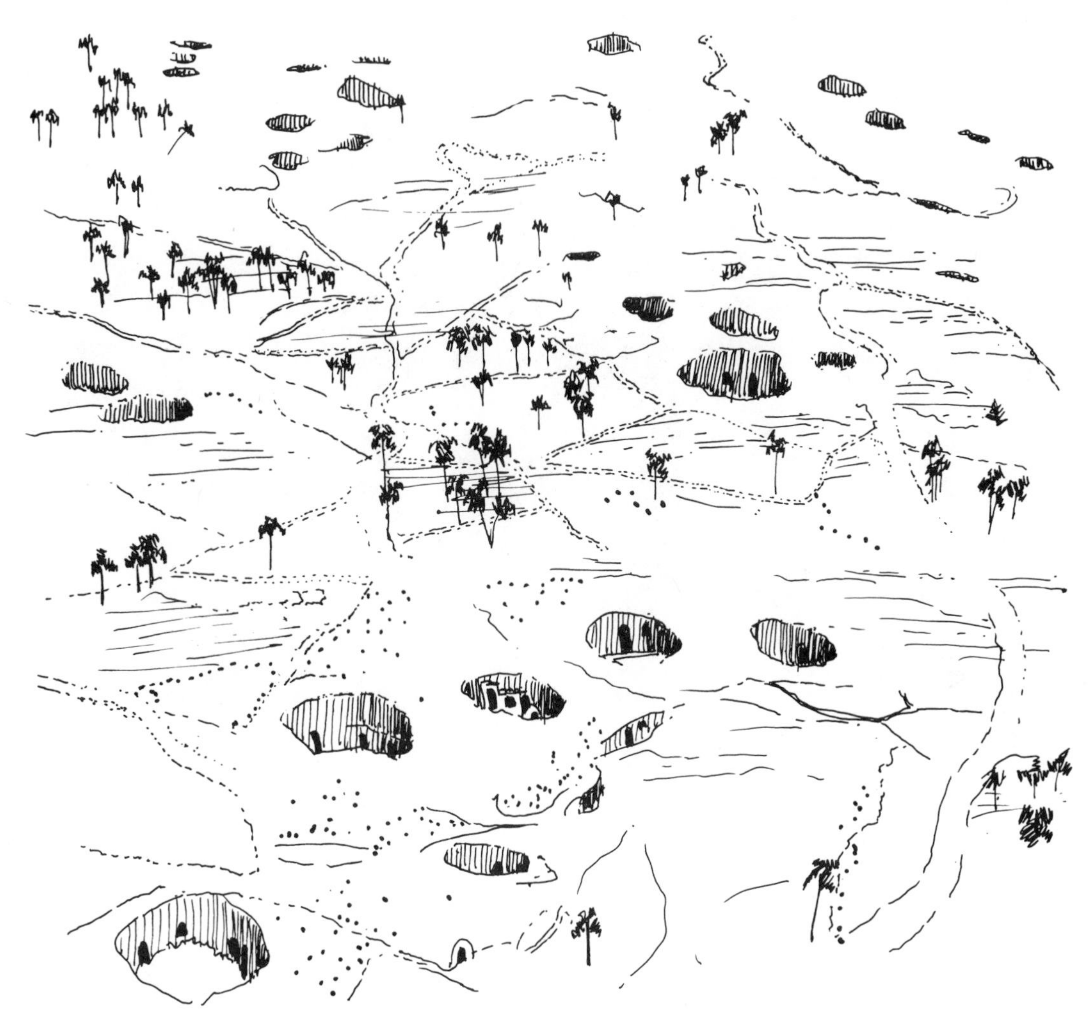

Fig. I-6. Overall view of Matmata, a subterranean settlement in Tunisia bordering the Sahara desert. Adapted from a publication of the Tunisian Tourist Information Office.

tion purposes, cooking space is located on the side of the patio. The lower rooms are cooler and suited the need of coping with the harsh, dry summer climate with temperatures exceeding 100°F.[6] Conversely, the rooms are warmer in winter, and the frequent wind and dust storms do not interfere with daily activities. Today, water for the villages comes primarily from deep wells dug more than 200 feet into the ground, as well as from the cisterns. The main employment is pasturing and gardening.

Chenini in southern Tunisia developed from troglodyte communities established nine centuries ago. Here again, the Berbers dug caves into the mountain slopes at the edge of the Sahara. The population of Chenini is about 1700 today. The Berbers also developed troglodyte villages in the northern mountains of Tunisia. Here the alternating soft and hard limestone strata did not exist. So the houses were built into the mountain rock with thick stone walls on the other sides. As in the south, these tribes built completely subterranean villages when they traveled to the valley and lowlands for water resources. Because of the Berber influence, the Roman colonies at Bulla Regia in northern Tunisia, built in the second century, had subterranean houses with uniform, straight walls. Such houses were built be-

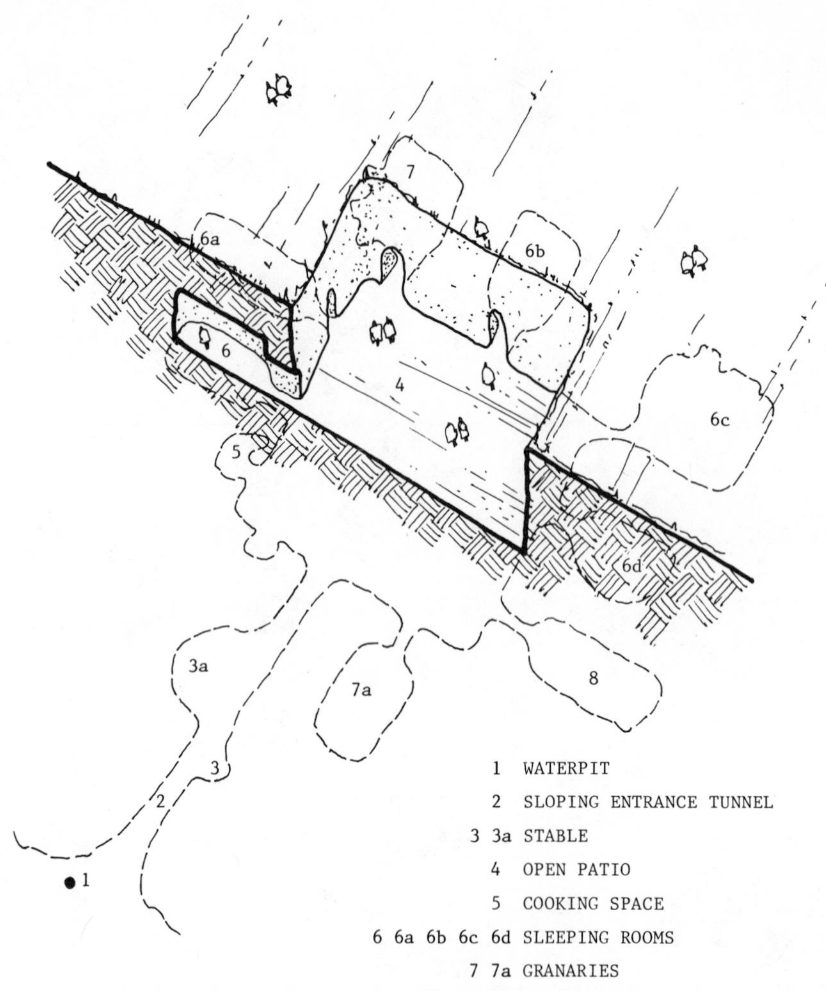

Fig. I-7. Plan of a Berber subterranean house at Matmata village. The central patio is used for family activities. Note that the tunnel leading to the patio has side spaces used for animal stables. Adapted from a number of sources.

lowground in pits and covered over with earth, except for small central patios. Indeed, the Berber tradition of subterranean usage can be traced back to the time when the Phoenicians controlled the North African coast.[7]

CHINA: SUBSURFACE DWELLINGS

Northern China, with its semiarid climate, has limited and uncertain rainfall. The area is subject to drought and flood. Therefore, irrigation is used locally for the 4- to 6-month growing season. Winters are dry and cold with temperatures dropping below zero, while summers are rainy with temperatures sometimes reaching 90°F.[8] In addition, the area is subject to earthquakes. The loess soil is fine and has high porosity; thus, it holds moisture. (Fig. I-9). The fine texture supports capillary movement of the moisture to the surface; the upper layer becomes dry because of the aridity of the climate and forms a hard crust while the interior parts are still humid. This supports living underground both by facilitating the digging and by retaining a cool microclimate when the outside is very dry and warm, and a warm pleasant atmosphere when it is cold and very windy. Also, the char-

Fig. I-8. Overview of the courtyard of a subterranean house in Matmata village, southern Tunisia. The courtyard is 9–12 meters (30–40 feet) wide and 12 meters (40 feet) deep. This depth keeps the court in shadow most of the day. Adapted from publications of the Tunisian Tourist Information Office.

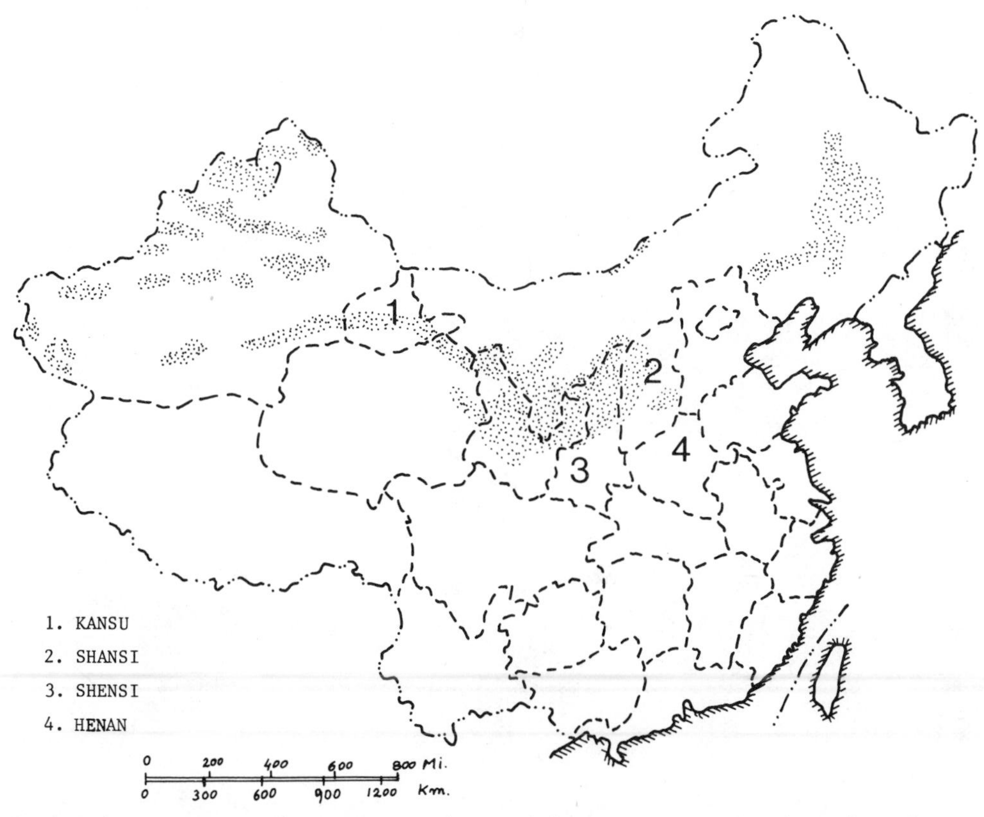

Fig. I-9. Chinese provinces of Henan, Kansu, Shansi and Shensi in eastern and northern China where 10 million people inhabit subterranean villages. Note that some are located in loess soil (dotted).

acter of the loess supports vertical straight walls and flat horizontal ceilings which again facilitates subterranean housing. The farmer retains the surface soil for agriculture while literally living below it.

The subterranean Chinese settlement is not seen from a distance; only close by can one see the sunken hole of the patio, the hole of the chimney and the smoke. There are no buildings aboveground (Fig. I-10). Chinese subterranean houses are well planned, and a great many are compact in design (Fig. I-10).

Fig. I-10. Overall view of Chinese farmers' subterranean dwellings. The houses are cool in summer and warm in winter.

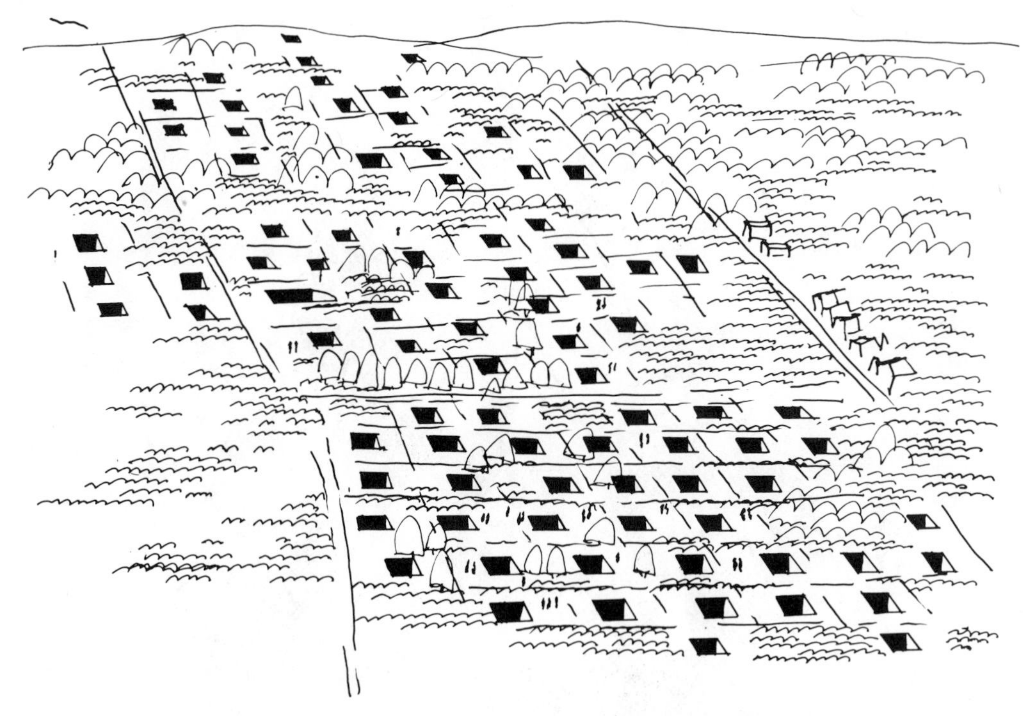

Fig. I-11. An overall view of a Chinese subterranean settlement built within loess soil. The spaces include dwellings, storage areas, schools, government offices and hotels. The surface is reserved for agricultural use.

The Chinese experience in living underground goes back to the fourth century B.C. or earlier and has included temples and shrines as well as settlements. In the provinces of Shensi, Shansi, Kansu and Henan there are a large number of towns and villages which accommodate more than 10 million inhabitants (Fig. I-11.)[9] Subterranean houses are limited to areas with loess soil, and they face south with their entrances through L-shaped stairways.[10]

Living space is more dense for the Chinese than for the Tunisians in the Matmata area. The need to use the land for agriculture has necessitated units being well organized spatially. Although there is much similarity between the Chinese pattern and that of North Africa, there are some differences:

1. The Chinese dwelling is more compact and better organized spatially.
2. Chinese rooms are built with angular corners, while the Tunisian rooms have curved corners.
3. Chinese entrances are through stairways leading to a tunnel perpendicular to the patio. Every entrance is marked by trees nearby. The Tunisian entrances were sloping.

CAPPADOCIA, TURKEY: UNDERGROUND SETTLEMENTS

Located 400 kilometers southeast of Ankara at Nevsehir in the Göreme Valley, the Cappadocia settlements are in a volcanic tufa. The morphology of

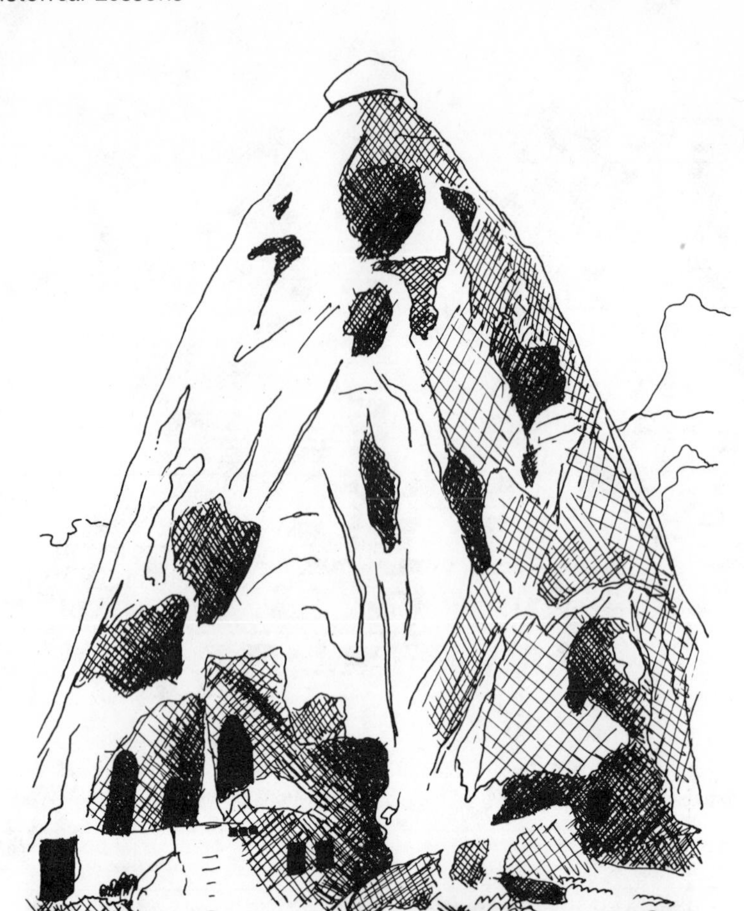

Fig. I-12. Complex of subterranean houses built into the tufa in Göreme Valley, Cappadocia, Turkey. The cone was formed by natural forces, and houses were integrated within it.

the area, especially conical forms which are soft and greatly eroded, has been used by the settlers (Fig. I-12).

In the southern part of the region, there are two subterranean cities, Derinkuyu and Kaymakli, where people still are living and which were inhabited by tens of thousands of people. The first settlements in the region were built c. 2300 B.C. Most of them were communal dwellings cut out of the rock of the hillsides.[11] The region was occupied later by the Persians, Romans, Byzantine Christians, and Islamics (Fig. I-13). The latter decimated the Christian population which was then forced in self-defense to develop the subterranean system of the area intensively. During the third century, these people built churches and monasteries; later they changed some parts to multistory apartments. It is estimated that by the seventh century A.D., the communities grew to 30,000 inhabitants, all of them living underground with wide communication networks. Some of the conical formations into which they cut living space are 100 feet high.

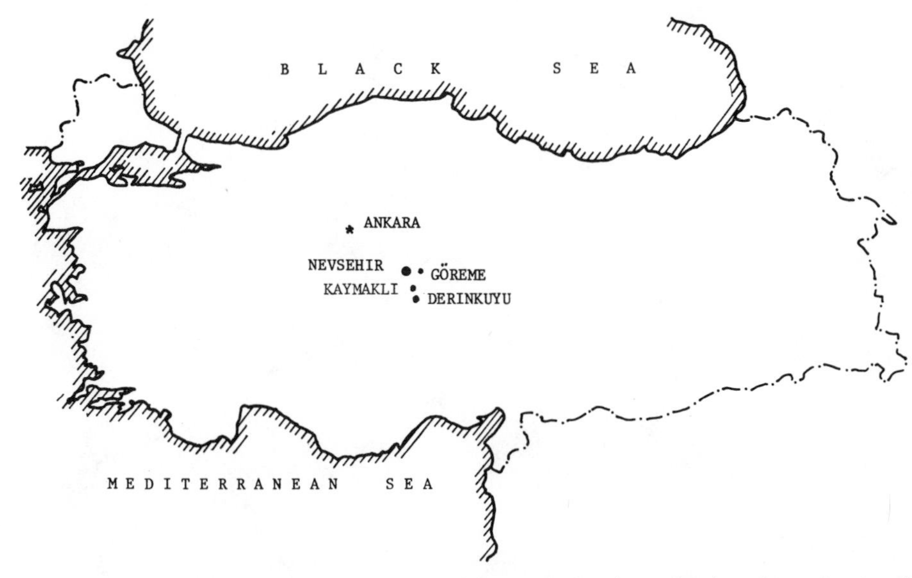

Fig. I-13. Nevsehir, 400 km southeast of Ankara (the capital of Turkey), developed in volcanic tufa. Derinkuyu and Kaymakli are two subterranean cities located in volcanic soil in Göreme Valley.

"In 1965," Koslof reports, "three entirely rock-cut towns were discovered in Cappadocia, a province of Turkey, one of which, penetrated through a single entrance, extended over an area of six kilometers."[12] These partially underground settlements included main urban centers, secondary villages, forts and watchtowers, monastic and eremitic communities, military and trade routes linking the region with the exterior environment, and networks of tracks linking the villages, fields and other places of work.[13] Giovannini gives a detailed description of underground cities in Cappadocia, 8 to 10 floors deep and several kilometers in length.[14] According to Hazer, there were 41 underground towns.[15] Housing units were built one above the other along the towers. These settlements are a remarkable example of man's ability to adapt to nature.

The communities of Cappadocia grew grapes for wine and other fruits and vegetables in the valley of Göreme. They used the subterranean space for food storage after the food was dehydrated in the dry surface climate.

APULIA PROVINCE, ITALY: LIMESTONE DEVELOPMENT

In Apulia Province, in southern Italy, is a limestone area which has been eroded by karst activity (Fig. I-14). The southern part of the region is covered by tufa. The subterranean settlements in the area were begun about 1500 B.C. and later developed by the Byzantines. The region has not been highly developed, however. There are man-made caves 1300 to 2000 feet above sea level and at as many as four levels.[16] The sites and surrounding area are described as:

> a great sponge of stone, for it has no ponds, lakes, streams, or rivers. It does not shed or hold water, but absorbs it. Its surface is a series of shallow, closed basins which funnel the rainfall into fissures in the thick underlying layer of white limestone, from which, by long underground labyrinths, it is eventually conducted beneath the surrounding coastal plains to emerge again from springs at the shore.[17]

Fig. I-14. Apulia Province in southern Italy. In this region there are numerous caves of eroded limestone which served as homes and churches in the past, possibly as a result of the Mediterranean climate (cold and wet in winter, warm and dry in summer).

In Matera, there were many subterranean dwellings which were extensively occupied until recently.

Both the Cappadocia region in Turkey and Apulia Province in Italy "share unusual climatic, geological and fauna characteristics, which are largely responsible for the unique architectural features of secular and religious structures of the most enduring type: underground cities and cave dwellings."[18]

UNITED STATES: INDIAN SUBTERRANEAN DWELLINGS

The indigenous population of the United States also used the earth-covered house in order to avoid the extreme heat of the desert. In the Southwest, the Indians used subterranean space called kivas for male religious ceremonial purposes as well as for dwellings. The kivas were circular chambers dug several feet into the ground and covered by semicircular domes. For air flow, the Indians used "external ventilator shafts to effect a natural cool air convective system."[19]

Cliff dwellings in Mesa Verde, an Indian village located in the southwest corner of Colorado, were built into the walls of the cliff and, as such, were sheltered on five sides either by earth or other buildings. All of the village was surrounded by the stone of the cliff itself (Fig. I-15). Montezuma Castle is another example of a cliff dwelling. Named by early European settlers who believed it was built by the Aztec king Montezuma, the structure was built c. 1250, next to the cliff at the edge of the valley. It faces south. In this case, the shadow of the cliff protects the structure from direct radiation by the high-angled summer sun; but the low-angled winter sun may penetrate. This structure was built of adobe.

Fig. I-15. Cliff dwellings at Mesa Verde, Colorado, where ancestors of the Pueblo Indians built semisubterranean structures for use as dwellings and kivas for ceremonial purposes. The structures, which were interconnected, are on the shaded cliffs which protects them from the harsh arid climate.

SERIPE, GHANA: SEMIRECESSED VILLAGE

In northeastern Ghana along the upper Volta River is the village of Seripe, a nearly circular shape with no conventional streets or alleys which forms one cohesive whole. It covers 20 to 30 acres. The houses are attached to one another from all sides, each house having a central patio surrounded by rooms. Thus, the village forms a solid perimeter wall which aids in its defense (Fig. I-16).[20]

Access to each house unit is by wooden ladder from the courtyard; the only pedestrian paths are over the flat roofs which are used for meetings, for drying clothes and food, and for children's play. Access to the village from the outside is by ladder, too. Each individual family compound is compact, as is the entire village; therefore, the climatic characteristics are almost the same as in subterranean communities because heat gain and loss is minimal.

The total population of Seripe today is 400 persons who live within five sectors with each sector made up of 15 to 50 family compounds. The village is agricultural. Its form has resulted from the desire to save agricultural land, to ease the effect of the semiarid climate and for defense.

In the construction of their dwellings, the people use such indigenous materials as wood, earth and cow excrement.[21] Walls are plastered with

Fig. I-16. The village of Seripe in northwest Ghana is a conglomerate of aboveground housing of single units without streets or alleys. Each house surrounds an open patio. The main pedestrian traffic is on the roofs. Adapted from a number of sources.

earth and cow dung mixed with sticky juice from ground creepers for waterproofing. The roofs, of course, transmit heat gain and heat loss since they are not covered with much earth. The patio also contributes to heat gain and loss within the house.

OTHER ANCIENT EXPERIENCES

The historical models introduced so far show patterns mainly determined by the topographical form, the soil type, some degree of technological experience, and, above all, the climatic necessity. Among the Chinese and Matmatans, for example, rooms surround the patio which becomes the house center. The Cappadocians developed underground walkway networks for pedestrians between the residential units and between the settlements. There are, however, numerous other examples of subterranean living throughout ancient times (Table I-1). At the northern corner of the western desert of Egypt in the oasis of Siwa, for instance, there are subterranean dwellings using former burial grounds. This is an area with a hot, dry climate (Fig. I-17).

The Eskimos have developed semisubterranean seasonal dwellings. In the igloo, the entrance through the tunnel is designed to keep out the wind. Several dwellings are connected for this purpose, too. The entrance is faced away from the prevailing wind and protected by a wall made of snow blocks. The tunnel itself is a transitional section to keep the inside warm; the living level is raised to be closer to the heated air. Seal oil lamps provide heat to the space in addition to the body radiation from the people themselves. The Eskimos change dwellings seasonally which necessitates a change of build-

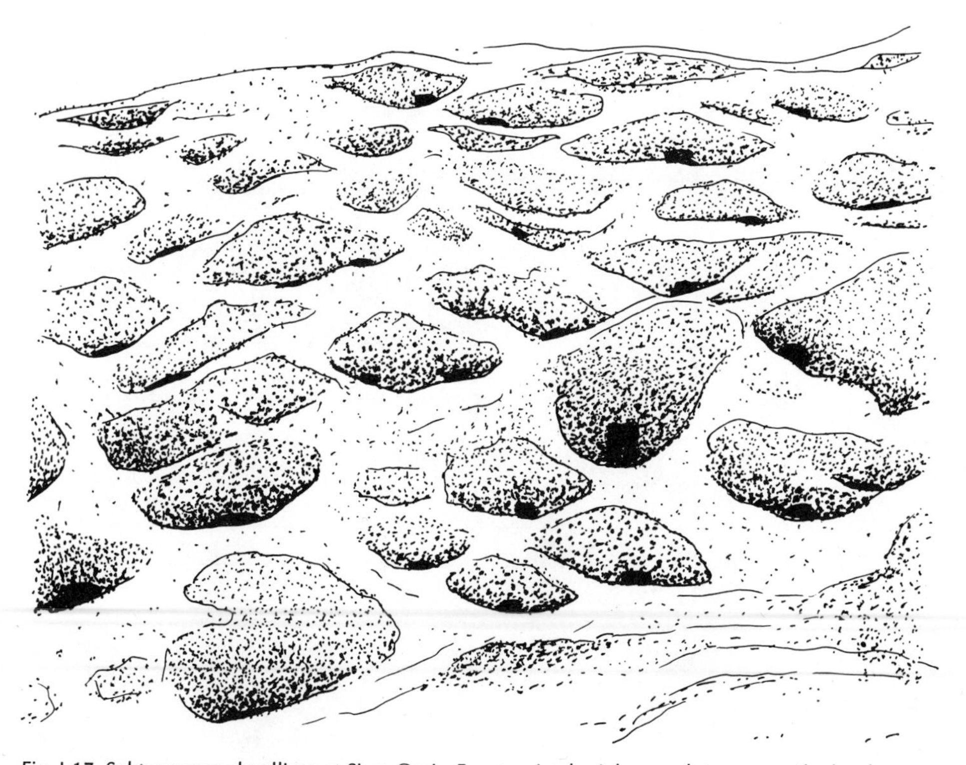

Fig. I-17. Subterranean dwellings at Siwa Oasis, Egypt, using burial grounds to escape the harsh, warm, dry climate of the Egyptian western desert.

Table I-1. Ancient Subterranean Structure Usage.

USAGE	PLACE	TIME	SOIL	CLIMATE	OTHER DATA
CAVE DWELLINGS	MIDDLE EAST, AFRICA, ASIA, AUSTRALIA, EUROPE AND AMERICA	PALEOLITHIC	NATURAL CAVES IN ROCK DEVELOPED BY EARTH MOVEMENT OR EROSION	PRIMARILY SEMIARID AND HUMID	
CAVE DWELLINGS	KOONALDA CAVE, AUSTRALIA	INHABITED 40,000 B.C.		ARID	
DWELLINGS	ESKIMO DENBIGH CULTURE, NORTH AMERICA	4000 B.C.		VERY COLD AND SNOWY IN WINTER, COOL IN SUMMER	THE ESKIMO DENBIGH CULTURE INHABITED SEMISUBTERRANEAN HOUSES WITH WALLS OF SOD AND STONE, OCCASIONALLY RE-INFORCED WITH WHOLE RIBS AND JAWS.
AGRICULTURAL VILLAGES & HOUSING	CHINA IN THE NORTH & NORTHEASTERN PROVINCES OF: HENAN, KANSU, SHANSI & SHENSI	4TH CENTURY B.C. TO DATE	LOESS. FERTILE. EASY TO CUT. RETAINS HARD CRUST WITH HUMIDITY IN LOWER LEVELS.	ARID AND SEMIARID	TEN MILLION PEOPLE LIVING IN THOSE SETTLEMENTS. ALL TYPES OF LAND USES ARE INCLUDED: RESIDENCES, SCHOOLS, FACTORIES, OFFICES AND STORAGE. BUILT TO SAVE AGRICULTURAL LAND AND COPE WITH THE STRESSED CLIMATE.
VILLAGES & HOUSING	SOUTHERN TUNISIA	BEFORE ROMAN OCCUPATION TO DATE	CLAY AND HARD ROCKS ALTERNATED.	ARID: DRY AND WARM	FULL SETTLEMENTS. BUILT TO COPE WITH THE STRESSED CLIMATE.
TOWNS & VILLAGES	CAPPADOCIA, CENTRAL TURKEY	FROM 4TH CEN-TURY B.C. TO 1100 A.D.	VOLCANIC TUFA IN CONIC FORMS (PLATEAU)	ARID: DRY, WARM IN THE SUMMER AND COLD IN WINTER	41 TOWNS AND VILLAGES. HOUSING WITH MORE THAN TEN STORIES. INTERCONNECTED BY LONG TUNNELS. EXTENDS SEVERAL MILES. THIS LIVING PROTECTED THE RESIDENTS AGAINST ATTACK BY MARAUDING TRIBES.
HOUSING	TRULLI REGION, SOUTHEAST ITALY	SECOND MILLEN-NIUM B.C.	SPONGE STONE. ROCK ERODED BY KARST	SEMIARID TO HUMID.	
CLIFF DWELLINGS	SOUTHWESTERN U.S.A.		LIMESTONE ROCKS	ARID: DRY AND WARM	ALSO INCLUDES KIVA, A CIRCLE CHAMBER PIT FOR RELIGIOUS CEREMONIES.
CATACOMBS	MALTA ISLAND	3000 B.C.	ROCKS	COLD AND RAINY IN WINTER, WARM AND DRY IN SUMMER (MEDITERRA-NEAN CLIMATE)	
CATACOMBS	NEAR ROME, ITALY	EXCAVATED IN 1ST-5TH A.D.	ROCKS	MEDITERRANEAN CLIMATE	600 ACRES AT DEPTHS RANGING FROM 22 TO 65 FEET BELOW GROUND
SEMIRECESSED VILLAGE	NORTHWESTERN GHANA ON THE BLACK VOLTA. SERIPE VILLAGE.	TO DATE		SEMIARID AND ARID	HOUSES CONGLOMERATED WITH NO ROADS OR ALLEYS. ACCESS THROUGH THE ROOFS TO THE HOUSE PATIO. COMMUNITY OF 20-30 ACRES. 400 PERSONS.
SEWAGE & WATER	ROMAN EMPIRE	ROMAN	VARIED		
SEWAGE SYSTEM	PARIS, FRANCE	MEDIEVAL		COLD AND HUMID	
MONASTERIES	METEORA, GREECE	GREEK	ROCKS	MEDITERRANEAN CLIMATE	MOLARLIKE ROCK OUTCROPPINGS FORMED INTO MASSIVE MESAS AND OBELISKS
TROGLODYTE	PENTALICA, ANAPO VALLEY, SICILY	ONE MILLEN-NIUM B.C.	ROCKS	MEDITERRANEAN CLIMATE	USED IN PREHISTORIC TIME, ONE MILLENNIUM B.C. USED ALSO AS BURIAL GROUNDS AND FOR LIVING DURING THE MIDDLE AGES.
SETTLEMENTS	BAMIYAN VALLEY, AFGHANISTAN	1ST CENTURY A.D. UNTIL 10TH CENTURY A.D.	SANDSTONE	ARID	BUDDHISTS. PRIMARILY HOUSES FOR THE MONKS WHICH WERE CURVED INTO THE SANDSTONE CLIFFS ALONG THE VALLEY.
CLIFF HOUSES	ADI KADO, ETHIOPA	FOUND AT THE 4TH CENTURY	ROCKS	ARID	USED BY THE CHRISTIANS
WATER TUNNEL	JERUSALEM,	ONE MILLENIUM B.C.	LIMESTONE ROCKS	SEMIARID	DEVELOPED BY KING HEZEKIAH TO SUPPLY WATER TO JERUSALEM FROM WELLS LOCATED OUTSIDE THE CITY.
WATER TUNNEL	MEGIDDO, ISRAEL	4TH MILLENNIUM B.C.	LIMESTONE ROCKS	SEMIARID	DEVELOPED TO SUPPLY WATER TO THE CITY FROM OUTSIDE WELLS.
WATER IRRIGA-TION SYSTEM (CANALS)	PERSIA (IRAN)	SINCE ANCIENT TIMES	VARIED	ARID	A VERY SOPHISTICATED SYSTEM DEVELOPED DEEP INTO THE GROUND TO AVOID LOSING WATER BY EVAPORATION
TRADE CITY	PETRA, JORDAN	SINCE THE GREEKS AND ROMANS TO THE NABATEAN	SANDSTONE	ARID: DRY AND WARM	DEVELOPED PRIMARILY BY THE NABATEANS AS CITY ON IN-TERNATIONAL TRADE ROUTE. FORTIFIED BY THE CRUSADERS UNTIL IT WAS ABANDONED AT THE TIME OF THE MOSLEM CON-QUEST. INCLUDES: DWELLINGS, TEMPLES AND PALACES.
TEMPLE	ROCK TEMPLE, BAHAJA, INDIA				
TOWN	AVDAT, ISRAEL	2ND CENTURY B.C. TO 6TH CENTURY A.D.	CHALKSTONE	ARID: DRY AND WARM	DEVELOPED BY THE NABATEANS AS AGRICULTURE AND TRADE CENTER.
SALT MINES	WIELICZKA, KRAKOW, POLAND	SINCE THE 11TH CENTURY	ROCKS	COLD AND SNOWY IN WINTER, MODERATE AND RAINY IN SUMMER	THIS IS THE MOST ELABORATE COMPLEX OF SUBTERRANEAN SPACE. DEPTH OF 800 FT. AND LENGTH OF MORE THAN 75 MILES. THERE ARE SEVEN LEVELS WITH MANY FACILITIES SUCH AS THEATRES, CHURCH AND BALLROOM. USED AS RESIDENCES IN WORLD WAR II.
BRONZE MINES	NEGEV (SOUTH), ISRAEL	11TH CENTURY B.C.	ROCKS	ARID: DRY AND WARM	DEVELOPED BY KING SOLOMON
HOUSES	PERSIA & SOUTHERN MESOPOTAMIA	ANCIENT	MOSTLY ALLUVIAL	ARID: DRY AND WARM	

ing materials. The summer dwelling is made of stone walls or sod, similar in construction to that of the igloo, and it is also built semisubterraneanly. Here, too, the entrance is narrow and the floor is raised.

In the southwestern part of the city of Beer Sheva, Israel, archaeologist Jan Perrot excavated Tels Mater and Sfr (Fig. I-18). In these tels, he discovered subterranean housing the oldest of which dates to the fourth millennium B.C. Tel Mater is located on a hill near the valley of Beer Sheva. Here, too, the soil is loess 2½ meters deep. The loess is deposited on a loam which is rich in sand. Below is a Miocene chalk. The shelters built here, which are approached from the cliff of the valley, were developed with a vertical tunnel leading to a large open rectangular space. The entrance to the buildings was through this vertical tunnel.

Subterranean space was used by the Nabateans in their city of Avdat in

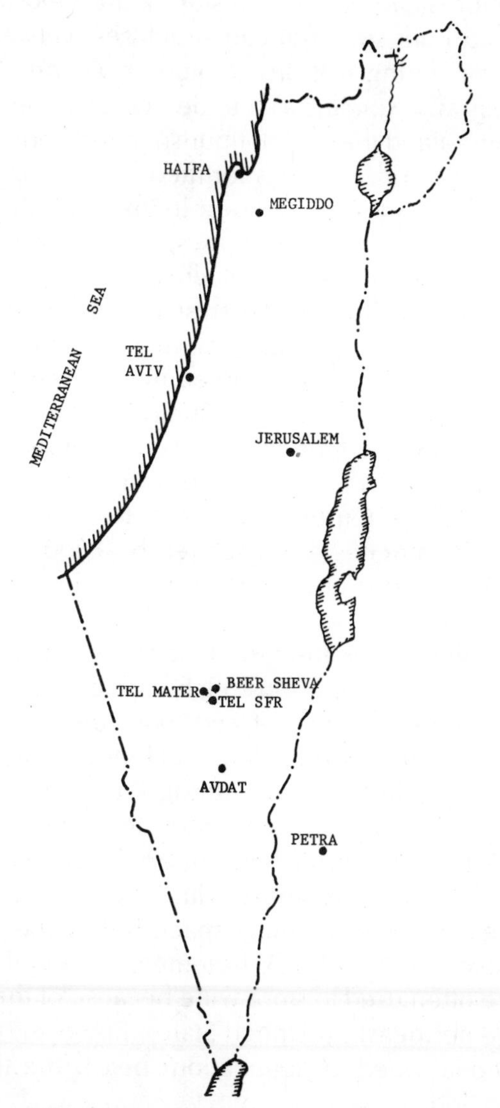

Fig. I-18. Avdat, Israel, and Petra, Jordan, where the two Nabatean cities used underground space extensively (2nd century B.C. to 6th century A.D.). The Jewish King Hordus' castle was erected south of Jerusalem. Underground tunnels for water supply were also used in ancient Jerusalem and Megiddo.

the Negev of modern Israel. According to Professor Avraham Negev, the Byzantines later used these unventilated subterranean structures in Avdat for agricultural storage from 400–700 A.D. (Fig. I-18). There is very little rain in the area, and it does not support dry farming. Therefore, a sophisticated underground water storage system and, consequently, agriculture were developed. The lithology of the area is alternating layers of limestone and chalk which are used for water storage and the development of water systems. Arch forms were used in the construction and were primarily cut out of the chalk. Houses are both aboveground and underground.

Also in Israel, King Hordus' castle was built 2000 years ago south of Jerusalem and used as a summer house. Later it was used by Bar-Kochba who dug underground tunnel systems for defense. The hill on which it is located is made of chalk.

Located in an area with a hot, dry climate at the eastern edge of the Arava Valley, near the border of southern Israel and Jordan, the Petra settlement contains structures carved into the side of the sandstone canyon (Fig. I-18). Nabateans also built subterranean structures with elaborate ventilation systems in Petra during the first century A.D. The city of Petra was known to the Greeks and Romans as a trade center along the international route between the west and the east. It flourished for many years, especially during the Nabatean period, and was fortified by the Crusaders but abandoned at the time of the Muslim conquest in the seventh century A.D. The city includes dwellings, temples or shrines, and a palace. Near Petra is another settlement cut out of rock, Al Madras.

In the fertile Bamiyan Valley of Afghanistan there were settlements developed by the Buddhists which date from the first century A.D. The chambers and houses for the monks were carved into the sandstone cliffs along the valley. The development of the community ended with the Islamic occupation at the end of the tenth century A.D. In Ethiopia, there are a few subterranean spaces in the cliffs of the remote mountains used by the Christians and discovered in the fourth century A.D. Located in the Wallo province, 11 ancient rock churches are inhabited by 1000 Coptic priests. This Lalibala area is now a tourist attraction with a modern hotel and air connections to Addis Ababa.

Finally, the salt mines near Krakow, Poland, which have been in use since the eleventh century, are a most elaborate complex of subterranean space. Placed at a depth of 800 feet and more than 75 miles long, they include seven levels with varied facilities such as a theater, a church, and a ballroom. People lived in the mines during World War II.[22]

The most commonly known semisubterranean spaces are the basements and the cellars used widely throughout history. A basement is defined as the space constructed below the house with 50 percent or less of its height below ground level. A cellar is similar space below the house with more than 50 percent below ground level. Basements and cellars in the Middle East and Europe are often used to store wine because of the relatively stable temperature. In the northeastern United States, however, they are built because of the additional need for foundations below the frost line to avoid freezing the inside floor.

Contemporary Achievement

Contemporary building of subterranean structures focuses on:

1. Housing, mainly earth covered as in the United States;
2. Shopping centers, as in Japan and Stockholm;
3. Military installations, as in the United States;
4. Oil storage space, as in Norway and Sweden;
5. Parking spaces for cars, as in many places throughout the world;
6. Atomic reactors, as in a few places in the world;
7. Other uses, such as libraries in Urbana (Illinois), the University of Minnesota, and in Baltimore; or manufacturing, storage and offices as in Kansas City.

Table I-2 lists by function some of the most important underground shelters built in modern times. Particular note should be made of the many designs that were developed in the 1950s and early 1960s for ground shelters to protect industry in the event of nuclear attack.[23] Other noteworthy underground designs include the Etnicon subsidiary of the Johnson & Johnson Corporation in San Angelo, Texas, and the Abo Elementary School and fallout shelter in Artesia, New Mexico, which was studied twice by Dr. Frank Lutz in an attempt to determine the effect of the windowless structure on the students.[24]

This author has noticed through his study that a new movement is emerging in the use of subterranean space in a variety of forms and in different countries throughout the world. This movement is receiving more and more energetic public support so that the average person today more readily accepts underground placement. There is recent understanding of the earth as an immense potential source for living, for storing, for refrigerating, for manufacturing, for environmental protection and most importantly, for energy conservation.

Many countries in the world are involved now, publicly or privately, in this new usage of subterranean space. Some of the leading participants are in the United States, Scandinavia, Japan, Canada, Australia, and Russia.

UNITED STATES: EARTH-COVERED SHELTERS

The American experience is diverse and pioneering. Contributing to this is the basically positive national attitude, the perception of the innovative concept and the practicality of the underground approach, as well as a

Table I-2. Contemporary Subterranean Structure Usage.

TYPE OF STRUCTURE	PLACE	TIME	SOIL	CLIMATE	OTHER DATA
RESIDENTIAL					
HOUSE	FRESNO, CA, USA	1908	ROCKS	SEMI ARID. HOT AND DRY, UP TO 120° F. OCCASIONALLY.	DEVELOPED BY BALDASARE FORESTIERE. 7 ACRES OF TUNNELS, PASSAGEWAYS, AND ROOMS 10 AND 20 FEET BELOW GROUND. ALL SPACES USE NATURAL LIGHTS. SOME HORTICULTURE. MAINTAIN 70° F.
DWELLING	KEYES HOUSE, ROCHESTER, MN, USA	1950-1		SNOWY AND COLD IN WINTER; RAINY AND WARM IN SUMMER.	DESIGNED BY FRANK LLOYD WRIGHT.
HOUSE	GEIER HOUSE, CINCINNATI, OH, USA	1965		TEMPERATE AND HUMID	DESIGNED BY ARCH. PHILIP JOHNSON. 4,000 SQ. FT. EARTH COVER: 15 IN. STRUCTURE: 6 IN. REINFORCED CONCRETE ROOF, 12 IN. REINFORCED CONCRETE WALLS. INSULATION: 2 IN. FOAMGLAS, ROOF AND WALLS. WATERPROOFING: ROOF - 5 PLY BUILT-UP ROOFING, WALLS - 3 PLY BUILT-UP WATERPROOFING, FLOOR - 3 PLY BUILT UP WATERPROOFING. HEATING SYSTEM: 4 AREA FURNACES.
HOUSE	CLARK-NELSON HOUSE. RIVER FALLS, WI, USA	1972		SNOWY AND COLD IN WINTER; RAINY AND WARM IN SUMMER.	DESIGNED BY ARCH. MICHAEL MCQUIRE. 2500 SQ. FT. EARTH COVER: VARIES FROM 6 IN. AT PEAK. STRUCTURE: STEEL CULVERTS WITH CONCRETE FLOOR SLABS. INSULATION: 2 IN. POLYURETHANE SPRAY ON INTERIOR OF SHELL. WATERPROOFING: ASPHALT DAMPPROOFING AND BITUTHAENE AND SHEET PLASTIC. HEATING: OIL, FORCED AIR - COOLING: NONE. ENERGY USE: REDUCED 25%.
DWELLING	WINSTON HOUSE, LYME, NH, USA	1972		SNOWY AND COLD IN WINTER; RAINY AND WARM IN SUMMER.	DESIGNED BY ARCH. DON METZ. 2800 SQ. FT. INCLUDING GARAGE. EARTH COVER: 12". BUILT OF CONCRETE BLOCK WALLS WITH TIMBER AND WOOD DECK ROOF. INSULATION: ROOF - 1 5/8 COMPRESSED FIBER GLASS, WALLS - ZONOLITE IN BLOCK CORES, FLOOR - 1 IN. URETHANE PERIMETER ONLY. WATERPROOFING: ROOF - 5 PLY COAL-TAR PITCH BUILT UP, WALLS - 2 PLY COAL-TAR PITCH BUILT UP.
HOUSE	"ECOLOGY HOUSE." OSTERVILLE, MA, USA	1973		SNOWY AND COLD IN WINTER; RAINY AND WARM IN SUMMER.	DESIGNED BY ARCH. JOHN BARNARD. 1200 SQ. FT. WITH EARTH COVER 10 IN. TO 16 IN. REINFORCED CONCRETE WALLS, PRECAST PLANK ROOF WITH STEEL BEAMS. INSULATION: 2 IN. STYROFOAM ON ROOF AND WALLS. WATERPROOFING: ROOF - 3 PLY BUILT-UP, WALLS - HOT MOPPED PITCH. HEATING: SOLAR COLLECTOR WITH FORCED AIR FURNACE. ENERGY USE: 25% OF NORMAL
COMMERCIAL					
SHOPPING CENTERS	JAPAN	CONTEMPORARY		HUMID AND COLD	VARIOUS UNDERGROUND SHOPPING CENTERS ON A LARGE SCALE. ONE OF THESE IS VISITED BY 800,000 SHOPPERS A DAY AND HAS 225 SHOPS.
SHOPPING CENTER	JERUSALEM, ISRAEL	CONTEMPORARY	ROCKS	SEMI ARID.	SEVEN FLOORS UNDERGROUND.
SHOPPING CENTER	MONTREAL	1956	ROCKS	SNOWY AND COLD IN WINTER; RAINY AND WARM IN SUMMER.	WHEN THE SHOPPING CENTER STARTED, IT USED THE CANADIAN NATIONAL RAILWAY EXCAVATION. THE SHOPPING CENTER CONSISTS OF OFFICES, SHOPS, RESTAURANTS, MALLS, THEATRES & OTHER SERVICES.
CAFETERIA	ST. LOUIS, MO, USA	COMPLETION-1968		TEMPERATE	MONSANTO COMPANY. 35,000 SQ. FT. DESICN TEAM: VINCENT KLING AND ASSOC. AND RAY MARTIN. DESIGNED TO SERVE 1,000 EMPLOYEES AS A PART OF A GROUP OF COMPANY OFFICE BUILDINGS. MAJOR CONSIDERATIONS WERE CENTRALITY AND EASE OF ACCESS FROM THE OFFICES. DINING AREA IS U-SHAPED. 1,300 PEOPLE CAN BE SEATED. SHOPS, RESTAURANTS & OTHER USES.
SHOPPING CENTER	STOCKHOLM, SWEDEN	CONTEMPORARY	GRANITE ROCKS	SNOWY AND COLD IN WINTER; RAINY AND COOL IN SUMMER.	SHOPS, RESTAURANTS & OTHER USES.
INDUSTRIAL					
RECORD STORAGE COMPLEX	NATIONAL STORAGE CO., INC., BOYERS, PA, USA	COMPLETED, 1956		SNOWY AND COLD IN WINTER; RAINY AND WARM IN SUMMER.	EXISTING MINE. CHAMBERS AND TUNNELS. PAINTED MINE. ALUMINUM WALLS.
SHELTER AND RECORD STORAGE	IRON MOUNTAIN, HUDSON, NY, USA		ROCKS	SNOWY AND COLD IN WINTER; RAINY AND WARM IN SUMMER.	700 STORAGE CLIENTS: MANUFACTURERS HANOVER TRUST CO. - CAPACITY, 24; SHELL OIL - CAPACITY, 44 (44,000 FT^2); STANDARD OIL - CAPACITY, 200 (20,000 FT^2). EXISTING MINE. CHAMBERS AND TUNNELS.
REFRIGERATED AND DRY FOOD STORAGE COMPLEX	SPACENTER, INC., KANSAS CITY, MO, USA		ROCKS	SNOWY AND COLD IN WINTER; RAINY AND WARM IN SUMMER.	EXISTING MINE. CHAMBER AND TUNNELS. STANDBY EQUIPMENT NOT REQUIRED UNDERGROUND. MINE TEMPERATURE RISES 1° C. PER DAY.
LABORATORY AND STORAGE COMPLEX	MEDUSA PORTLAND CEMENT COMPANY, WAMPUM, PA, USA			SNOWY AND COLD IN WINTER; RAINY AND WARM IN SUMMER.	LABORATORY - 18,000 FT^2. NO VIBRATION. EXISTING MINE. CHAMBERS AND TUNNELS. CAPACITY: 75 PERSONS. STORAGE - PAGE AIRWAYS, WEYERHAUSER CO., U. S. GENERAL SERVICE ADMINISTRATION. GROUND TEMPERATURE: 54° F.
PRECISION INSTRUMENT MANUFACTURING FACILITY	BRUNSON INSTRUMENT CO. KANSAS CITY, MO, USA		ROCKS	SNOWY AND COLD IN WINTER; RAINY AND WARM IN SUMMER.	140,000 FT^2. EXISTING MINE. CHAMBERS AND TUNNELS. CAPACITY: 200 TONS/FT^2. NO VIBRATION. CAPACITY: 125 PERSONS. GROUND TEMPERATURE: 54°F.
HYDROELECTRIC POWER	CHURCHILL FALLS HYDRO-ELECTRIC POWER PLANT		ROCKS		
LIQUID STORAGE	SWEDEN, USA & OTHER COUNTRIES	SINCE WORLD WAR II	ROCKS, MOSTLY GRANITE	VARIED	MAINLY FOR OIL. SOME FOR WATER.

TRANSPORTATION and INFRASTRUCTURAL

SHELTER AND GARAGE	MALMO, SWEDEN		ROCKS	SNOWY AND COLD IN WINTER; RAINY AND COOL IN SUMMER.	CAPACITY: 4,300 PERSONS.
GARAGE	STOCKHOLM, SWEDEN		ROCKS	SNOWY AND COLD IN WINTER; RAINY AND COOL IN SUMMER.	CAPACITY: 500 CARS.
MANNED COMMUNICATION CENTERS	26 LOCATIONS BETWEEN BOSTON AND MIAMI, OPERATED BY AMERICAN TELEPHONE & TELEGRAPH COMPANY			DEPENDENT UPON SITE.	OPEN EXCAVATION WALLS/ROOFS - 7 OZ. COPPER SHEETED.
ARTERIAL NETWORK (LES HALLES)	PARIS, FRANCE	CONTEMPORARY	ROCKS	SNOWY AND COLD IN WINTER; RAINY AND WARM IN SUMMER.	MULTI-LEVEL SYSTEM INCLUDING RAIL SUBWAY, CAR HIGHWAYS, PARKING, OFFICES, COMMERCIAL AND RECREATIONAL COMPLEXES.
PUMPING STATION	SEA OF GALILEE, ISRAEL	CONTEMPORARY	ROCKS	RAINY AND COLD IN WINTER; WARM AND DRY IN SUMMER.	STATION PUMPING WATER FROM SEA OF GALILEE TO THE WATERFALL AND HYDROELECTRIC POWER STATION.
SEWAGE SYSTEM	CHICAGO AND BOSTON, USA	CONTEMPORARY	DIFFERS WITH SITE	SNOWY AND COLD IN WINTER; RAINY AND WARM IN SUMMER.	UNDERGROUND STORAGE OF SEWAGE DURING THE PEAK FLOW AND SUBSEQUENT TREATMENT DURING LOW FLOW TIME. (STOCKHOLM ALSO HAS UNDERGROUND SEWAGE STORAGE.) GREATER CHICAGO METROPOLITAN SANITARY DISTRICT WILL HAVE STORAGE CAPACITY OF 11,300 CUBIC METERS AT A DEPTH OF 215 METERS AND A COST 25% OF COMPARABLE ABOVEGROUND FACILITY.
GARAGE	GENEVA, SWITZERLAND	CONTEMPORARY		SNOWY AND COLD IN WINTER; RAINY AND WARM IN SUMMER.	THE LATEST GENEVA AIRPORT CAR PARK IS ENTIRELY BELOW GROUND, CONSERVING OPERATION AND CONSTRUCTION COSTS.
SUBWAYS	LONDON, PARIS, MOSCOW, WASHINGTON, DC, NEW YORK & MONTREAL	VARIED	ROCKS	SNOWY OR RAINY AND COLD IN WINTER; RAINY AND WARM IN SUMMER.	
UTILITIES & INFRASTRUCTURE	ALL OVER THE WORLD	CONTEMPORARY	VARIED	ALL TYPES	THIS INCLUDES SYSTEMS FOR SEWAGE, WATER SUPPLY, ELECTRICITY, TELEPHONE, TV CABLE, ETC.

EDUCATIONAL and OFFICES

ELEMENTARY SCHOOL	ARTESIA, NM, USA	COMPLETED 1962		ARID: HOT AND DRY.	28,000 FT^2. ONE FLOOR. OPEN EXCAVATION. ROOF SLAB - 21"; WALLS - 12"; FLOOR - 4". CAPACITY: 540 PERSONS.
EISENHOWER LIBRARY, JOHNS HOPKINS UNIVERSITY	BALTIMORE, MD, USA	EARLY 1960'S		SNOWY AND COLD IN WINTER; RAINY AND WARM IN SUMMER.	
HIGH SCHOOL	ROSWELL, NM,USA	COMPLETED, 1965		ARID: HOT AND DRY.	182,000 FT^2; 82,000 FT^2 BELOW GRADE. 2 FLOORS. OPEN EXCAVATION. BASEMENT ROOF SLAB - 13". CAPACITY: 2,000 PERSONS.
HIGH SCHOOL	LAREDO, TX, USA	COMPLETED 1964		ARID: HOT AND DRY.	CAPACITY: 540 PERSONS.
UNDERGRADUATE LIBRARY	UNIVERSITY OF ILL, URBANA, ILL, USA	COMPLETED, 1969		SNOWY AND COLD IN WINTER; RAINY AND WARM IN SUMMER.	98,689 FT^2. 2 FLOORS. OPEN EXCAVATION. FLOOR 6" THICK; WALLS 16" (LOWER) AND 12" (UPPER). HEATING - CENTRAL STEAM; CAPACITY - 150,000 VOLUMES; REFRIGERATION - CENTRAL; SEATS FOR 1,905 READERS.
LIBRARY	HENDRIX COLLEGE, CONWAY, AR, USA			SNOWY AND COLD IN WINTER; RAINY AND WARM IN SUMMER.	BEAMED. CAPACITY: 110,000 VOLUMES. WALLS - CONCRETE; ROOF - WAFFLE SLAB.
ELEMENTARY SCHOOL	TERRASET ELEMENTARY SCHOOL, RESTON, VA, USA	COMPLETED, 1977		SNOWY AND COLD IN WINTER; RAINY AND WARM IN SUMMER.	TO SERVE AS SCHOOL AND COMMUNITY NEEDS. MULTI-PURPOSE HALL, KITCHEN, MULTI-MEDIA THEATER, CLASSROOMS ACCOMMODATING 990 STUDENTS IN 66,000 NET FT^2. DESIGNERS: DAVIS, SMITH AND CARTER, INC. COST PER SQ. FT. $38.
PUBLIC GATHERING SPACE	ALL OVER THE WORLD	CONTEMPORARY	VARIED	VARIED	THEATERS, ASSEMBLY HALLS, CONCERT HALLS, MUSEUMS, EXHIBITION HALLS, AND EDUCATIONAL AND SPORT FACILITIES.
LABORATORY	WAMPUM, PA, USA			SNOWY AND COLD IN WINTER; RAINY AND WARM IN SUMMER.	18,000 SQ. FT. WORK REQUIRES PRECISION.
COMPUTER CENTER AND ADMINISTRATIVE OFFICE	UNIVERSITY OF TEXAS, AUSTIN, TX, USA			ARID: HOT AND DRY.	
BOOKSTORE AND ADMISSIONS OFFICE	UNIVERSITY OF MINNESOTA, MINNEAPOLIS, MN, USA	COMPLETION-1977		SNOWY AND COLD IN WINTER; RAINY AND WARM IN SUMMER.	83,480 SQ. FT. THIS IS A COURTYARD TYPE SUBSURFACE BUILDING. 200 X 200 FT. THREE LEVELS BELOW GRADE. 95% OF THE FLOOR AREA IS SUBTERRANEAN. 21,000 SQ. FT. ARE FOR BOOKSTORE USE AND 35,660 SQ. FT. FOR OFFICE USE. DESIGNED FOR ENERGY EFFICIENCY. ENERGY SAVING IS 80% DURING THE HEATING PERIOD AND APPROX. 45% DURING THE COOLING PERIOD. DESIGNERS: MAYERS AND BENNETT. COST CONSTRUCTION PER SQ. FT. WAS $51.00. THE DESIGN AIMED TO PRESERVE THE VIEW TO TWO HISTORIC CAMPUS BUILDINGS.
LIBRARY	PUSEY LIBRARY, HARVARD YARD, HARVARD UNIVERSITY, CAMBRIDGE, MA, USA	1975		SNOWY AND COLD IN WINTER; RAINY AND WARM IN SUMMER.	THREE STORIES. 8,700 SQ. FT. COST PER SQ. FT. - $96.55. DESIGN TEAM ARCH., HUGH STUBBINS AND ASSOCIATES, INC. THE HARVARD UNIVERSITY LIBRARY IS THE LARGEST IN THE U.S.
OFFICES	UNIVERSITY UNION, UNIVERSITY OF NORTHERN IOWA, CEDAR FALLS, IA			TEMPERATE	
BOOKSTORE	CAMPUS BOOKSTORE, CORNELL UNIVERSITY, ITHACA, NY, USA	COMPLETION-1977		SNOWY AND COLD IN WINTER; RAINY AND WARM IN SUMMER.	33,000 SQ. FT. THE THREE-LEVEL STORE ALSO SHAPED TO AVOID THE TREES. NATURAL LIGHT IS BROUGHT TO THE CENTER OF THE STORE BY A COURTYARD. DESIGN TEAM: EARL FLANSBURGH AND ASSOCIATES, MANSON AND GREY & OTHERS. THE DESIGN AVOIDS DESTRUCTION OF THE MAGNIFICENT VIEW OF THE VALLEY AND IS CLOSE TO THE STUDENT UNION BUILDING.
ELEMENTARY SCHOOL	FREMONT ELEMENTARY SCHOOL, SANTA ANA, CA, USA	COMPLETION-1974		ARID: DRY AND WARM	46,600 SQ. FT. 2.8 ACRES. AN AMPHITHEATER IN FRONT. EARTH-COVERED ROOF WHICH IS USED FOR PLAYGROUND. POURED-IN-PLACE CONCRETE WITH ALL COLUMNS SPACED 30 AND 40 FT. CENTERS. EXTERIOR WALLS ARE BOARD-FORMED, SAND BLASTED CONCRETE WITH SNOW ROCK AGGREGATE. THE ROOF LAMINATION CONSISTS OF A COAL-TAR PITCH WATERPROOFING SYSTEM, COVERED BY INSULATION AND 3 INCH THICK SLAB OF CONCRETE. DESIGNERS: ALLEN & MILLER. COST PER SQ. FT. - $34.23. 850 STUDENTS.

Table I-2. Contemporary Subterranean Structure Usage. (*Continued*)

RECREATIONAL and RELIGIOUS

CEREMONIAL STRUCTURES (KIVAS)	ARIZONA AND NEW MEXICO, USA	ANCIENT		ARID: HOT AND DRY.	USED BY INDIAN TRIBES.
THEATER	PARIS, FRANCE			SNOWY AND COLD IN WINTER; RAINY AND WARM IN SUMMER.	ONE FLOOR. 65 FEET IN DIAMETER. FLOOR SLAB - 20" THICK. CAPACITY: 1,000 PERSONS. OPEN EXCAVATION.
ART GALLERY	PHILIP JOHNSON GALLERY, NEW CANAAN, CT, USA			SNOWY AND COLD IN WINTER; RAINY AND WARM IN SUMMER.	BEAMED.
STUDENT CENTER	UNIVERSITY OF HOUSTON, HOUSTON, TX, USA			COOL AND RAINY IN WINTER; HOT AND DRY IN SUMMER.	RECREATION, SALES, FOOD SERVICE, CHECK CASHING, POSTAL SERVICE, BARBER SHOP, STUDENT ORGANIZATION OFFICES, LOCKER ROOMS, AND OUTDOOR CONCERT FACILITIES.
CHURCH	ST. BENEDICT'S ABBEY CHURCH, BENET LAKE, WI	COMPLETION-1972		SNOWY AND COLD IN WINTER; RAINY AND WARM IN SUMMER.	8,950 SQ. FT. DIMENSIONS ARE 68' X 68'. DESIGNERS: STANLEY TIGERMAN DESIGN ASSOCIATES. COST PER SQ. FT. - $37.98. DESIGNED TO PRESERVE THE OPEN SPACE AND THE LANDSCAPE. EARTH-COVERED.
CHURCH	TAIVALLAHTI CHURCH, TEMPPELIAUKIO, HELSINKI, FINLAND			COLD AND SNOWY IN WINTER; COOL AND RAINY IN SUMMER.	ARCHITECT: TIMO AND TUOMO SUOMALAINEN.

DEFENSE

COMBAT OPERATIONS HEADQUARTERS	NORTH AMERICAN AIR DEFENSE COMMAND, COLORADO SPRINGS, CO, USA	COMPLETED-1965	ROCKS	COLD AND RAINY IN WINTER; HOT AND DRY IN SUMMER.	11 BUILDINGS. 200,000 FT^2. 1 TO 3 FLOORS. TUNNELS AND CHAMBERS. STEEL FRAME. INDEPENDENT STRUCTURES ON SPRINGS. CAPACITY: 450 PERSONS. BUILT INTO THE GRANITE OF THE CHEYENNE MOUNTAINS.
NAVAL BASE	MUSCO ISLANDS, SWEDEN	COMPLETED, 1968-69	ROCKS	SNOWY AND COLD IN WINTER; RAINY AND COOL IN SUMMER.	
SHELTER	STOCKHOLM, SWEDEN			SNOWY AND COLD IN WINTER; RAINY AND COOL IN SUMMER.	CAPACITY: 10,000
FALLOUT SHELTER	TENNESSEE VALLEY, USA	CONTEMPORARY		HUMID	
DEFENSE LINE	MAGINOT LINE, NORTHERN BORDER OF FRANCE	BEFORE WORLD WAR II		COLD AND SNOWY IN WINTER; RAINY AND WARM IN SUMMER.	AN ELABORATED DEFENSE NETWORK WHICH SURVIVED SHORT TIME DURING WORLD WAR II.
SHELTERS	ALL OVER THE WORLD	CONTEMPORARY	VARIED	ALL TYPES	THIS INCLUDES SHELTERS AGAINST ATOMIC ATTACK.

MIXED USE

COMMERCIAL COMPLEX (PLACE VILLE-MARIE AND PLACE BONAVENTURE)	MONTREAL, CANADA	1956	ROCKS	VERY SNOWY AND COLD IN WINTER; RAINY AND WARM IN SUMMER.	SHOPS, OFFICES, RESTAURANTS, CINEMAS, RAILROAD STATION.
INDUSTRIAL PARK COMPLEX	DOWNTOWN INDUSTRIAL PARK CO., KANSAS CITY, MO, USA		ROCKS	SNOWY AND COLD IN WINTER; RAINY AND WARM IN SUMMER.	2,000,000 FT^2 ULTIMATE SIZE. EXCAVATED. CHAMBERS AND TUNNELS. WALLS SPRAYED WITH CEMENT.
CITY CENTER	MOSCOW, USSR	IN PLANNING	ROCKS	VERY SNOWY AND COLD IN WINTER; RAINY AND WARM IN SUMMER.	RESTAURANTS, THEATERS, STORES, EXHIBITION HALLS, PARKING GARAGES, SWIMMING POOLS, MARKETS, WAREHOUSES, REPAIR SHOPS, AND HAIRDRESSING SALONS. 2 1/2 - 3 MILES OF ONE-WAY TUNNELS THROUGH THE CENTER.

combination of the availability of advanced technology and resources. Since the energy crisis in 1973 there has been construction in the United States of a large number of earth-covered homes, schools, public buildings, business offices, industrial spaces, refrigeration spaces, military installations, etc. (Fig. I-19).

The oldest "contemporary" use of earth-covered space is the sod house built by the pioneers as they moved westward. Though some dwellings were also made of hay and timber, the soddy became the true house for the prairie because of its effective insulating capability and low cost. It was warm in winter and cool in summer, it stood against strong winds and it was more effective than the frame house as protection against tornadoes. In addition, the settlers could water the grass roofs thus protecting them from fires. Sod houses were built in canyons and in hollows, but mostly in the slopes; they called the latter dugouts. Such a site was subject to flood. The walls were built of sod bricks, around a simple wood frame. The bricks were laid with the grass side down, and mortar and loose soil was applied to any holes or cracks in the sod. Before being plastered, the inner and outer walls were shaved. The roof, because it was heavy, demanded strong support. Roofs had some leaks during rainy times; but in spite of this, they had high insulating quality.

Other more contemporary projects include structures which alleviate climatic stress by reducing the negative ecological impact of conventional building forms.[25] One of the most interesting examples of such structures is the underground tunneled shelter built by an Italian, Baldasare Forestiere, in 1908 at Fresno, California, in an almost barren desert where temperatures are usually 120°F. The shelter, which furnishes protection against the hot, dry stressful climate, maintains a constant temperature of 70°F. It is composed of tunneled passageways and rooms (65 rooms and gardens and grottoes) covering 7 acres, 10 to 20 feet below the surface. All the spaces are lighted by natural sunlight. Underground gardens in this shelter can produce a variety of lemons, grapefruit and oranges.[26]

Another example of a building design which meliorates the effects of climate—this time a nonarid one—is in Marston Mills, Massachusetts. It is an underground house (called Ecology House) featuring a central patio open to the sky. The house, designed by John E. Barnard, costs approximately 25 percent less to build than a usual frame dwelling; and—more significantly—its heating bill during the winter is one third of that for a conventional house.[27]

In the United States, beginning in the 1970s, a new movement developed which began to encourage the building of underground housing in order to save energy. Many of the houses which were constructed as a result of this movement were primarily built in extreme climates: cold, such as in Minnesota, or warm and dry, such as in the Southwest (Table I-3).

In Kansas City, mining space began to be used for warehouses in the 1950s. These underground horizontal limestone mines form a wide network in several valleys in the Kansas City area. Recently quarries have been dug with plans for secondary use of their spaces for warehouses, factories and offices. According to Stauffer, the construction of mines is now adjusted to the needs of this second use, with consideration given to ceiling heights and pillar distances.[28] Stauffer reports that 1641 employees, earning a

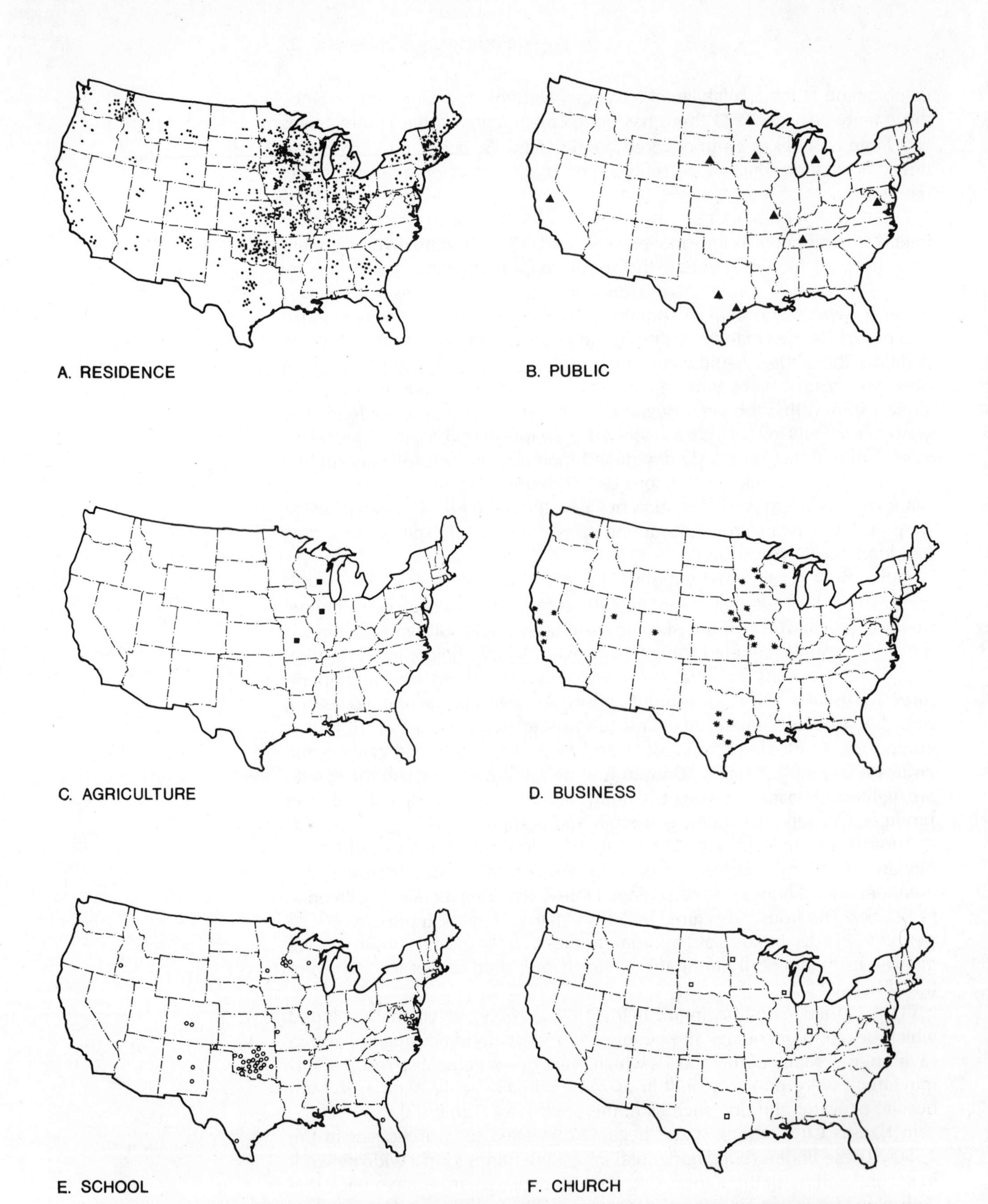

Fig. I-19. Recently developed earth shelters and subterranean structures in the U.S. Adapted from: *Earth Shelter Digest & Energy Report,* No. 11 September/October, 1980; *Earth Shelter Digest and Energy Report,* No. 2 March/April, 1979; *Underground Space,* 4/2 May/June 1980; and Kim Cunningham, "Underground Architecture: An Alternative Approach to Architecture and Energy Conservation," Stillwater: Oklahoma State University, 1977 (Thesis for Master of Architecture, 1975).

Table I-3. Selected list of recently Constructed Earth-Covered Houses in the United States (after 1973).

NAME	PLACE	TIME	CLIMATE	ARCHITECT	OTHER DATA
BALDTOP DUGOUT	LYME, NEW HAMPSHIRE	1977	SNOWY AND COLD IN WINTER. WARM AND RAINY IN SUMMER.	DON METZ	2,200 SQ. FT. EARTH COVER: 8 INCHES. STRUCTURE: CONCRETE BLOCK WALLS, TIMBER AND WOOD DECK ROOF. INSULATION: ROOF- 3 INCHES URETHANE, WALLS - 2 INCHES URETHANE. WATERPROOFING: ROOF - 5 PLY ASPHALT PITCH, WALL - 2 PLY ASPHALT PITCH. HEATING SYSTEM: WOOD STOVES, OIL FURNACE BACKUP. CODING SYSTEM: NONE.
JONES HOUSE	NEAR STILLWATER, WISCONSIN	1977	SNOWY AND COLD IN WINTER. WARM AND RAINY IN SUMMER.	CARMODY AND ELLISON. DERRICK JONES AND JOHN CARMODY	1,500 SQ. FT. PLUS 800 SQ. FT. ENTRY/GARAGE. EARTH COVER: 12 INCHES. STRUCTURE: CAST-IN-PLACE REINFORCED CONCRETE WALLS, PRECAST CONCRETE PLANK ROOF, WOOD JOIST FLOOR. INSULATION: ROOF - 8 INCHES STYROFOAM, WALLS - 4 INCHES STYROFOAM. WATERPROOFING: BENTONIZE. HEATING SYSTEM: ELECTRIC FORCED AIR, CRAWL SPACE USED AS PLENUM, AND WOOD STOVE.
KARSKY HOUSE	ST. CROIX FALLS, WISCONSIN	1979	SNOWY AND COLD IN WINTER. WARM AND RAINY IN SUMMER.	MR. AND MRS. DONALD KARSKY	2,000 SQ. FT. STRUCTURE: SILO BLOCK WALLS, CAST-IN-PLACE CONCRETE ROOF AND FLOORS. SELF-BUILT HOUSE.
TOPIC HOUSE	SHAKOPEE, MINNESOTA	1977	SNOWY AND COLD IN WINTER. WARM AND RAINY IN SUMMER.	JOE TOPIC	2,400 SQ. FT. EARTH COVER: 16 INCHES. STRUCTURE: PRECAST-CONCRETE WALLS, PRECAST-CONCRETE PLANK ROOF. INSULATION: ROOF - 6 INCHES STYROFOAM, WALLS - 4-1/2 INCHES STYROFOAM. WATERPROOFING: BENTONIZE - ROOF AND WALLS. HEATING SYSTEM: HEAT PUMP.
DAVIS CAVE	ARMINGTON, ILLINOIS	1975	SNOWY AND COLD IN WINTER. WARM AND RAINY IN SUMMER.	ANDY DAVIS AND FAMILY	1,200 SQ. FT. PLUS 800 SQ. FT. EARTH COVER: 3 FT. TO 4 FT. STRUCTURE: ROOF, WALL AND FLOOR - CAST-IN-PLACE, REINFORCED CONCRETE. INSULATION: ROOF - 1 INCH STYROFOAM, WALLS - 1 INCH STYROFOAM EXTENDING DOWN 8 FT. FROM SURFACE. WATERPROOFING: WALLS AND ROOF - ASPHALT COATING. HEATING SYSTEM: FIREPLACE. COOLING: NONE REQUIRED.
SOLARIA	VINCENTOWN, NEW JERSEY	1975	SNOWY AND COLD IN WINTER. WARM AND RAINY IN SUMMER.	MALCOLM WELLS	2,800 SQ. FT. EARTH COVER: 24 INCHES (12 INCHES SUBSOIL, 8 INCHES MULCH, 4 INCHES TOP SOIL). STRUCTURE: CONCRETE BLOCK WALLS, TIMBER AND WOOD DECK ROOF. INSULATION: 2-1/2 INCHES STYROFOAM ROOF, WALLS AND FLOOR EDGE. WATERPROOFING: ROOF- 1/8 INCH BUTYL RUBBER, WALLS - FOUNDATION COATING ON PARGETTED CONCRETE BLOCK. HEATING SYSTEM: SOLAR COLLECTORS (THOMPSON TYPE) WITH OIL BACKUP. COOLING SYSTEM: CONVENTIONAL A/C WITH ROCK STORAGE.
BORDIE RESIDENCE	AUSTIN, TEXAS	1975	ARID: DRY AND WARM	COFFEE AND CRIER. ENGINEER: GEORGE MAXWELL	2,000 SQ. FT. EARTH COVER 14 INCHES TO 18 INCHES. STRUCTURE 8 INCHES REINFORCED CONCRETE WALLS, PRECAST CONCRETE ROOF. INSULATION: ROOF - 3 INCHES URETHANE FOAM. WATERPROOFING: ROOF - SHEET MEMBRANE, WALLS - MASTIC AND FIBERGLAS, MOPPED ON, UNDER FLOOR - PLASTIC SHEET. HEATING SYSTEM: SOLAR HOT WATER WITH ELECTRIC FURNACE FORCED AIR BACKUP. COOLING SYSTEM: DIRECT EXPANSION SPLIT SYSTEM, ELECTRIC POWERED, 2 TON. ENERGY USE: ABOUT 50% OF TYPICAL CONSTRUCTION.
DUNE HOUSE	ATLANTIC BEACH, FLORIDA	1974	WARM AND HUMID	WILLIAM MORGAN AND WILLIAM MORRIS. ENGINEER: GEIGER BERGER ASSOC.	750 SQ. FT. EACH UNIT (1500 TOTAL). EARTH COVER: 22 INCHES MINIMUM. STRUCTURE: REINFORCED GUNITE SHELL. INSULATION: NONE REQUIRED. WATERPROOFING: LIQUID BITUMINOUS, BRUSHED ON. HEATING AND COOLING: 1-1/2 TON WATER COOLED REVERSE CYCLE HEAT PUMP.
SUNDOWN HOUSE	SEA RANCH, CALIFORNIA	1976	WARM AND HUMID	DAVID WRIGHT, ENVIRONMENTAL ARCHITECT	1200 SQ. FT. PLUS 400 SQ. FT. EARTH COVER: 6 INCHES WITH NATIVE GRASS. STRUCTURE: REINFORCED CONCRETE BLOCK WALLS, WOOD RAFTERS AND PLYWOOD ROOF BRICK ON SAND FLOOR. INSULATION: ROOF 2 INCHES STYROFOAM, WALLS - 2 INCHES STYROFOAM, FLOOR - 1 INCH STYROFOAM. HEATING SYSTEM: 95% PASSIVE SOLAR. COOLING SYSTEM: NATURALLY INDUCED VENTILATION. ENERGY USE: 1% to 5% OF NORMAL.
ALEXANDER HOUSE	MONTECITO, CALIFORNIA	1974	SEMIARID WARM AND DRY IN SUMMER, COOL AND RAINY IN WINTER	RONALD COATE	7,000 SQ. FT. EARTH COVER: VARIES. STRUCTURE: REINFORCED CAST-IN-PLACE CONCRETE FOR WALLS, ROOF, FLOOR. HEATING SYSTEM: GAS; FIRED, FORCED AIR. COOLING SYSTEM: INSTALLED BUT NOT NEEDED.
HILL TOP HOUSE	CENTRAL FLORIDA	1975	WARM AND HUMID	WILLIAM MORGAN ARCHITECTS	3,300 SQ. FT. IT HAS PANORAMIC VIEWS OF THE CITRUS GROVES IN THE ROLLING TERRAIN. BUILT AT THE TOP OF THE HILL.
DAVIDSBURG SUBDIVISION	DAVIDSBURG, MICHIGAN	1979	SNOWY AND COLD IN WINTER. RAINY AND WARM IN SUMMER.	LOUIS A. DI GERONIMO	4,700 SQ. FT. COST PER SQ. FT. $25.00. IT IS A SUBDIVISION OF 160 ACRES OF LAND IN THE COMMUNITY. 92 UNITS PLANNED FOR THE SITE. THE DEVELOPER WANTED TO DEVELOP THE LAND WITH A MINIMUM DISTURBANCE TO THE NATURAL CHARACTERISTICS OF THE SITE, AND TO SAVE ENERGY CONSUMPTION UP TO 50%.
ROUSELLOT RESIDENCE	TAOS, NEW MEXICO	1971	SEMIARID: WARM AND DRY	ARCHITECT: RON MC CLURE. ENG: BUCK ROGERS	3,500 SQ.FT. MATERIALS: STUCCO ON CONCRETE BLOCKS WITH STEEL FRAME ROOF. SOIL: CALICHE CLAY. WATERPROOFING: BITUMINOUS ON PARGETTING. EARTH COVER: 12"-16." ENERGY SAVINGS 20-25%. HEATING: UNDERFLOOR PLENUM. NO COOLING NEEDED.
SOLAR/EARTH SHELTERED HOUSE	WILD RIVER STATE PARK, CHISAGO COUNTY, MINNESOTA	1979/80	SNOWY AND COLD IN WINTER. WARM AND RAINY IN SUMMER.	MC GUIRE/ENGLER/DAVIS ARCHITECTS. ENG: BREDOW ASSOCIATES	1,950 SQ. FT. PLUS DECK AND GARAGE. REINFORCED CONCRETE BLOCK AND WOOD FRAME. EXPOSED AGGREGATE FLOORS. SLOPING ROOF WITH 18" EARTH COVER. HEATING: PASSIVE SOLAR SYSTEM SUPPLEMENTED BY WOOD/FIRED FORCED AIR HEATING SYSTEM. COOLING: NONE. THE HOUSE OPENINGS ARE FACING 15° EAST OF SOUTH. ALL OTHER SIDES COVERED WITH EARTH. SOIL IS SAND-GRAVEL MIX. THE UNIVERSITY OF MINNESOTA UNDERGROUND SPACE CENTER HAS INSTALLED AN ENERGY MONITORING SYSTEM TO MONITOR ALL ENERGY USED IN THE HOUSE.

combined total of $13 million annually, have been working in nine sites of warehouses, industries and offices some 50 to 100 feet underground. The ultimate goal is to develop an underground industrial park with cooperation between the quarry operators and the developers. The total subsurface area is 45 million square feet. One seventh of Kansas City's warehouses are already underground. There is, however, more than a 34-million-cubic-foot capacity for frozen food storage underground, a capacity which makes up one tenth of the entire United States capacity. In addition to the vast space that is available, the floor has high bearing capacity for industries. The reduced fluctuations in temperature and humidity are ideal for some industries such as printing shops and producers of lacquers and glues. Stauffer also lists some of the advantages of subterranean land uses: low cost for purchase or lease, reduced investment in roofs and foundations, high capacity of the floor to support heavy loads for industry and storage, the possibility of noise and vibration control, fireproof space which leads to low insurance rates, low cost in heating, airconditioning or freezing, and a high degree of security for equipment.[29]

Missouri is the leading state to make secondary use of subsurface areas for storage. Stauffer concludes that the value of secondary usage of space is greater than the original use of limestone.[30]

Subterranean space can be efficiently used for storage of documents or other materials which require stable humidity. For example, one of the world's largest microfilm collections is located in space specially mined for this purpose near Salt Lake City, Utah, in the Wasatch mountain range. The controlled environment is ideal for housing a million rolls of microfilm records of important genealogical data from all over the world.[31]

One of the most technologically impressive subterranean spaces of this century is the large combat operations center of the North American Air Defense Command (NORAD) situated at Cheyenne Mountain, southwest of Colorado Springs, which includes a three-story-high space of 200,000 square feet excavated in granite.[32] NORAD was begun in the early 1960s to provide warning of a ballistic missile attack on North America. NORAD's inside temperature is a constant 58°F throughout the year, important because of the need for air-cooling systems in order to dissipate heat from rooms occupied by heat-generating computers. Further, the inside air is extremely clean.[33]

Another important subterranean project built in recent years is the University of Minnesota Bookstore and Admissions Office in Minneapolis. Lack of space and a need to preserve historically valuable buildings nearby, led the University of Minnesota to build a subterranean bookstore combining a successful design with environmental considerations. Both the bookstore and the admissions and records office have different needs and characteristics. A wall of glass at a 45° angle introduces a view and sunlight. To allow sunlight penetration in winter and reduce it in the summer, louvers holding plants are used to block the intense rays at the peak hour. The sun's angle is at a maximum of 68° in summer; in the winters the sun is at its minimum angle of 22° and penetrates the space easily. The building is pleasant physically and visually. One aspect of its attractiveness is that the roofs are used as gardens (Fig- I-20).

Many other universities in the United States have also built their libraries or bookstores below the ground, primarily to preserve open space or adja-

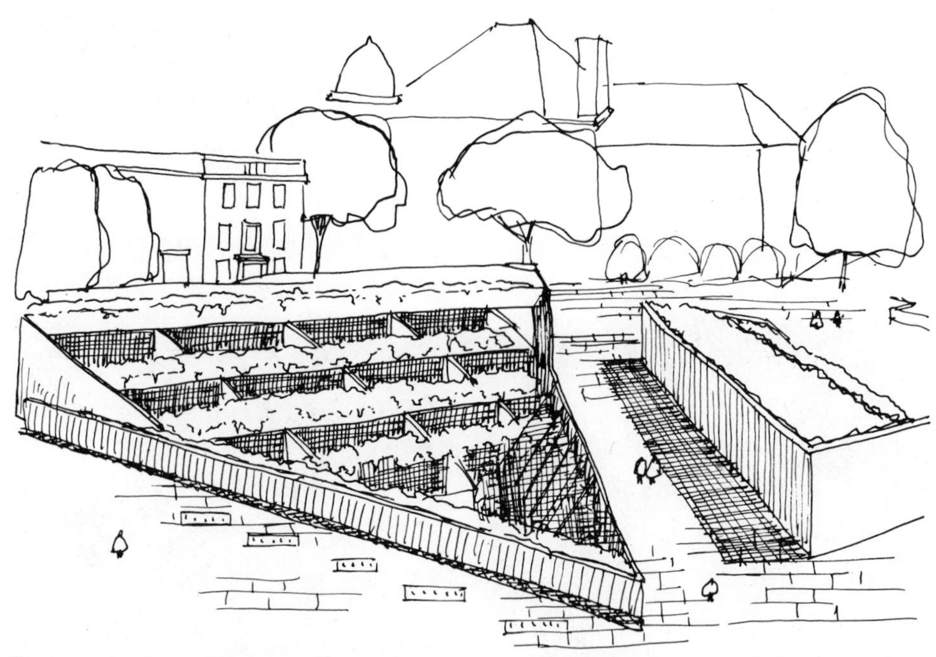

Fig. I-20. The University of Minnesota Bookstore and Admissions Office, built underground in Minneapolis. The research effort behind its planning and design have made it a landmark in the new earth-covered movement in the United States. This subterranean building is a result of the need to preserve the open space and the adjacent historic buildings and to provide proximity to other land uses.

cent historical buildings as in Minnesota, to retain proximity to other buildings, or to expand existing libraries (Table I-2). The university at Urbana, Illinois, built a two-story subterraean building for the undergraduate library (Fig- I-25), similar to the Pusey Library at Harvard and the bookstore at Cornell University (Table I-2).

Oil storage is another important underground use. In the mid-1970s, approximately 150 million barrels of liquid petroleum and nearly 5½ billion cubic feet of natural gas were stored underground[34] in the United States. See Figs. I-21–24 for other contemporary underground houses.

SCANDINAVIA: STORAGE SPACES

Sweden is considered to have the largest subterranean space in Europe, including all types of uses: oil storage, subways, garages, shopping centers, restaurants, power plants, military installations and factories. Two billion dollars were invested in subterranean space as of 1967.

The subterranean facilities for military and civilian purposes are very comprehensive. The country spends $28 million a year for its sophisticated underground shelter system. In addition, Swedes have developed a giant naval base inside the rock of Muskö, an island 25 miles south of Stockholm.[35] Other shelters can accommodate more than 20,000 people.

All the underground space in Stockholm is cut into granite or gneiss, usually by means of drill and blast systems. The subway in Stockholm is very extensive, and it is reported that it cost several million dollars per kilometer. Further, an underground garage is used for Stockholm garbage trucks. One of the main motivations of the Swedish people for underground space is defense. For this reason, they have developed a large number of

Fig. I-21. Earth-covered house, Rousselot Residence, Taos, New Mexico, designed by Architect Ron McClure. Buck Rogers, Engineer, 1970. Materials: Stucco on concrete block with steel frame roof. Courtesy of Architect Ron McClure.

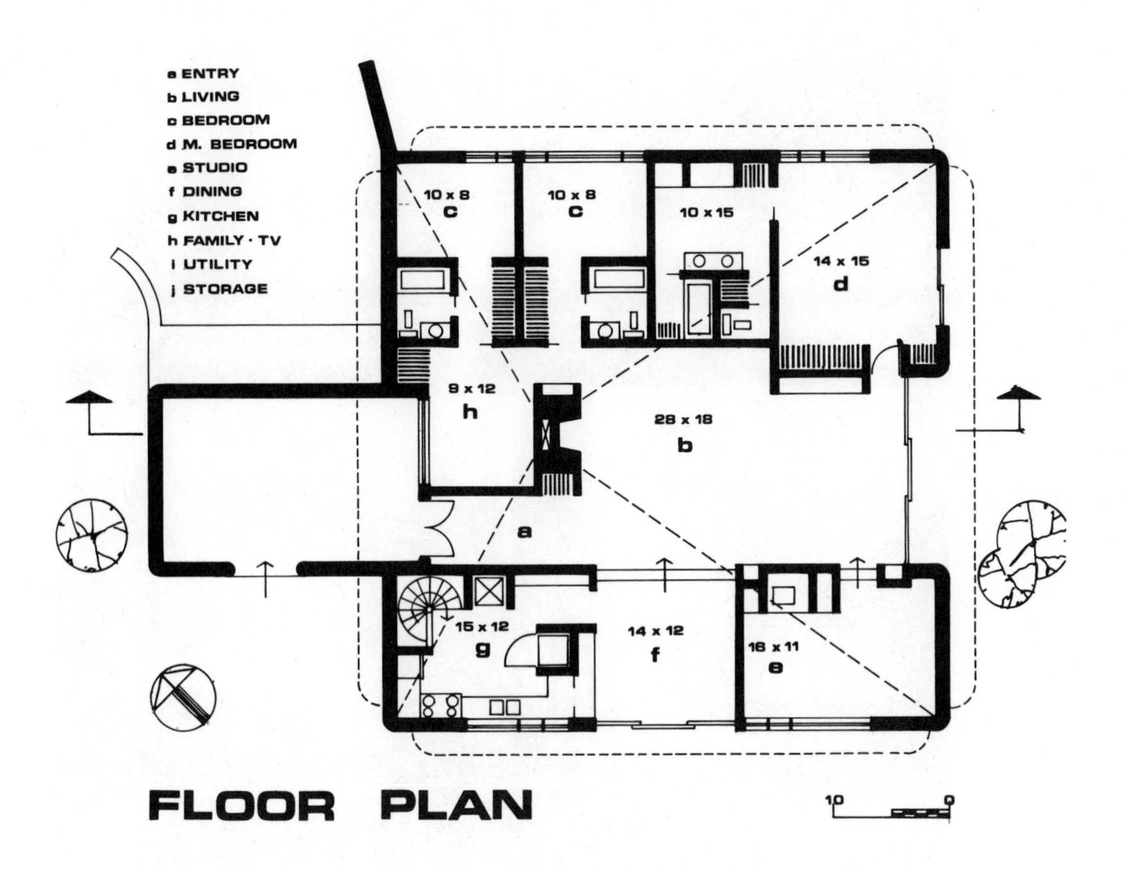

FLOOR PLAN

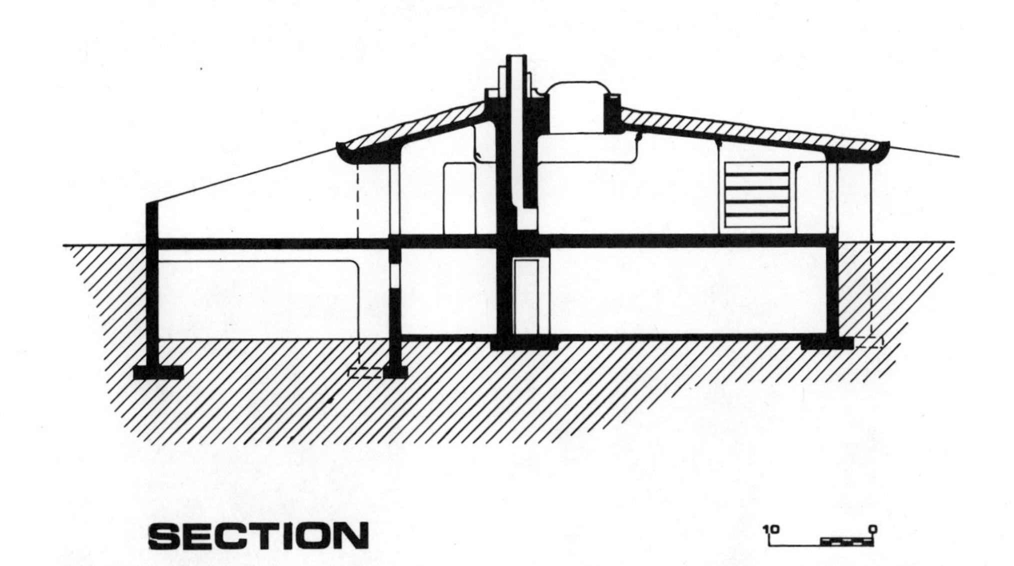

SECTION

Fig. I-22. The Southern Indiana House, designed by Architect John E. Barnard and constructed recently, a combination of an earth-covered atrium house built with an elevational home. When outside temperature was below 0°F, the atrium temperatures never went below 40°F. Courtesy of John E. Barnard, Architect.

Fig. I-23. Solar/earth-sheltered house in Wild River State Park, Chisago County, Minnesota. Designed by McGuire/Engler/Davis Architects in 1978 and constructed in 1979–80 as Park Manager's Residence. Courtesy of McGuire/Engler/Davis Architects.

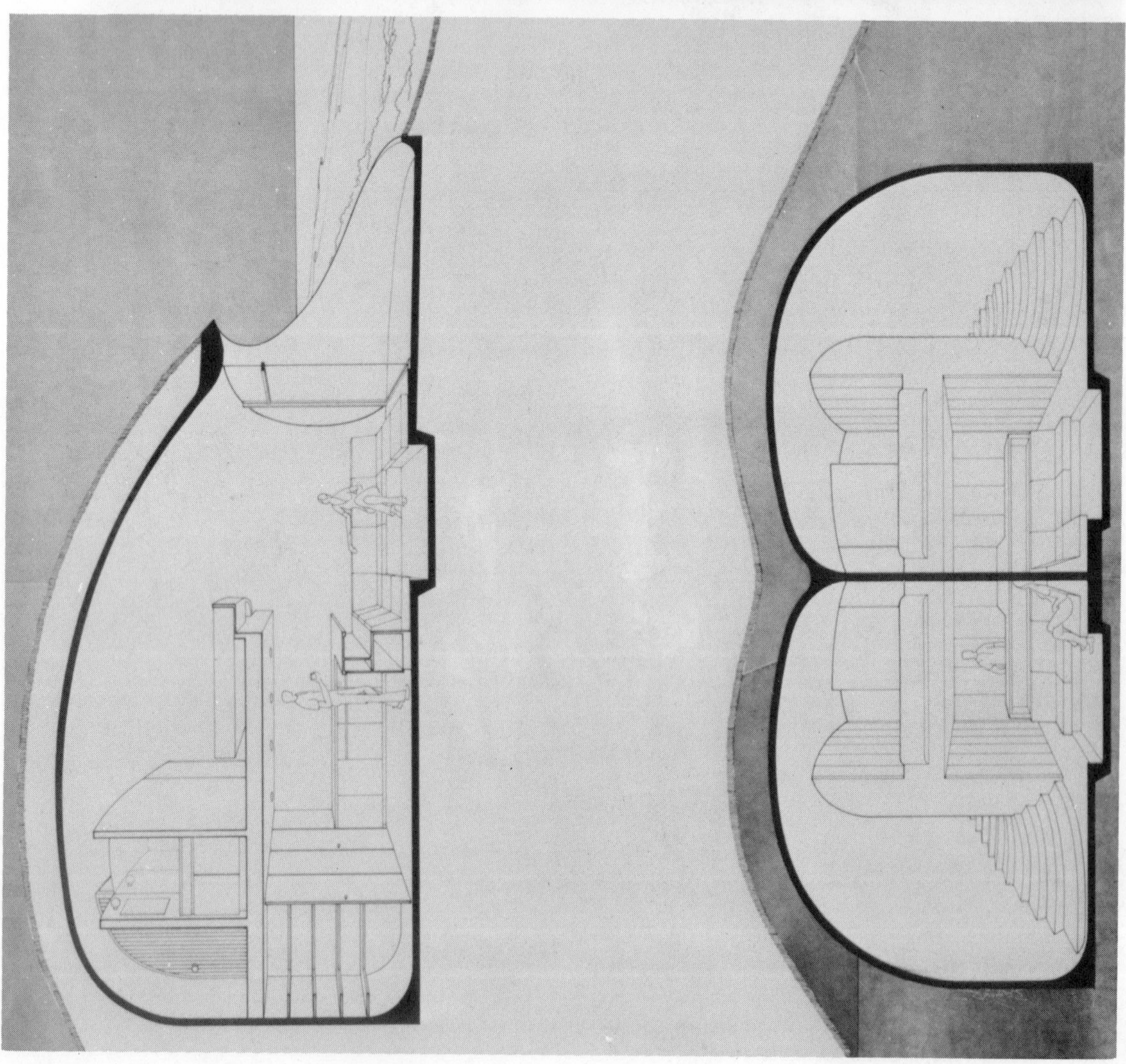

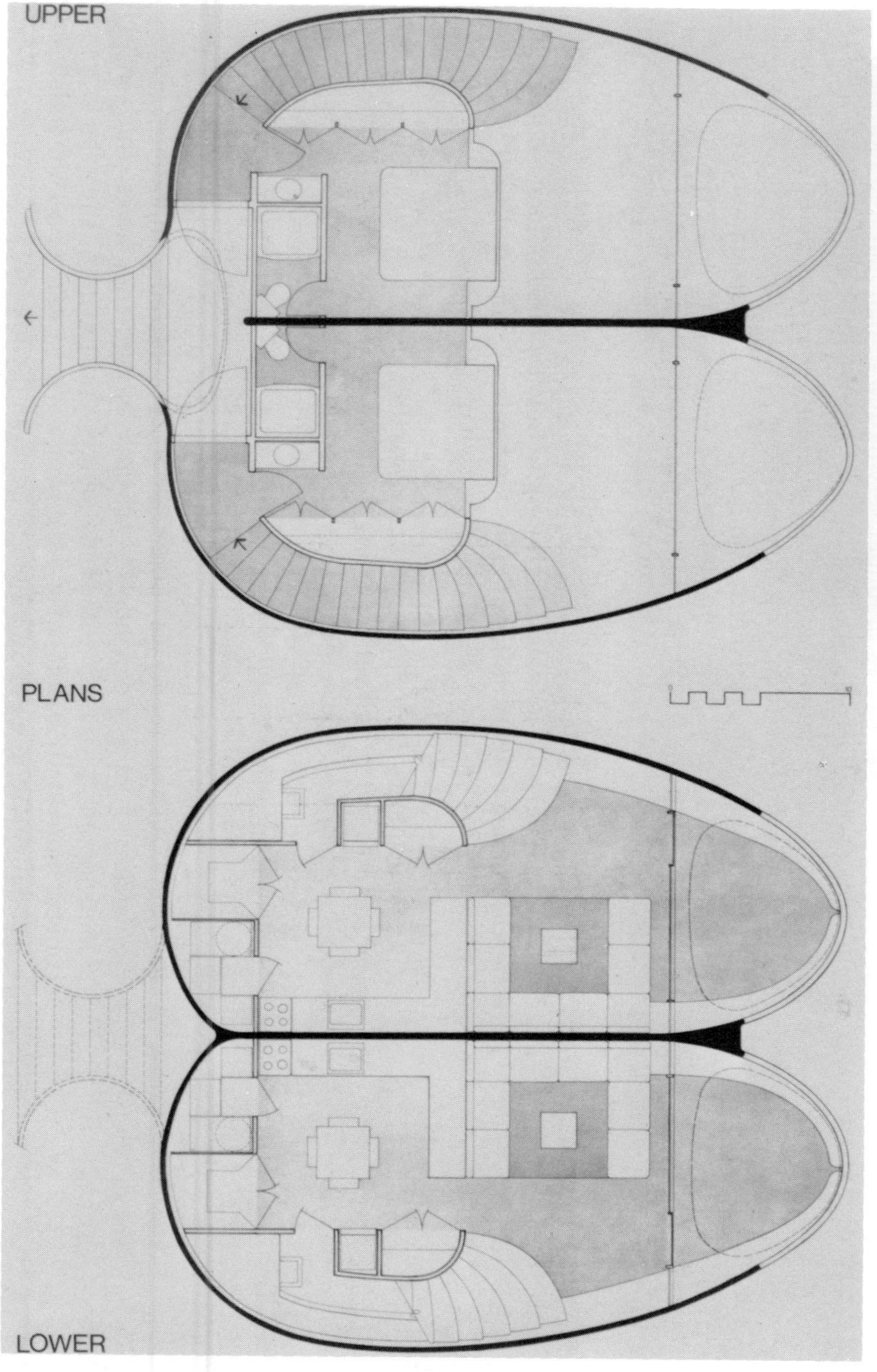

Fig. I-24. Atlantic Beach Dunehouses, Florida, designed by Architect William Morgan in 1974 and constructed in 1975. The two apartments are set into the existing sand dunes to preserve the original site's environment for neighboring structures built above grade to the north and south. Courtesy of William Morgan Architects.

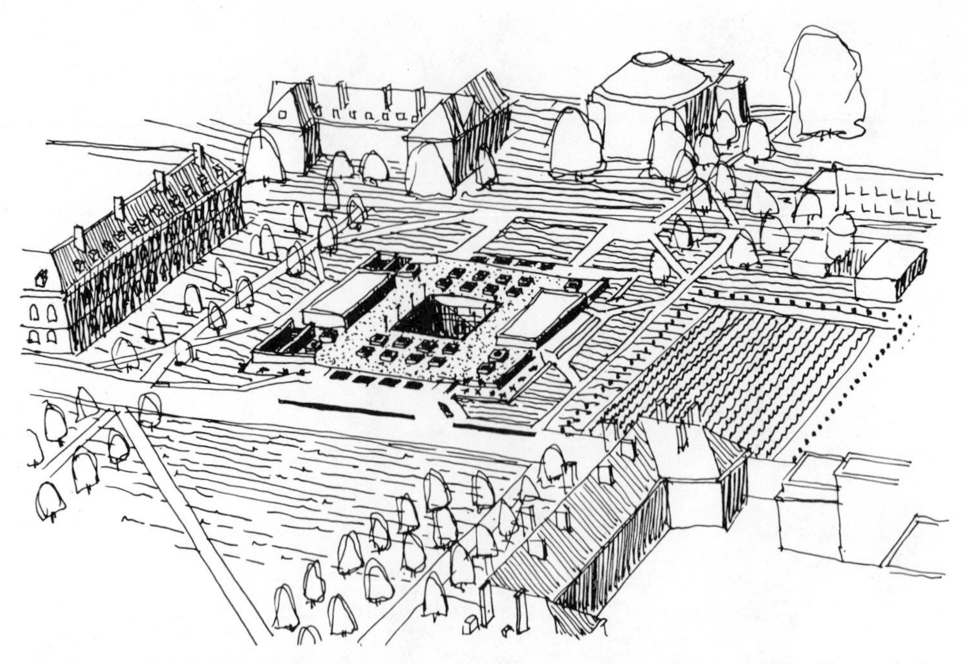

Fig. I-25. Undergraduate Library built belowground at the University of Illinois, Urbana. The expansion of the old library was necessary, but the area was already congested. The new two-story library preserved the open space and provided proximity to other uses.

underground spaces for use in case of war. Also, they have built a repair shop in an underground garage and a sewage treatment plant, Kappala, which consists of 37 miles of tunnel serving more than half a million persons in ten suburban communities. The organizers have obtained the right by law to dig underground without the necessity of getting permission from the people who own homes on the surface.[36]

In Helsinki, Finland, architects Timo and Tuomo Suomalainen designed an underground church in a neighborhood where space was short (Fig. I-26). The Taivallahti church is well integrated with its environs.

JAPAN AND CANADA: SHOPPING CENTERS

Japan has become a leading country in the development of large subterranean shopping centers. In Japan there are twelve "Shopping Cities" called Chikagai. One of those shopping centers has 225 shops which 800,000 shoppers visit every day.

Construction of the shopping center in Montreal began in 1956 using the Canadian National Railway's existing excavation, and it consists of offices, shops, restaurants, malls, theaters and other services.

Underground Montreal, especially Place Ville Marie, is one of the largest belowground urban spaces in the world. As a result of Expo 67 and construction of an artificial island in the St. Lawrence River, the underground metro was constructed in the space left by the dirt which was taken to build the island. Another metro was built later and connected with the first, and the whole system was expanded to six underground complexes which include shopping facilities, restaurants, offices and theaters. The plan is to connect more than 50 percent of the downtown area so as to accommodate a half-million people at one time in a controlled climate.

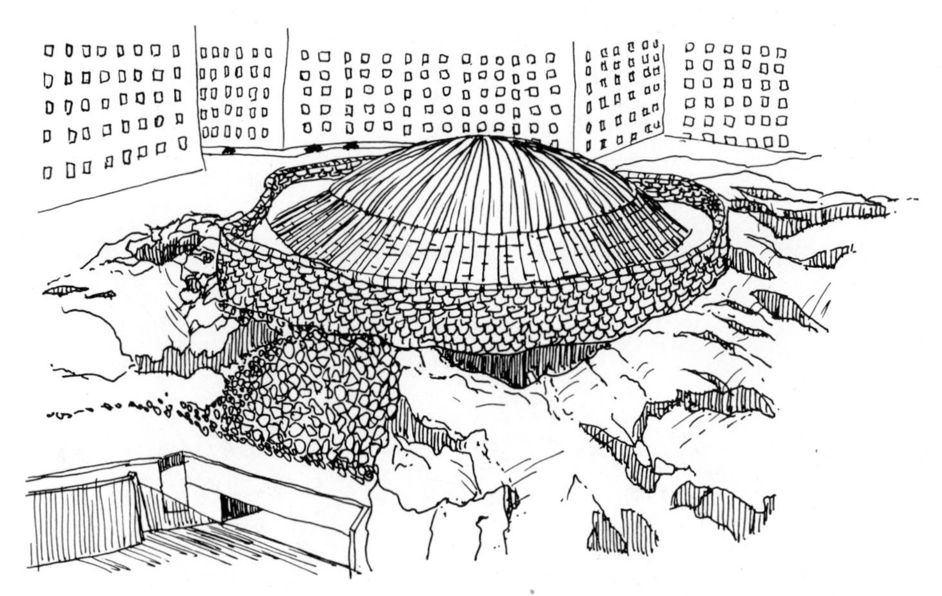

Fig. I-26. Taivallahti Temppeliaukio, Helsinki, Finland, built belowground to retain the open space of the neighborhood. Adapted from personal picture.

Designed by I. M. Pei, Place Ville Marie is a whole new multilevel underground transportation and pedestrian center with stores, restaurants and miles of pedestrian and subway networks. The square on the surface is enclosed by a tower and small buildings bordering the subterranean complex. Noisy traffic and harsh weather aboveground do not affect the underground complex. This subterranean multiuse center will be connected with six subway stops, 9000 parking spaces, eight skyscrapers, three department stores, two railroad stations, four luxury hotels, eight theaters, forty first-class restaurants and scores of shops. In Winnipeg, developers and planners have successfully joined an underground shopping mall with a belowground pedestrian crossing at a busy intersection complete with additional stores and services.

AUSTRALIA: DWELLINGS FOR DESERT CLIMATE

According to Sydney Baggs, the only English-speaking underground communities in the world are in Australia. These are Coober Pedy and White Cliffs (Fig. I-27). Coober Pedy, with a population of 1903 (1976 census), of whom only 125 live belowground, is located in an arid zone at the edge of the Simpson Desert. The White Cliffs population is 3000. In Coober Pedy, underground living has existed for more than 60 years, and in White Cliffs for more than 85 years.[37] The White Cliffs underground dwellings are made in easily excavated sandstone and claystone sediments.

Coober Pedy is a mining town 960 kilometers north of Adelaide, and 200 kilometers north of Tarcoola. Rain is rare, with a mean of 141 millimeters per year which cannot support farming; desalinated water is used for drinking. Temperatures range between 0°C and 47°C. Digging for the subterranean houses through horizontal layers of sandstone is accomplished with explosives, and walls are finished later. Also in use are air compressors or rock drills. Tunneling machines are heavy for maneuvering, and, thus, they

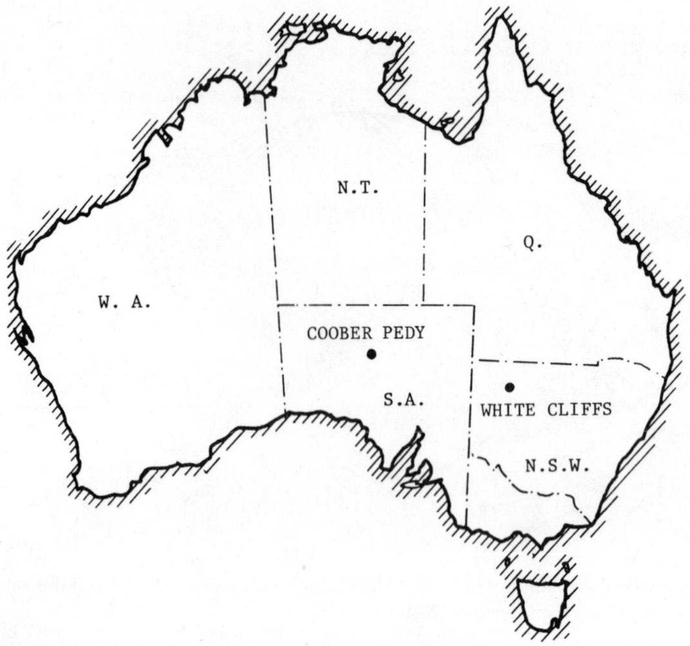

Fig. I-27. Existing underground communities in Australia are Coober Pedy in Southern Australia and White Cliffs in New South Wales. They are both mining towns and are located in arid, warm, dry regions.

are expensive. When digging is finished, three layers of epoxy-based varnish is plastered around the walls to hold back the water and to avoid flaking of the rock. Floors are made of concrete poured on plastic sheets because of the extreme dryness and porosity of the sandstone base. Warm air enters through vents, is cooled and rolls out through the doors providing circulation. Air temperature remains constant, around 21°C.[38] (Fig. I-28).

RUSSIA: MULTIUSE OF UNDERGROUND SPACE

It has been reported that the Russians have a large-scale development underground in Moscow where eighteen Moscow institutes contributed to a plan for multipurpose uses in underground space including stores, theaters, exhibit areas and a network of tunnels and parking garages. The goal of this project is to create new space underground in order to free the space aboveground for parks and recreational and sport activities. The plan should take 15 to 20 years to reach completion, and it is integrated with the city's master plan. According to the architects, the plan will save 18,000 acres of surface area and will include 447 miles of streets 100 yards wide. The plan also suggests the expansion of the Moscow subway from 90 miles to 198 miles in length. Currently, 5 million passengers use the subway daily.[39] There also have been many conjectures that Russia has moved extensively into the underground for military installations, protection of critical industries and civil defense.

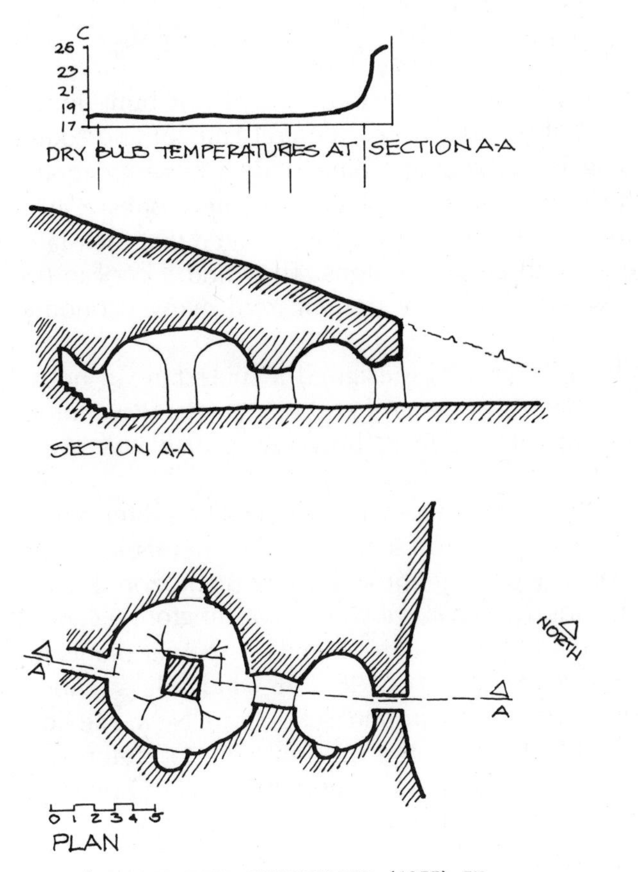

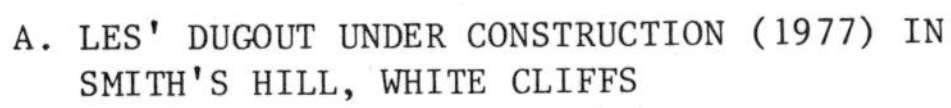
A. LES' DUGOUT UNDER CONSTRUCTION (1977) IN SMITH'S HILL, WHITE CLIFFS

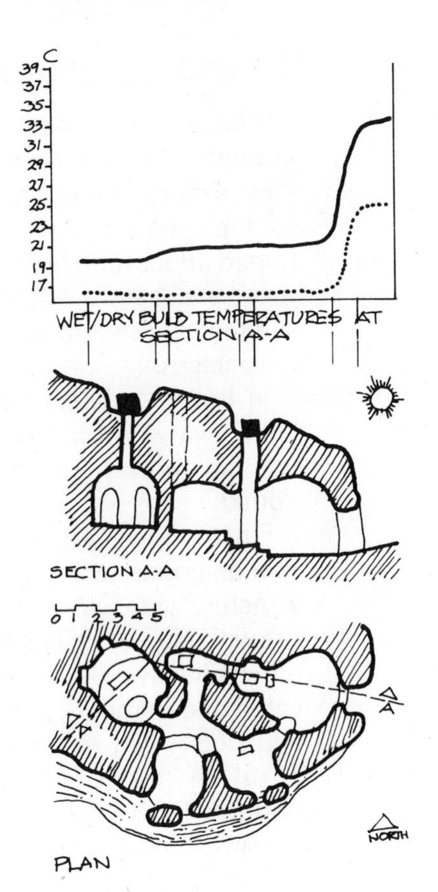

B. WARREN'S DUGOUT, THE BLOCKS, WHITE CLIFFS

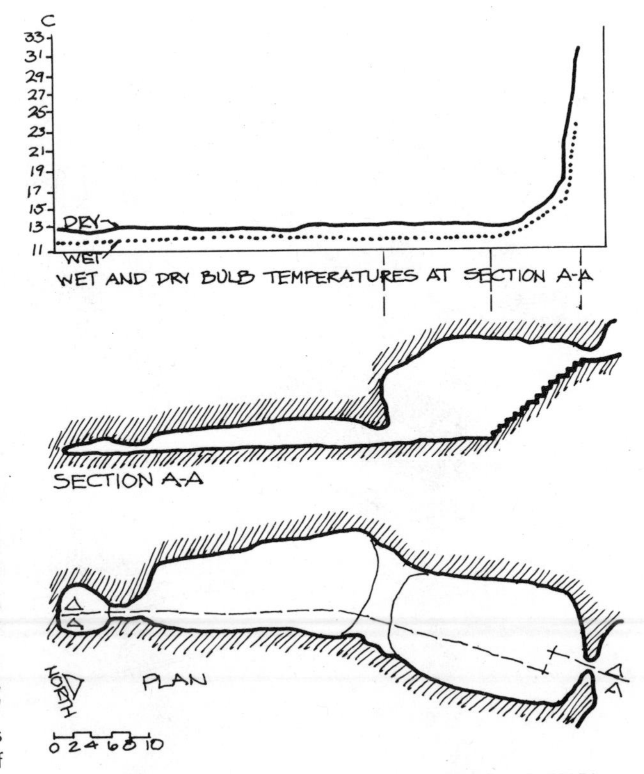

C. A SACK FORM OF STATIC CAVE, MT. ECCLES, VICTORIA

Fig. I-28. Dugout in Australia. Note the sharp drop in temperature in the three cases immediately after entering the subterranean space, especially in case C. In case B, according to Baggs ". . . the recessed shaft orifices induce increased viscous sucking. The vent shafts have the effect of raising the dry bulb temperature and dropping the vapor pressure immediately at the entrance to the shaft." Adapted from Sydney A. Baggs, "The Dugout Dwellings of an Outback Opal Mining Town in Australia," in *Underground Utilization: A Reference Manual of Selected Works.* 8 Vols. Truman Stauffer, ed. Volume IV, *Human Response and Social Acceptance of Underground Space.* Kansas City: Department of Geoscience, University of Missouri, 1978, pp. 582, 584 and 586.

EXAMPLES FROM OTHER COUNTRIES

In France, an estimated 2000 people live in housing which is built underground. The main regions are the Loire Valley (central France) and southwest France. Other housing is located at Château de Cheverny in the southern part of Paris. These are historical places, not new ones. Questioned about such living, the families from the area indicated that their families have been living there for three generations. They find it cool in the summer and warm in the winter, and they benefit from many economic advantages.[40]

In Israel, architect Avraham Tshareniak designed a subterranean house which was constructed in the 1950s in the arid port city of Eilat. Psychological bias prevented people from living in it, however, and it was abandoned.

Finally, there are many other contemporary examples of subterranean space, among which are the transportation tunnel, 12 kilometers long and 7 meters wide, which allows easy passage through Mont Blanc from France to Switzerland; the London, Paris, Moscow, Montreal, Washington, D.C., and other subway systems; more than 50 percent of Sweden's power plants located underground; in the United States as of 1967, 96 windowless schools had been built in 23 states, with underground schools in New Mexico, Texas, Maryland, Virginia and Oklahoma; the Maginot Line for defense in northern France developed before World War II, and the UNESCO building in Paris, built around six sunken courts (Fig. I-29).

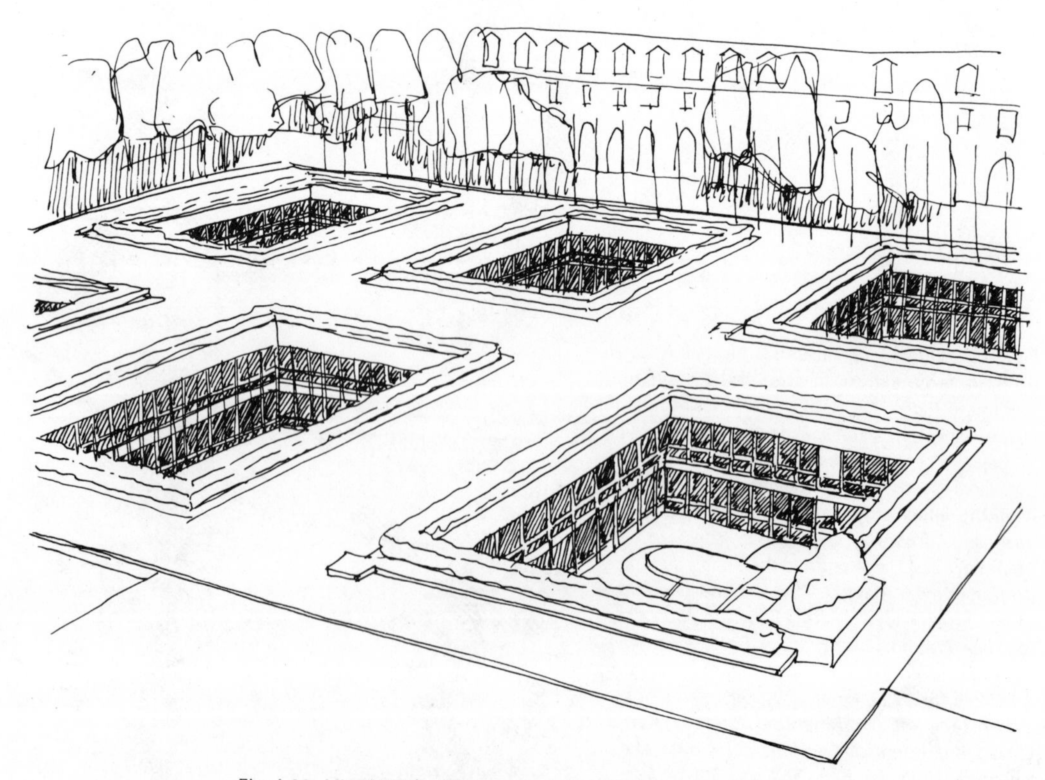

Fig. I-29. UNESCO building in Paris built below ground level.

In conclusion, subterranean space has been used for diversified purposes throughout history. Today, with the availability of high technology, subterranean space is used for military installations, defense shelters, housing, storage, refrigeration, industry and manfacturing, liquid storage (especially fuel), garages, offices, transportation and exploitation of natural resources. Gasoline stations, for example, use subterranean space for fuel storage to minimize hazards, especially in urban areas.

Underground Development: An Evaluation

The prime limitation of underground living is psychological bias. The image of windowless subterranean living, obviously a negative one, has been associated with cave societies and primitive cultures. Indeed, living underground has been, especially in our time, associated with poverty, as in North Africa, Iran and Turkey, and related to low socioeconomic status and isolation. Bechtel, for example, noticed that 2 percent of the Iranian people still living in caves are from the poorest economic level.[41] On the other hand, in the developed countries such as Scandinavia, the United States or Japan, middle-income people are seeking underground shelters as a means of saving energy. The feeling of isolation which results from the lack of windows and the unnatural absence of noise are a psychological condition which can be overcome by technology. Bechtel concluded that the problem of negative images will be solved through the passage of time, especially when more underground houses are built.[42]

Most people have objections to living underground. In addition to the low status image previously mentioned, such housing is usually associated with darkness, dampness, bad construction and design, graves, demons, death and unhealthy conditions. In any case, this bias and the misunderstanding of the potentialities offered by properly designed underground living necessitate a wide dissemination of knowledge about this renewed endeavor.

One possible benefit of underground housing is operation cost, which can be divided into two parts: (1) maintenance, and (2) heating and cooling. Maintenance cost underground is less than for the conventional supraterranean building for several reasons. Painting is limited to only those parts of the house which are not covered by soil. There are few or no windows to be maintained, and there is no need for new roofs every 10 to 15 years. Finally, damage by wind, hail, rain, snow or other natural causes is not a consideration.

As for heating and cooling, it has already been found that most underground houses save more than half to two thirds of conventional house expenses for the same purposes. Possibly the savings could go even higher if house design and insulation are improved still further, and if the house is operating on passive energy-conserving methods as described in Section II of this book.

In addition, the underground house is safer than its aboveground counterpart. For example, as already mentioned storm effect is minimal or absent. The underground house is likewise fireproof with minimal potential for a fire to spread to neighboring houses. Accordingly, insurance may be expected to be lower than for the conventional house. The underground house is safe against earthquake if properly designed and located, safe against cyclones or tornadoes, and also safe against radioactive fallout (if planned to meet this requirement). In very cold climates underground houses do not have the problems of water pipes freezing and cracking.

The underground house saves land, too, making aboveground space useful for different purposes such as playgrounds, recreation or even for other buildings. Such a house provides a very quiet environment, much quieter than that aboveground. Thus, we can build underground houses near airports and utilize the space which is usually left vacant between the residential area and the airport for housing. Also, the underground house provides privacy for the people who live in it. For some people, who are engaged in certain types of work such as art, painting, music, writing and sculpture and who may desire a very quiet isolated environment, the underground house would be an ideal space for their work.

Some scientists have found that new schools constructed without windows produced no negative results. Research findings also show that labor in underground structures results in no health or production problems.[43] Claustrophobia—a pathological fear of confined space—can develop among some people. Confined space can, on the other hand, stimulate creativity—among writers, for example.

In summary, then, there are many advantages to underground construction. For example, traffic noise and noise from neighbors, a problem in apartment units but also in unattached housing, are entirely eliminated in the underground house. Fire cannot spread easily to other houses, for winds are virtually nonexistent underground. Furthermore, an underground structure can save money. Figure I-30 compares heating/cooling and operating costs for an underground and aboveground manufacturing plant; Figure I-31 compares installation and operating costs for underground and aboveground dry storage and refrigeration facilities. Figure I-32 compares life-cycle costs for typical underground and aboveground structures, and finds

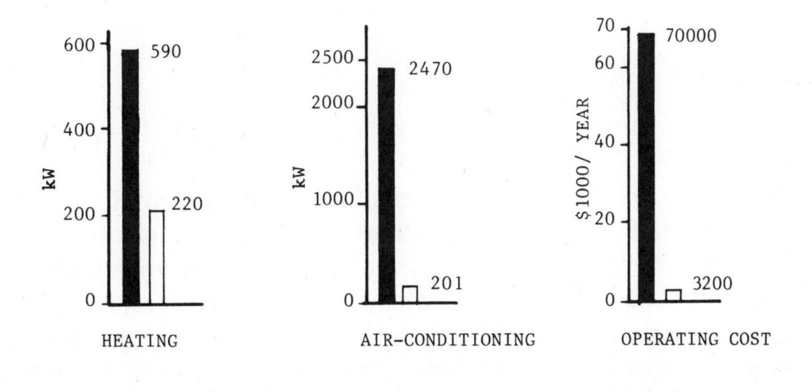

Fig. I-30. Comparison of heating/cooling and operating costs for an underground and an aboveground manufacturing plant. From Thomas Bligh, "A Comparison of Energy Consumption in Earth Covered vs. Non-Earth Covered Buildings," in *Alternatives in Energy Conservation: The Use of Earth Covered Buildings, Proceedings of a Conference,* Fort Worth, Texas, July 9–12, 1975. Washington, D.C.: National Science Foundation, 1975, p. 95.

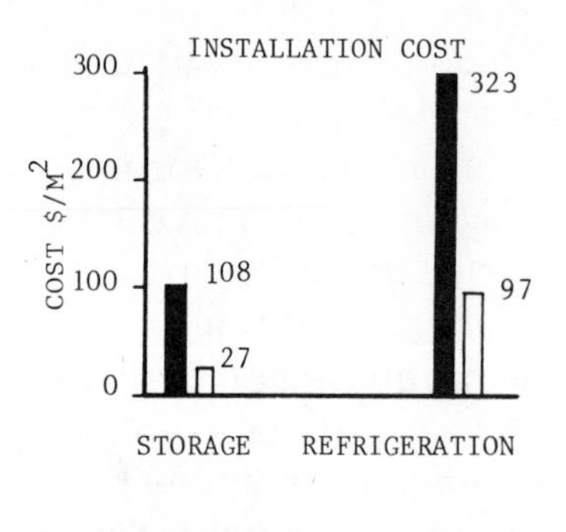

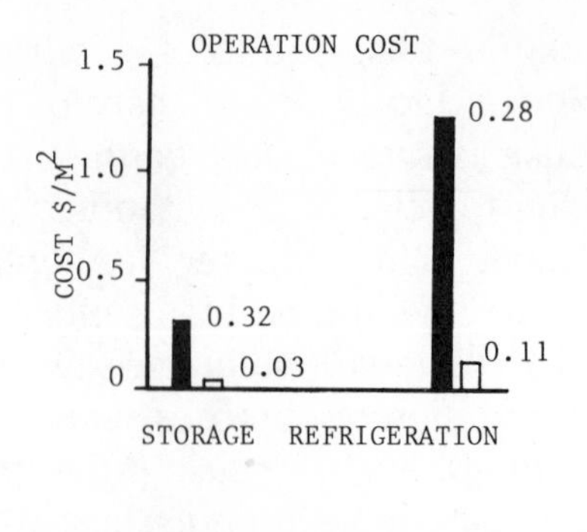

SUBSURFACE ABOVEGROUND

Fig. I-31. Comparison of installation and operating costs of underground and aboveground dry storage and refrigeration facilities. Source: From Bligh, "A Comparison of Energy Consumption in Earth Covered vs. Non-Earth Covered Buildings," p. 94.

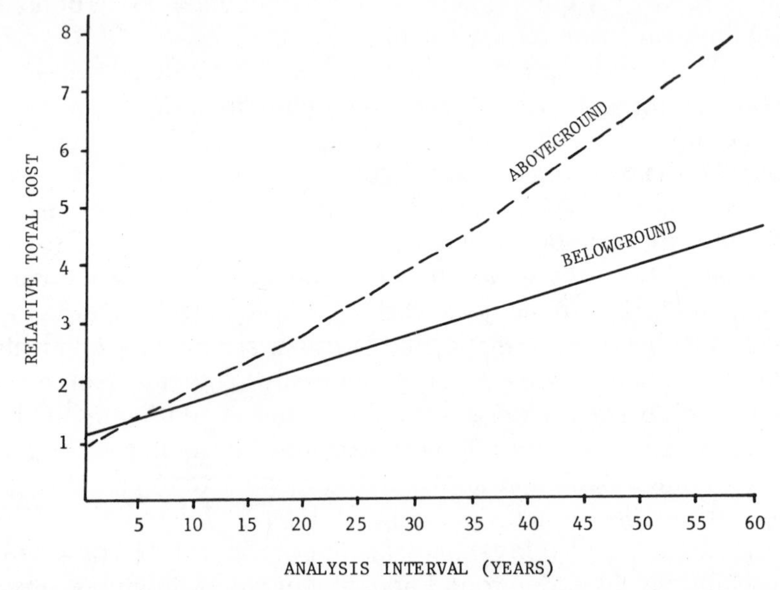

Fig. I-32. Comparison of life-cycle costs of typical underground and aboveground structures. From John E. Williams, "Comparative Life Cycle Costs," in *Alternatives in Energy Conservation: The Use of Earth Covered Buildings,* p. 59.

the cost of the former progressively lower throughout the life of the structures.

The excavation required for underground construction, on the other hand, may require a relatively larger investment in rocky ground than in more "soily" ground. However, excavated materials such as limestone can be sold. If these materials are not excavated and sold, excavation will increase the price of the structures to the point where they become uneconomical. Also, other geological conditions may make construction very expensive. A thorough geological study of the site is thus required before any excavation. Generally speaking, hillside development may be preferable to that in a flat area, for excavated dirt can then be used to reshape the hill.

Another issue is the water table and any water movement within the ground. In arid zones, however, planners will not be especially involved with such issues. Still, a detailed geological map including such matters is necessary. Table I-4 summarizes these and other issues affecting subterranean development, noting both the advantages and the problems.

Table I-4. Summary of Underground Construction

PROS	CONS
PROTECTION FROM THE CLIMATE	
*WEATHERPROOFING: WEATHERPROOFING THE STRUCTURE AGAINST EXTREMES SUCH AS SOLAR RADIATION, COLD OR WARM TEMPERATURES, WIND AND DUST STORMS, TORNADOES, ETC., IS BENEFICIAL. *TEMPERATURE FLUCTUATION: THE USE OF SUBTERRANEAN STRUCTURES MINIMIZES DAILY TEMPERATURE FLUCTUATION AND MAKES A STABLE TEMPERATURE POSSIBLE. *COMFORT: CREATES A COMFORTABLE CLIMATE THROUGHOUT THE YEAR. *MODERATE: UNDERGROUND DWELLINGS WILL HAVE A MODERATE MICROCLIMATE COMPARED WITH THE OUTSIDE MACROCLIMATE. *SURVIVAL: PROTECTION AND SURVIVAL WHEN ELECTRICITY BREAKDOWNS OCCUR IN AN EXTREME COLD WEATHER CONDITION. *STRESSFUL CLIMATE: MOST SUITABLE WITH STRESSFUL CLIMATES OF HOT OR COLD,DRY.	*DUST COVER: DURING DUST STORMS IN DESERTS, AN UNDERGROUND STRUCTURE CAN BE BURIED IN SAND IF SPECIAL DESIGN ARRANGEMENTS ARE NOT CONSIDERED.
ENERGY COSTS	
*ENERGY CONSUMPTION: THE USE OF UNDERGROUND STRUCTURES CAN CUT ENERGY CONSUMPTION THROUGHOUT THE DAY AND WINTER SEASON BY ELIMINATING AIR-CONDITIONING AND REDUCING HEATING NEEDS. THIS REDUCTION MAY BE 50%-80% OR MORE. *HEAT LOSS: HEAT GAIN AND LOSS IS MINIMAL OR NONE. ALSO, SUCH USE CAN CUT ELECTRICAL COSTS BECAUSE THERE IS NO HEAT LOSS THROUGH WINDOWS AND BECAUSE IT REDUCES FLUCTUATION OF DAILY TEMPERATURES. *REFRIGERATION: REQUIRES LESS ENERGY FOR REFRIGERATION.	*LIGHTING: UNDERGROUND STRUCTURES MAY REQUIRE MORE ENERGY FOR LIGHTING DURING BOTH DAY AND NIGHT AND FOR VENTILATION.
CONSTRUCTION COSTS	
*LAND COST: SAVINGS WILL RESULT FROM MINIMAL LAND COST BECAUSE OF THE INTENSIVE OR DUAL USE OF THE LAND SITE. *DESIGN COST: THE COSTS OF DESIGN WILL BE LOW BECAUSE OF THE SIMPLICITY. *BUILDING MATERIALS: LESS MATERIALS WILL BE USED FOR EXTERIOR FINISH, WINDOWS, OR LANDSCAPING. *FLOOR LOAD: FLOORS WILL BEAR A HIGHER LOAD THAN FLOORS IN ABOVEGROUND STRUCTURES.	*BLASTING: COSTS CAN BE HIGH WHEN THE BLASTING OF ROCK IS NECESSARY OR SUBSTANTIAL EXCAVATION IS REQUIRED. (THIS CAN BE ECONOMICAL WHEN LARGE-SCALE DEVELOPMENT JUSTIFIES THE SYSTEM ECONOMICALLY AND OPERATIONALLY.) *EXCAVATION: EXTENSIVE EXCAVATION MAY REQUIRE COSTLY RECLAMATION OF SURFACE LAND TO REFIT IT FOR AGRICULTURE OR OTHER USES. *MAPPING: DETAILED GEOLOGICAL MAPPING AND SOIL STUDIES WILL BE REQUIRED.
MAINTENANCE COSTS	
*MAINTENANCE: THERE MAY BE MINIMAL OR NO EXTERIOR MAINTENANCE COSTS, SUCH AS FOR PAINTING, REPAIRS, REMODELING, WINDOW OR ROOF REPLACEMENTS, ETC. *DURABILITY: LONG DURABILITY OF THE BUILDING MATERIALS ENVELOPING THE STRUCTURE BECAUSE OF THE MINIMIZED INFLUENCE OF WEATHER ON THE HOUSE'S MATERIALS. *HOUSEKEEPING: THERE WILL BE LESS HOUSEKEEPING SINCE A MINIMAL AMOUNT OF OUTSIDE DUST WILL ENTER THE HOUSE. *INSURANCE: FIRE INSURANCE RATES SHOULD BE LESS.	*UTILITY REPAIRS: AN UNDERGROUND STRUCTURE MAY REQUIRE HIGH EXPENDITURES FOR REPAIRS TO UTILITY SYSTEMS.
LAND USE	
*PRESERVATION: PRESERVES THE LAND AS POSSIBLE OPEN SPACE. *SPACE SAVING: PROVIDES PLENTY OF GREEN OR OPEN SPACE. *COMPACT: MAKES LAND USES COMPACT AND, THEREFORE, REDUCES ENERGY FOR TRANSPORTATION. *UTILITIES: REDUCES LENGTH OF UTILITIES IF COMBINED WITH COMPACT LAND USE. *LANDSCAPE: PRESERVES BEAUTY OF NATURE. *DUAL USES: UNDERGROUND CONSTRUCTION WILL SAVE LAND AND MAKE SURFACE GROUND AVAILABLE FOR OTHER USES. *PROXIMITY: UNDERGROUND CONSTRUCTION WILL BRING LAND USES INTO CLOSE PROXIMITY, SUPPORTING SOCIAL INTERACTION AND ENCOURAGING PEDESTRIANS.	*DENSITY: THE USE OF UNDERGROUND CONSTRUCTION MAY INCREASE SETTLEMENT DENSITY.
ENVIRONMENTAL IMPACT	
*PRIVACY: PROVIDES PRIVACY. *ENVIRONMENTAL IMPACT: UNDERGROUND CONSTRUCTION WILL HAVE A MINIMAL IMPACT ON THE ENVIRONMENT AND ON THE ECOSYSTEM OF THE ARID ZONE. *NOISE: SOUNDPROOF. MAKES LAND USE NEAR AIRPORTS POSSIBLE. SUCH CONSTRUCTION WILL REDUCE OUTSIDE NOISE AND CREATE A QUIET ATMOSPHERE WITHIN THE STRUCTURE WHICH WILL BE SOUNDPROOFED AGAINST THE OUTSIDE VIBRATION CAUSED BY ROADS, VEHICLES, OR A NEARBY AIRPORT. *PRODUCTIVITY: MORE PRODUCTIVE ENVIRONMENT, ESPECIALLY FOR LABOR AND CREATIVE ACTIVITIES. *WIND IMPACT: SUCH CONSTRUCTION WILL MINIMIZE THE IMPACT OF THE WIND.	*WATER TABLE: REQUIRES SPECIAL TREATMENT FOR SITES WITH A HIGH WATER TABLE. *TRANSPORTATION VIBRATION: THE STRONG VIBRATION OF RAILROAD SYSTEMS AND OTHER HEAVY TRANSPORTATION FORMS MAY BE TRANSFERRED ACROSS LARGE DISTANCES THROUGH THE GROUND.
SAFETY	
*PROTECTION: PROVIDES SAFETY FROM VANDALS, NOISE, AIR POLLUTION AND DUST AS WELL AS RADIOACTIVE FALLOUT (IF PLANNED TO MEET THIS REQUIREMENT). *SHELTER USE: UNDERGROUND SPACE, ESPECIALLY THAT USED FOR MANUFACTURING, CAN ACCOMMODATE LARGE NUMBERS OF PEOPLE DURING WARS OR NATURAL DISASTERS. *NUCLEAR: NUCLEAR POWER PLANTS ARE POSSIBLY SAFER UNDERGROUND THAN ABOVE IT. *FREEZING: NO WATER PIPES WILL FREEZE. *FIRE: GREATER FIRE RESISTANCE AND LOWER INSURANCE RATES WILL RESULT BECAUSE FIRE GROWTH IS LIMITED, AND EXTENDED COVERAGE RISKS ARE REDUCED. *EARTHQUAKES: UNDERGROUND CONSTRUCTION COULD BE SAFE IN REGIONS SUBJECT TO EARTHQUAKES, IF THE SITE IS PROPERLY SELECTED.	*FIRE EVACUATION: THERE CAN BE EVACUATION PROBLEMS IF FIRE OCCURS. *FLOODING: UNDERGROUND STRUCTURES CAN BE SUBJECT TO FLOODING WHEN UTILITIES FUNCTION INEFFICIENTLY. *FAULT: SUCH STRUCTURES CAN BE DANGEROUS WHEN LOCATED ON A GEOLOGICAL FAULT.
HEALTH	
*RELAXATION: UNDERGROUND STRUCTURES CAN PROVIDE A COMFORTABLE CLIMATE AND AN ISOLATED AND QUIET ENVIRONMENT, AND CAN, THEREFORE, BE HEALTHFUL. *QUIET: CAN STIMULATE CREATIVITY, ESPECIALLY AMONG WRITERS AND ARTISTS; A QUIET ENVIRONMENT CAN CONTRIBUTE TO MENTAL HEALTH. *DAMPNESS: THE RISK OF DAMPNESS DOES NOT EXIST IN THE ARID ZONE AS IT DOES IN THE HUMID ZONE. *POLLUTION: REDUCES POLLEN AND DUST.	*EXTENSIVE VENTILATION: UNDERGROUND STRUCTURES REQUIRE SPECIAL CONSTANT VENTILATION TO BRING IN REQUISITE FRESH AIR. *DAMPNESS: UNDERGROUND STRUCTURES CAN BE DAMP WHEN THEIR ENVIRONS ARE OVERWATERED. *CLAUSTROPHOBIA: A PATHOLOGICAL FEAR OF CONFINED SPACE CAN DEVELOP AMONG SOME PEOPLE, ESPECIALLY THE ELDERLY; SUCH PROBLEMS REQUIRE SPECIAL DESIGN TREATMENT.

Positive conclusions are to be drawn from ancient and contemporary experiences with subterranean structures for the climatically stressful zone: First, subterranean structures can have a microclimate of their own, independent of the exterior; second, their microclimates are almost stable on both a daily and a seasonal basis; third, the stability of their microclimate is primarily dependent on their depth in the ground and less influenced by the fluctuation of the outside climate; and fourth, underground structures can save substantial amounts of energy when extreme climatic conditions exist outside. Because of these positive features, underground structures are suitable for living, working, refrigerating or storing in climatically stressful zones.

One major—and expected—obstacle to the concept of underground shelters in general is the preoccupation of planners, architects, engineers and other professionals with conventional abovesurface construction. Fortunately, we can expect those professionals to become more receptive to the underground concept when it is introduced for stressed climate zones. We can also expect the underground settlement in arid zones to disrupt the ecosystem less than would the aboveground community. Thus, underground settlements can have a positive environmental impact by relieving the load on the delicate natural ecosystem of the arid zone. Moreover, in countries where land is scarce (Israel, for example), where the environment is sensitive and security is important, the underground settlement can provide a remarkable solution to design problems. In nonarid zones, underground settlements offer noticeable relief to the problems of increasing energy consumption and of decreasing agricultural land which is being swallowed up by urban expansion.

As a result of this historical and contemporary review, we can outline a few lessons:

1. Subterranean living is a remarkable adjustment to the environment, a good example for proper use of local resources and of building materials, and less harmful to the natural environment, to the landscape and to the ecosystem than supraterrestrial construction.
2. The development of subterranean space to meet our modern norms and standards requires the combination of knowledge, advanced technology, proper physical-environmental conditions and, above all, an imaginative design and proper understanding of the solutions offered to us by ancient civilizations.
3. Finally and most important, the most effective and optimal year-round performances of the subterranean houses are in regions of stressful climate, especially that which is dry and cold or dry and warm.

This is certainly evident from study of the historical lessons. The effectiveness of the subterranean houses in climatically stressful regions is far more impressive than in moderate climate zones.

Notes

1. Kenneth Labs, "The Architectural Underground," in *Underground Space,* 1 (May/June 1976), pp. 2–3; Spiro Kostof, *Caves of God: The Monastic Environment of Byzan Capatindonia* (Cambridge, Mass: The MIT Press, 1962); Christopher H. L. Owen, "Lalibala Rock-cut Churches in the Highland of Ethiopia," *Architecture Plus* 2/6 (Nov./Dec. 1974); and George Woodcock, "Cave Temples of Western India," *Arts* (May/June, 1962).
2. The term "geotecture" is used for "the concept or practice of subterranean special construction of any kind, at any depth, providing service or accommodation for any purpose," in Patrick Horsbrugh, "Urban Geotecture: The Invisible Features of the Civic Profile," in *Alternatives in Energy Conservation: The Use of Earth Covered Buildings.* Proceedings of a Conference, Ft. Worth, Texas, July 9–12, 1975 (Washington, D.C.: National Science Foundation, 1975), pp. 152–153.
3. M. Goldfinger, *Villages in the Sun* (London: Lund Humphries, 1969); Labs, "The Architectural Underground," p. 2; Amos Rapoport, *House Form and Culture* (Englewood Cliffs, N.J.: Prentice-Hall, 1969); Royce LaNier, *Geotecture, Subterranean Accommodation and the Architectural Potential of Earthworks* (South Bend, Ind: Royce LaNier, 1970).
4. Stanley Hallet, "Mountain Villages of Southern Tunisia," *Journal of Architectural Education,* 29, 2 (1975), p. 23.
5. Ibid, p. 24.
6. Ibid, pp. 22–25.
7. LaNier, pp. 31–32.
8. George B. Cressey, *Land of the 500 Million* (New York: McGraw-Hill, 1955), p. 263.
9. Labs, "The Architectural Underground," 2–3; Cressey, p. 263; James Marston, *American Building: The Environmental Forces That Shape It* (New York: Houghton-Mifflin, 1972); Bernard Rudofsky, *Architecture Without Architects* (Garden City, N.Y.: Doubleday, 1964); Andrew Boyd, *Chinese Architecture and Town Planning 1500 B.C.–A.D. 1911* (Chicago: University of Chicago Press, 1962).
10. Rudofsky, "Troglodytes," *Horizon,* Spring, 1967; and Cressey, p. 263.
11. Fahriye Hazer, "Cultural-Ecological Interpretation of the Historic Underground Cities of Göreme, Turkey: In *Alternatives in Energy Conservation: The Use of Earth Covered Buildings,* p. 25.
12. Kostof.
13. Hazer, p. 24.
14. L. Giovannini, *Arts of Cappadocia* (Geneva, Switzerland: Nagel Publishers, 1971).
15. Hazer, p. 25.
16. Ibid; and Edward Allen, *Stone Shelters* (Cambridge, Mass.: The MIT Press, 1977), p. 1.
17. Allen, pp. 1–2.
18. Hazer, p. 21.
19. Labs, "The Use of Earth Covered Buildings Through History," *Alternatives in Energy Conservation: The Use of Earth Covered Buildings,* p. 8; George J. Gumerman, *Black Mesa: Survey and Excavation in Northeast Arizona* (Prescott, Ariz.: Prescott College Press, 1970); and Watson Smith, *Prehistoric Kivas of Antelope Mesa.* (Cambridge, Mass.: Peabody Museum, Harvard University, 1972).

20. Paul Oliver, *Shelter in Africa* (New York: Praeger Publishers, 1971), pp. 4–46; idem, *Shelter and Society* (New York, Praeger Publishers, 1969), pp. 80–90.
21. Labs, "The Use of Earth Covered Buildings," p. 8; Oliver, *Shelter in Africa*, pp. 4–46; idem, *Shelter and Society*, pp. 80–90; and Labelle Prussin, *Architecture in Northern Ghana* (Berkeley: University of California Press, 1969).
22. LaNier, p. 14.
23. *Underground Plants for Industry* (Washington, D.C.: United States Department of Defense, 1956).
24. Lutz, Frank W., "Studies of Children in Underground School," in *Alternatives in Energy Conservation: The Use of Earth Covered Buildings.*
25. *Industrial Design*, (July, 1964), pp. 56–61; Mart and Eleanor Karf, "The Ecological City," *Landscape* (Autumn, 1963), pp. 4–8; and Malcolm Wells, "Nowhere to Go But Down," *Progressive Architecture* (Feb. 1965), 174–179.
26. Henry Sanoff, "Seven Acres of Underground Shelter," *AIA Journal*, 47 (February, 1967), pp. 67–69.
27. "Saving by Going Underground," *AIA Journal*, 61 (February, 1974), p. 49.
28. Truman Stauffer, "Guidebook to the Occupancy and Use of Underground Space in the Greater Kansas City Area," in *Underground Utilization: A Reference Manual of Selected Works*, 8 vols., Truman Stauffer, ed. Vol. I: *Historical Perspective* (Kansas City: University of Missouri, Department of Geoscience, 1978), p. 1.
29. Ibid, p. 3.
30. Ibid, p. 19.
31. Lloyd S. Jones, "Thinking down through the Earth," in *Underground Utilization*, Vol. I, p. 46.
32. "Going Underground," in *Underground Utilization*, Vol. I, p. 50.
33. Elaine K. Lally, "The Strange Language of Cheyenne Mountain: The NORAD Underground," in *Underground Utilization*, Vol. VII: *The Future of Underground Development.*
34. Jones, p. 46.
35. "Going Underground," p. 50.
36. Stauffer, "Subsurface Uses in Sweden and France: A Report," in *Underground Utilization*, Vol. IV, *Human Response and Social Acceptance of Underground Space*, p. 671.
37. Sydney A. Baggs, "The Dugout Dwellings of an Outback Opal Mining Town in Australia," in *Underground Utilization*, Vol. IV, p. 573.
38. John Gelder, "Underground Desert Shelters," (Adelaide, Australia: University of Adelaide, 1977). Thesis for M. Arch., unpublished, pp. 164–172. Courtesy of Professor Zig.
39. John D. Rockaway and N. B. Aughenbaugh, "Go Underground for Low Cost Housing," in *Underground Utilization*, Vol. IV, p. 614.
40. Stauffer, "Subsurface Uses in Sweden and France," pp. 673–675.
41. Robert B. Bechtel, "Psychological Aspects of Earth Covered Buildings," in *Earth Covered Buildings and Settlements* (CONF-7805138-P2), Frank L. Moreland, ed. Vol. II (Springfield, Virginia: National Technical Information Service, 1978), pp. 71–72.
42. Ibid, pp. 71–77.
43. N. B. Aughenbaugh and John D. Rockaway, "Go Underground for Low Cost Housing," in *Proceedings* of IAHS International Symposium for Housing Problems, Parviz F. Rad, Herbert W. Busching, J. Karl Johnson, and Oktay Ural, eds. 2 vols. (Atlanta: International Association for Housing Science, 1976), p. 1233.

Section II—Subterranean House Design

In this section we focus on the ways and means of designing subterranean or earth-covered structures, especially houses, and we discuss most of the elements which constitute this design. We further believe that the design should be conceived in a comprehensive form.

It is our premise that the ancient experiences in city and house design are remarkable; there is a great deal we can learn from them, especially because such experiences have evolved through generations of practical experimentation and not merely as ideas on the drafting table. However, with regard to our discussion in this book, the lessons to be learned from these historical experiences can be summarized by the following three points:

1. Subterranean living is an effective way to cope with harsh climatic conditions. It offers a comfortable microclimate and convenience.
2. Ventilation can be passive, which is not only healthy but also saves resources, especially if combined with other energy-saving systems.
3. Cooling by evaporation is an effective way of changing the ambient climate with immediately noticeable results.

Our role here is not only to study these ancient experiences, but also to combine these three aspects of traditional subterranean dwellings to create one unified system, and then augment it with modern technological and design innovations so that it can meet our modern building codes and cultural expectations, including those of cost, which is of concern to low-income groups. This new, combined system is thoroughly discussed in a part of this section, Innovation Design of an Energy-Free Cooled House.

Site Selection Considerations

Site selection is a critical stage in the development of the subterranean or earth-covered house since it affects energy consumption. A proper choice will have an economic as well as a social return. Variables to be considered throughout this phase are certainly multiple and comprehensive; every site has its own unique characteristics resulting from the nature of each of its factors, the relation of one to the other, and the combination of all factors together. Therefore, each factor must be studied both

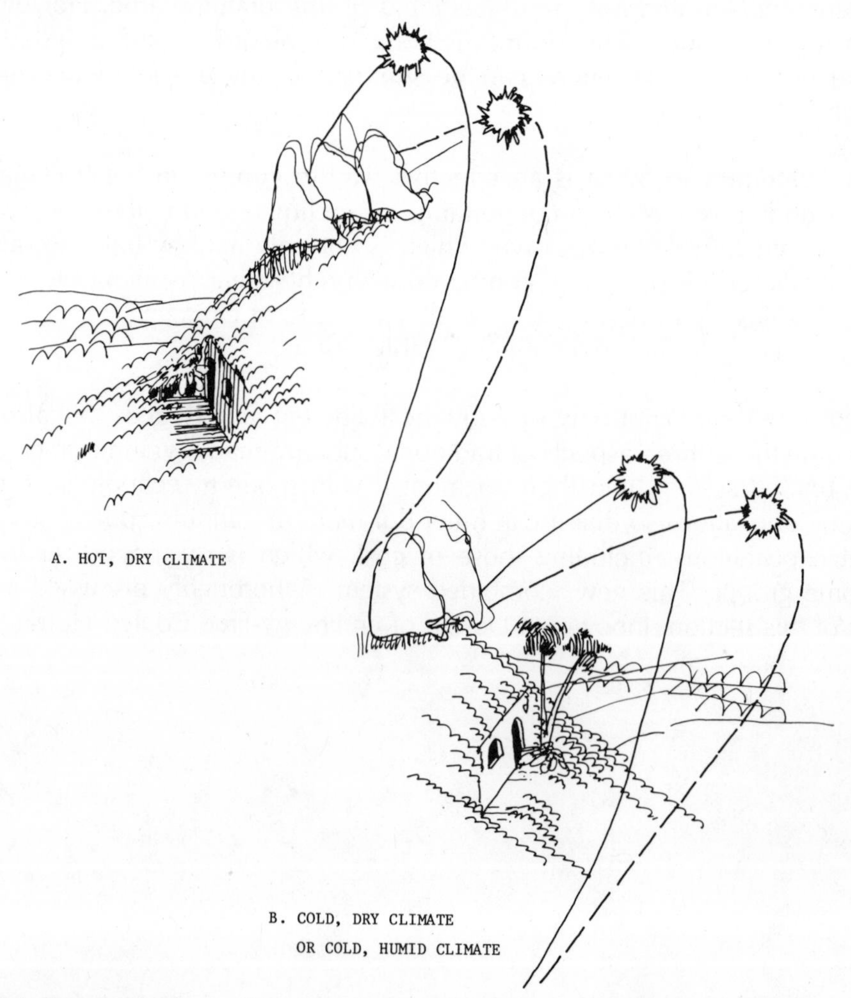

Fig. II-1. Site selection for a subterranean house with regard to solar radiation orientation in two different climatic conditions.

Table II-1. Basic Site Selection Considerations for a Subterranean House

MAJOR CRITERIA	SECONDARY CRITERIA
PHYSIOGRAPHY	-TOPOGRAPHICAL CONFIGURATIONS, ELEVATIONS (ABSOLUTE AND RELATIVE), CLOSED OR OPEN FORMS (VALLEY OR PLAIN, CRATERS, HILLTOPS, SLOPES, MESA), VISUAL VALUES, ACCESSIBILITY. -ORIENTATION: TO WIND VELOCITY, TO RADIATION, TO LIGHT, TO POLLUTION, TO NOISE. -PROXIMITY: TO SEA OR OTHER WATER BODIES, TO RESOURCES, AND TO MAIN SERVICES. -SOIL SUITABILITY: COMPOSITION: GRAIN AND PERMEABILITY, LANDSLIDES AND DEGREE OF STABILITY, SOIL TEMPERATURE AT DIFFERENT DEPTHS, THERMAL BEHAVIOR UNDER TEMPERATURE AND HUMIDITY CHANGES, HEAT GAIN AND LOSS PROCESSES, HUMIDITY AND MOISTURE AND WATER TABLE CONDITIONS, RESISTABILITY TO EROSION AND EOLIAN DEPLETION, SALINITY OF THE SOIL. -GEOLOGICAL CONSTRAINTS: FAULTS AND JOINTS, LAND AND ROCK SLIDES, EARTHQUAKES OR VOLCANOS, CAVITIES (NATURAL AND ARTIFICIAL), CAVES, AND SUBTERRANEAN TUNNELS.
CLIMATIC CONDITIONS	-TEMPERATURE: SEASONAL AND DIURNAL FLUCTUATIONS (DRY AND WET TEMPERATURES), EVAPORATION AND EVAPORTRANSPIRATION, SEASONAL AND DAILY SUNCYCLE, ILLUMINATION. -AERODYNAMICS: VELOCITY AND DIRECTION, BREEZES, TURBULENCE, DUST STORMS, AIR QUALITY, INVERSION. -PRECIPITATION: RAIN, SNOW, DEW, FOG, HAIL AND RELATIVE HUMIDITY.
HYDROLOGY	-PRECIPITATION PATTERNS AND TYPES. -RUNOFF BEHAVIOR AND PATTERNS OF DRAINAGE, EROSION, FLOOD HAZARD, GRAVEL AND LANDSLIDES, ERODED DEPOSITS, POTENTIAL RUNOFF CAPTURINGS, LOCAL DRAINAGE SYSTEM CONTROLS, WATERSHEDS, WATER TABLE CHANGES. -WATER RESOURCES: QUALITY, QUANTITY AND PROXIMITY.
ENVIRONMENTAL QUALITY	-ECOLOGY OF FLORA AND FAUNA. -POLLUTION SOURCES, REGIONAL INDUSTRIES, AIRPORT AND MAJOR TRAFFIC ARTERIES, SOIL POLLUTION, ODORS. -EOLIAN DEPOSITS, DUNES AND DUST RESOURCES (QUARRIES, UNPAVED ROADS, PLAINS) UNSTABLE SOIL, VEGETATION COVER. -RESIDUALS, ODORS, GARBAGE DUMP RESOURCES. -NOISE RESOURCES, INDUSTRY, TRAFFIC. -POLLUTED LAND, INDUSTRIAL WASTE, POORLY DRAINED AREAS. -SAFETY AGAINST HAZARDS.
ACCESSIBILITY AND PROXIMITY	-UTILITIES: ROAD, WATER SUPPLY, SEWAGE SYSTEM, TELEPHONE, ELECTRICITY, TV CABLE, WATER PUMPING COST. -YEAR-ROUND ACCESSIBILITY AND TERRAIN CONDITIONS, BUILDING MATERIAL SUPPLY AND ACCESSIBILITY. -SPACE REQUIREMENTS AND SITE DESIGN POTENTIALITIES.
RESTRICTIONS AND LIMITATIONS	-LEGAL ZONING AND CODE ENFORCEMENT, PSYCHOLOGICAL OR CULTURAL PRESERVATIONS, ENVIRONMENTAL AND PHYSICAL CONSTRAINTS.

individually and as a part of the whole. Flood plains are to be avoided, and lowlands and foot-of-the-slope sites are not optimal locations.

The criteria to be considered for the site selection are to be treated comprehensively (Table II-1). For instance, decisions on orientation to sun and to light are determined by the type of climate existing in the region (Fig. II-1). In addition, site selection and site analysis are both related directly to the structure's function. In a house which is to be located under the surface, it is essential to have a scenic panorama to add new and important value to the house, bring a new dimension to it and ease any feelings of claustrophobia which may exist.

Most important is the selection of the site in relation to the topography. In general, we can group the topographical configurations in their major forms: the lowland (flat), the slope and the top of the hill. Each one of these has its own characteristics which bring advantages and disadvantages (Fig.

II-2 and Table II-2). In analyzing each form, we conclude that the optimal form is the slope. A subterranean house located on a slope, if properly designed, can eliminate feelings of claustrophobia; provide light and sunshine as they are desired; establish a direct, wide view of the surrounding area; introduce ascending and descending entrances; offer good ventilation; permit potential house expansion vertically and horizontally; provide for easy sewage disposal, flowing by gravity; and be in terrain with a low water table (Fig. II-3).

On the other hand, although building earth-covered or subterranean houses on slopes has many advantages, it requires careful consideration in

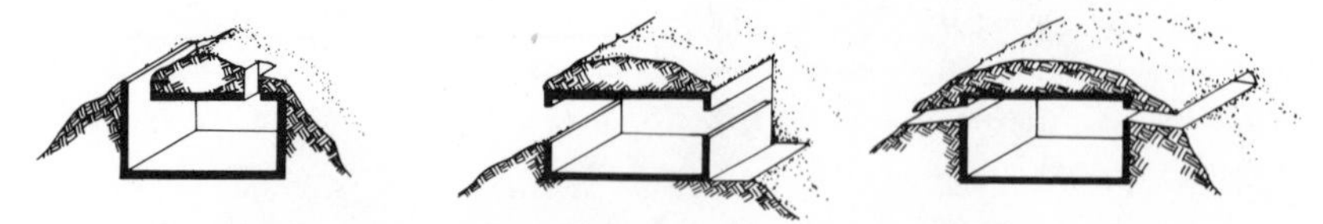

A. HILLTOP : WIDE VIEW TO THE ENVIRONS AND TO THE LOW LAND: POSSIBILITY OF PLENTIFUL LIGHT AND SUNSHINE, GOOD VENTILATION. POSSIBLE WATER TABLE.

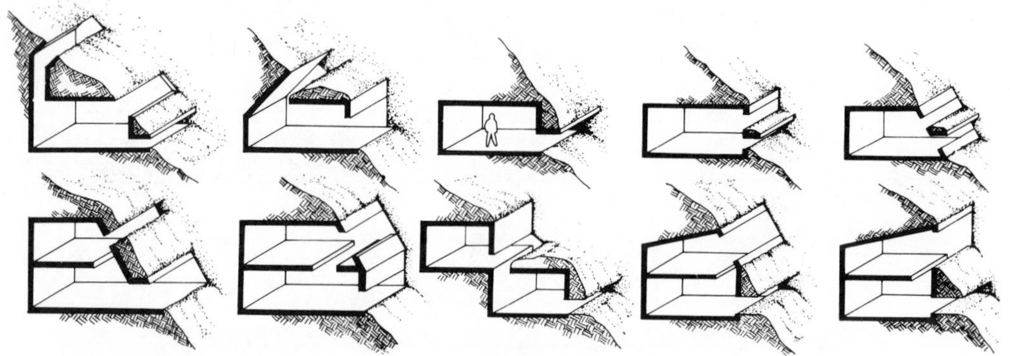

B. SLOPE TOPOGRAPHY: VIEW TO THE LOW LAND OF THE VALLEY OR THE PLAIN. HOUSE EXPANSION IS POSSIBLE SIDEWAYS, UPWARD AND DOWNWARD AND IS UNLIMITED. EASY SEWAGE. LOW WATER TABLE.

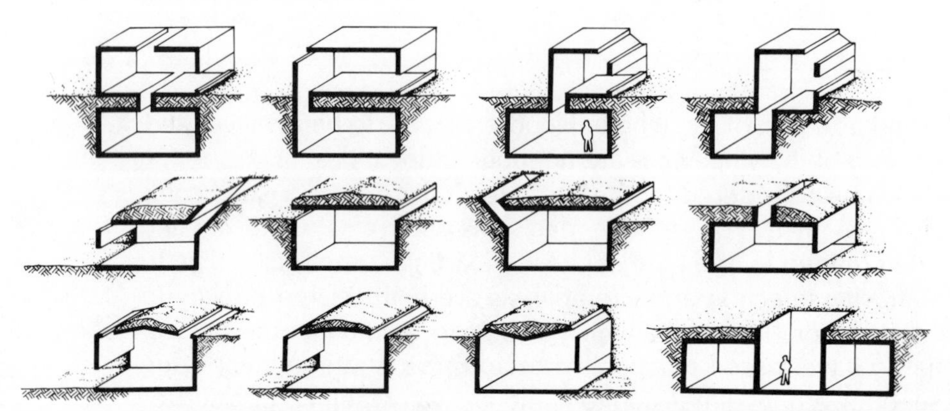

C. FLAT TOPOGRAPHY: BLOCKS THE VIEW TO THE SURROUNDING ENVIRONS, ALTHOUGH THE PATIO IS OPEN TO THE SKY. HOUSE EXPANSION IS LIMITED AND NOT DESIRED. RISK OF HIGH WATER TABLE.

Fig. II-2. Cross section of subterranean housing forms as related to the three major topographical configurations. Sites on the slope have many advantages compared with the other sites.

Table II-2. Evaluating Problems in Site Selection for a Subterranean House on Three Basic Topographical Forms

FACTOR	ON FLAT	ON SLOPE	ON TOP OF THE HILL
1. Water supply	Easy	May require pumping	Requires pumping
2. Sewage	Requires pumping	By gravity	By gravity
3. Drainage	Subject to flood	Good potential	Very good potential
4. View to environs	To the sky only	Good potential	Very good potential
5. Accessibility	Easy	Requires initial investment	Requires much initial investment
6. Excavation	Relatively easy	May require blasting	Requires blasting
7. Dust	Very high	Little or none	Nearly none
8. Ventilation	Very low	Good	Very good
9. Temperature (relative)	High	Lower	Reasonably lower
10. Water table	Potentially high and close to the surface	Very low, if any	None

site selection, design and construction. Slopes are associated with heavy landslides, rockfalls and erosion. Moreover, any alteration and grading of the slope will change its old balance and increase the landslide potential if special precautions are not taken. Also, the risk of flood on the slope is high when careful grading is not considered. Special ditches to divert water or protective walls may be necessary. In the arid zone, air movement coupled with the absence of vegetation can support eolian movement of the soil to the house or cover its entrance.

At any site, there are some facts that cannot be overlooked. We have mentioned already that rock excavation is costly. Also, blasting in rock can be a problem since it may affect nearby structures in addition to producing noise. Humidity increases as elevation increases, but air temperature decreases as the elevation increases by 1C° for every 100 meters. At the same time, elevation improves the aesthetic quality and ventilation, supports vegetation and reduces dust infiltration.

Flat sites have more limitations than sloped sites. A flat site limits the view (except to the sky if the house has a patio); it is subject to flood espe-

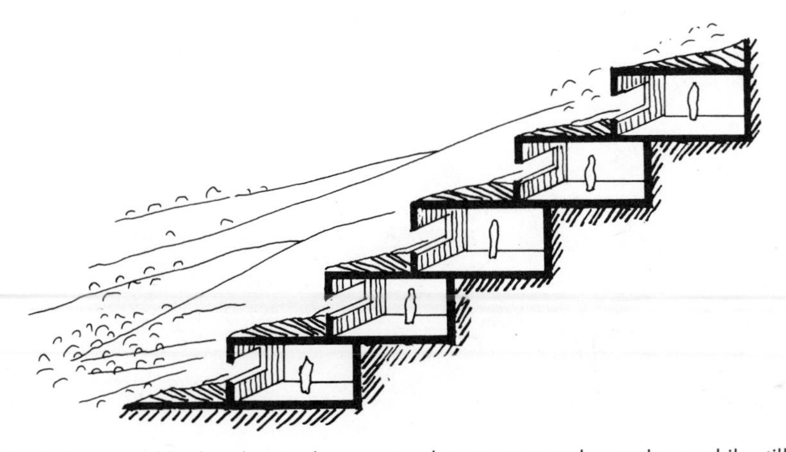

Fig. II-3. A site for a neighborhood on a slope opens the structure to the outdoors while still in the confined underground space. Light, sunshine, view and ventilation are accessible.

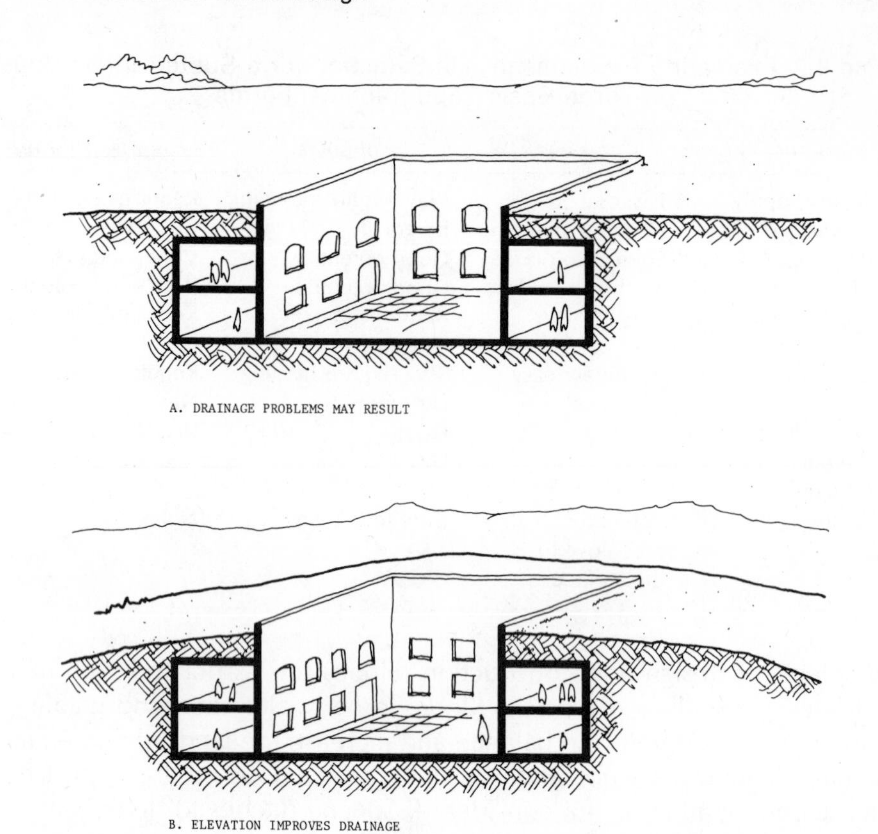

Fig. II-4. Development on a flat site. A subterranean dwelling unit in a flat area requires land adjustment (case B) in order to provide efficient drainage, to avoid floods and to reduce dust accumulation in the patio. Such development requires sewage pumping, depending upon availability of electricity.

cially if the water table is high; it has serious drainage problems (especially if the water or sewage system breaks down); it requires sewage pumping; it is subject to dust and litter accumulation; and it may also encourage feelings of claustrophobia (Fig. II-4). Also sewage disposal, construction and maintenance on a flat area are more expensive and more disruptive to the environment. House expansion is limited when units are close to each other.

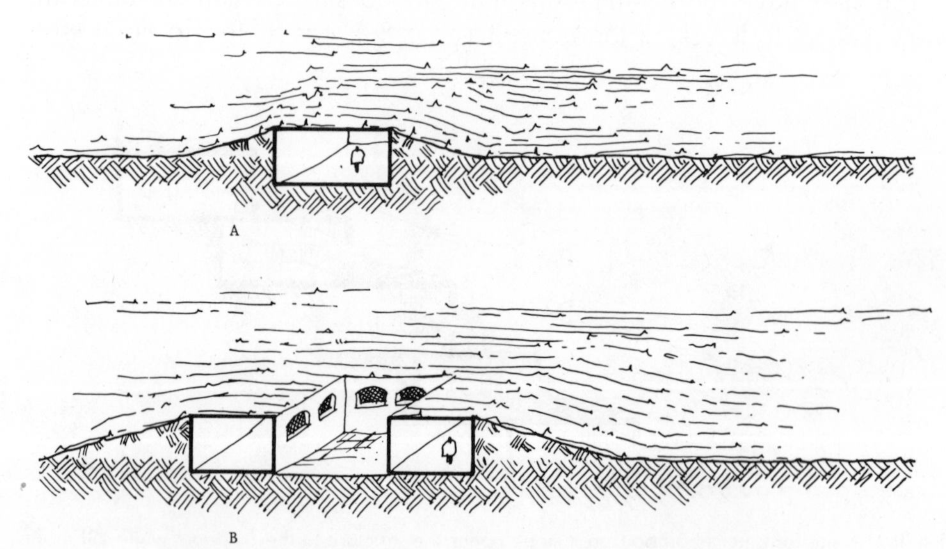

Fig. II-5. Grading adjustment for subterranean houses on flat land.

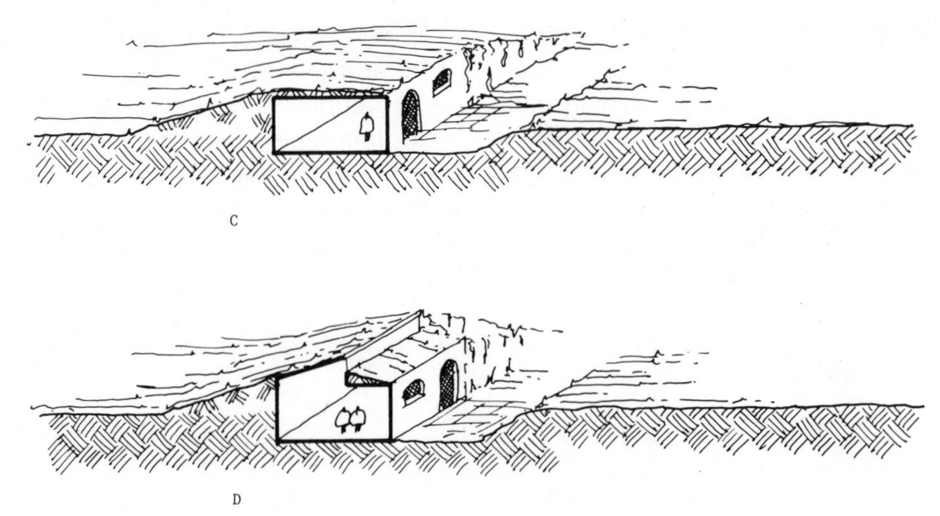

Fig. II-5. Continued.

However, house construction is fairly easy because dirt can usually be removed with little or no blasting (Fig. II-5); but high water tables necessitate expensive waterproofing. Finally, flat sites may limit the privacy of a residence located in a developed area (Fig. II-6).

Sites on the top of a hill also have their special characteristics: Such sites have a wide view of the surrounding area, with the possibility of plentiful light and sunshine if it is desired; and they offer good ventilation,

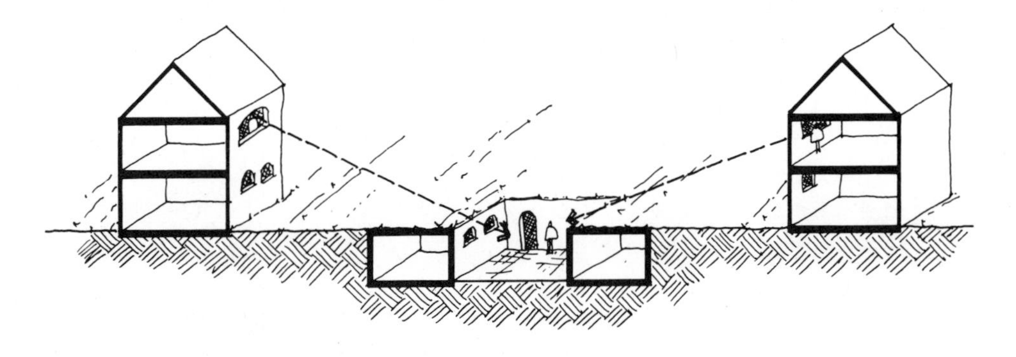

B. INTEGRATED MIX OF SUBTERRANEAN AND SEMISUBTERRANEAN DWELLINGS

Fig. II-6. Two examples of subterranean house site selection. The selection and the design may result in the loss of privacy (case A). An integrated mixture can provide privacy (case B).

though there may be the risk of strong, sometimes very cold, winds. There also is the potential for a low water table, and this location may require extensive blasting, which will make the development of the house more expensive. To achieve accessibility, road investment may be essential. Finally, while there would be good drainage and easy sewage disposal, bringing water to the site could necessitate a large investment.

The Soil's Influence on Subterranean Construction

Soil character, along with its density, the shape and size of its particles and its moisture content, determines its thermal conductivity, heat retention and resistance to freezing. Our major interests should be in the ability of a specific soil to gain or lose heat and in the associated water issues. Factors which affect heat conduction in the soil are varied: intensity of solar radiation, air temperature, humidity in the air, moisture or water within the soil and type of soil coverage.

Damp or wet soil can consume a large quantity of energy in heating and evaporating the water. However, additional water to further saturate the soil may well conduct the heat downward and therefore carry the temperature to the lower soil sections so that they later may have a higher temperature than that of the surface.[1] Usually subsurface soil temperatures differ from those above the surface. In general, subsurface temperatures are cooler than above-surface temperatures in the summer and warmer in the winter. This fundamental rule is the basis for the development of earth-covered or subterranean habitats (Fig. II-7).

Although solar radiation is the source of the soil's heat at the depth with which we are concerned, as a result of seasonal and diurnal solar position and other conditions energy received by the soil is not uniform. Other factors include:

1. The degree of slope angle in relation to sun radiation.
2. Diffuse radiation in the air which depends on the degree of humidity and other particles in the air.
3. The type of soil, its texture and surface. For example, rough surfaces absorb less energy than smooth surfaces.
4. The coverage of the soil by sod, grass, trees, etc.; or its light color. These determine the albedo or the amount of radiation absorbed by the soil. Soil covered by sod, for example, heats less than bare soil during the day, but gets cooler at night (Fig. II-8).
5. The moisture content of the soil. Soil with high moisture content absorbs large quantities of radiation and, therefore, albedo is reduced. If soil becomes extremely dry, heat is transported by conduction from one particle to another when the soil is compressed. If arid-zone soil is watered occasionally, then heat conduction will be moderated (Fig. II-9).
6. Wind clouds, rain, irrigation and other elements influence the soil temperature as well.

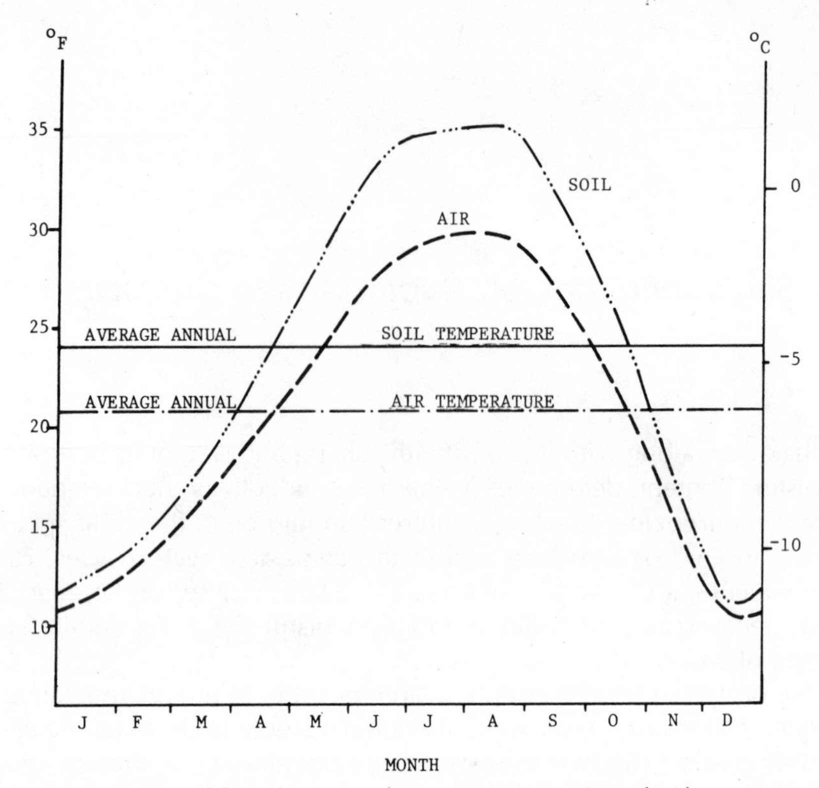

Fig. II-7. The significance of the soil mass as a heat retainer. Comparison of arid-zone air temperature with soil temperature at a depth of 2 inches in an arid zone, College Station, Texas (1951–1955). Source: B. J. Fluker, "Soil Temperatures," *Soil Science,* Vol. 86, 1958, 38.

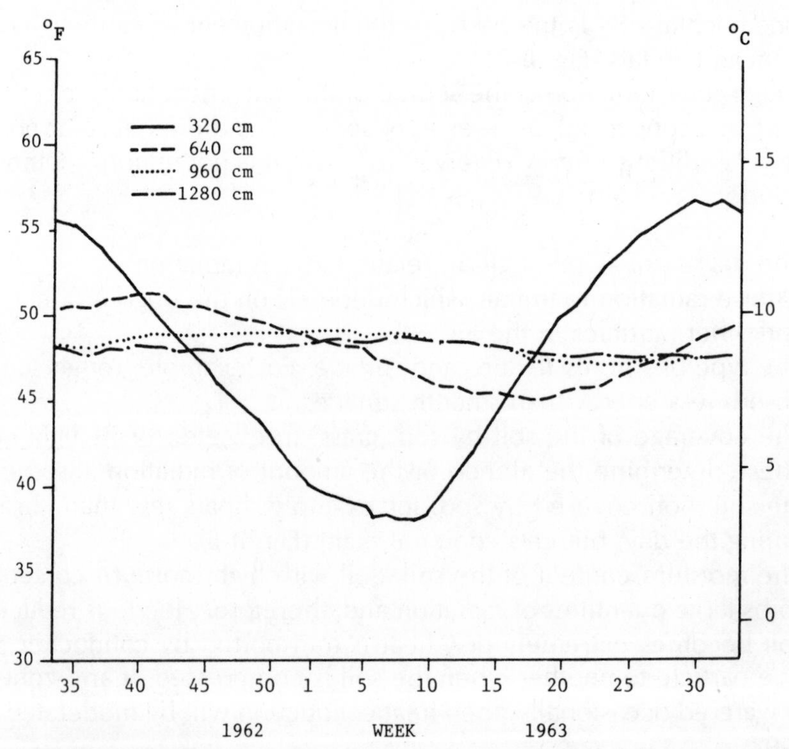

Fig. II-8. Weekly means of the soil temperature wave at various depths beneath sod. Source: Robert Kingsley Maxwell, "Temperature Measurements and the Calculated Heat Flux in the Soil." Minneapolis: University of Minnesota, August, 1964. Thesis, Master of Science, p. 77.

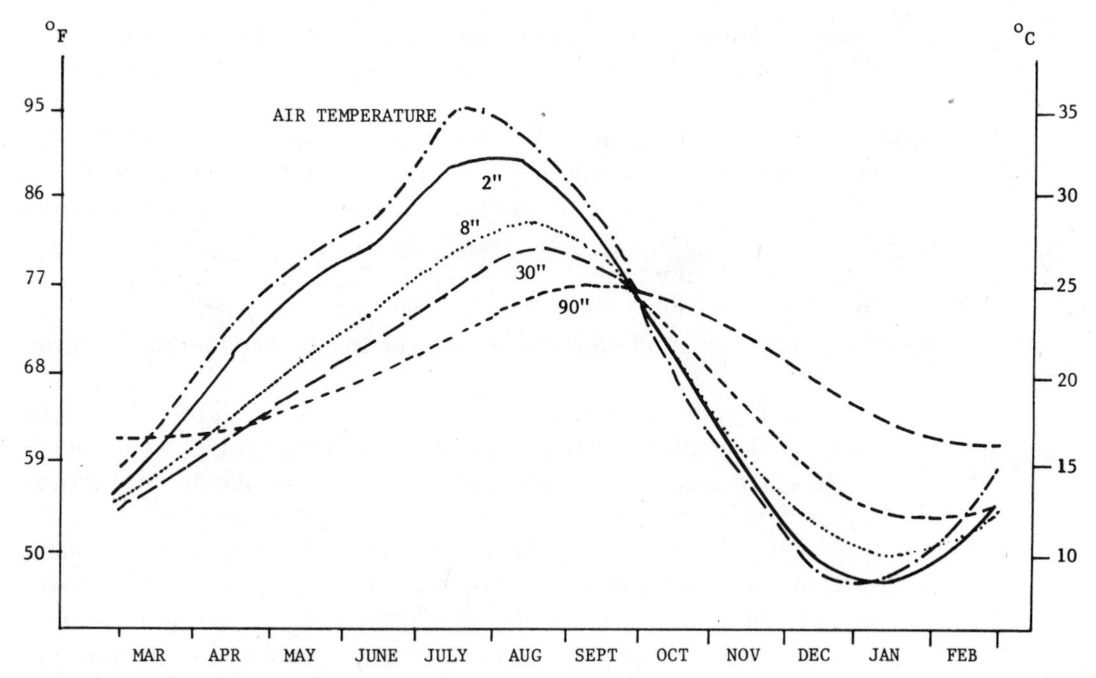

Fig. II-9. A year's cycle of soil temperature fluctuation at different depths in the hot dry climate of an arid zone, Tempe, Arizona. This data was collected by Dr. Robert H. Hilgeman, a horticulturist at the University of Arizona Citrus Experimental Farm in Tempe. The record was taken daily over an 8-year period in an orange grove with sandy, gravelly loam soil which was irrigated every 2 weeks in the summer with a total yearly irrigation of 14–16 days (excluding heavy rainfalls). The soil had an apparent specific gravity of 1.5 and held approximately 17 percent water. Source: James W. Scalise, editor, *Earth Integrated Architecture*. Tempe, Arizona: The Architecture Foundation, 1975, p. A-14.

Heat conduction within the soil is a slow process. For example, one way to decrease the effect of outside temperature on the soil around the subterranean house is to reduce the thermal conductivity of the soil by sprinkling water on the upper part of it at night. This water must penetrate deeply into the soil (40 centimeters or more) so that its cooling effect is less affected by the diurnal temperature fluctuation which usually only influences soil a few centimeters into the surface.[2]

SOIL: ITS PROPERTIES AND THERMAL BEHAVIOR

We should now be convinced that soil is a feature which must be carefully examined before any investment is made for purchasing a particular piece of land. Aspects of soil that must be studied are:

- Structure and composition and resultant internal pressures.
- Presence of water: possibility of impurities, height of water table in peak season and its impact on the soil and on the house. (This is an especially important consideration in humid regions where rain is heavy.)
- Heat conduction, possible contraction and expansion, effect of decreasing or increasing humidity and of watering, presence of impurities or permeability.
- Effect of planting trees and grass.
- Resistance to earthquake, cracks, joints or faults.
- Potential for landslide from adjacent area if earth movement occurs be-

cause of intensive soil vibration (by heavy traffic, for example), earthquake, runoff water or loose soil (such as sand).

Even if other side considerations do not make construction too costly, the study of temperature conditions in the soil is vital for the calculation of heat gain and loss and for determination of the degree and type of insulation to be used.

Therefore, let us now further consider basic aspects of the soil's thermal behavior. Careful analysis of this thermal behavior should enable us to correlate the depth of the underground house with the desired thermal comfort.

1. Seasonal temperatures of the soil fluctuate to the depth of 8 to 10 meters; below this level, the temperature becomes stable. As the depth reaches 3 meters, the temperature becomes similar to the aboveground yearly average. The seasonally stable temperature is usually around 9°C (48°F) (Fig. II-10) depending on climate and soil water content. Fluctuation of soil temperature follows the seasonal aboveground changes. Therefore, the general rule is: the deeper the construction (in the range of zero to eight meters), the smaller the fluctuation; the shallower the construction, the greater the temperature fluctuation.

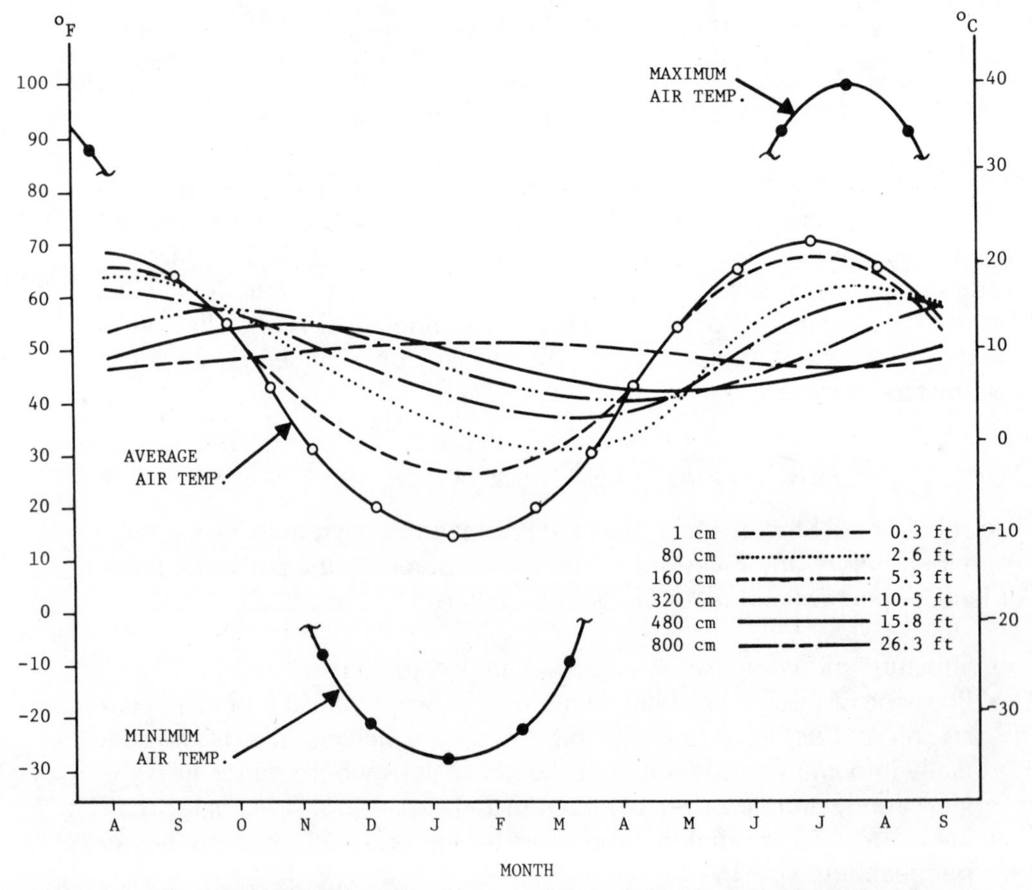

Fig. II-10. Average monthly soil temperature changes at different depths, St. Paul, Minnesota. Adapted from: Thomas Bligh, "A Comparison of Energy Consumption in Earth Covered vs. Non-Earth Covered Buildings," in *Alternatives in Energy Conservations: The Use of Earth Covered Buildings. Proceedings of a Conference,* Fort Worth, Texas, July 9–12, 1975. Washington, D.C.: National Science Foundation, 1975, p. 92.

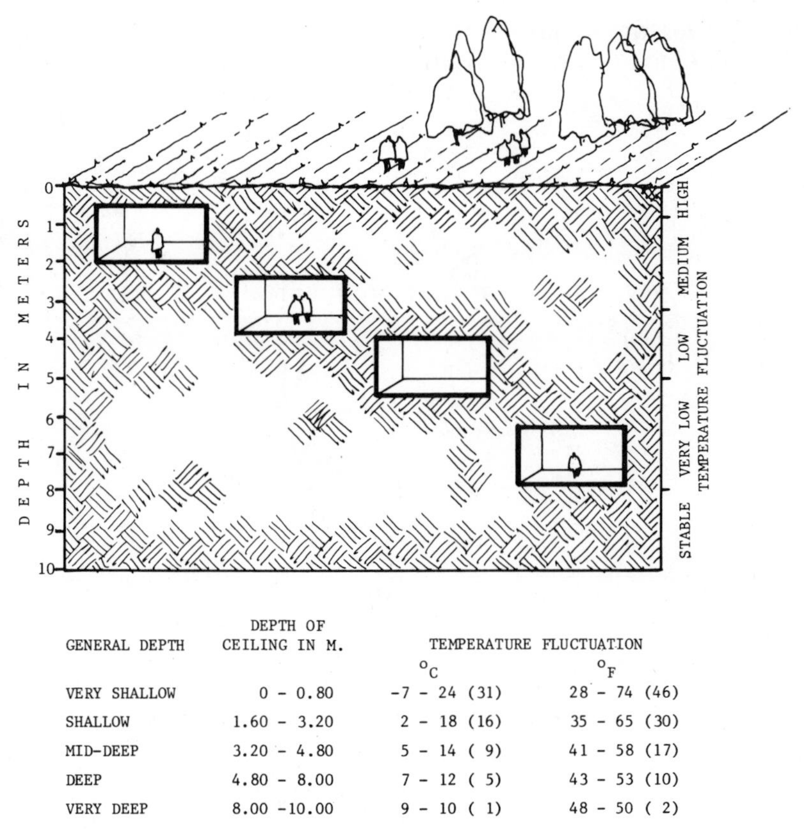

GENERAL DEPTH	DEPTH OF CEILING IN M.	TEMPERATURE FLUCTUATION °C	TEMPERATURE FLUCTUATION °F
VERY SHALLOW	0 - 0.80	-7 - 24 (31)	28 - 74 (46)
SHALLOW	1.60 - 3.20	2 - 18 (16)	35 - 65 (30)
MID-DEEP	3.20 - 4.80	5 - 14 (9)	41 - 58 (17)
DEEP	4.80 - 8.00	7 - 12 (5)	43 - 53 (10)
VERY DEEP	8.00 -10.00	9 - 10 (1)	48 - 50 (2)

Fig. II-11. Schematic illustration of soil heat gain fluctuation in the subterranean house at different depths. Note that the maximums of temperature show the lag which affects the house during the winter, while the minimums of the temperature affect the house in summer. Minnesota temperature calculated after Maxwell, "Temperature Measurements and the Calculated Heat Flux in the Soil."

2. In general, we can state that the greater the depth in the soil, the longer the lag in time for soil temperature to reach air temperature. The lesser the depth, the shorter this lag of time will be. This time-lag can be calculated to determine the necessary depth for placement of the subterranean house in order to achieve the desired temperature for proper thermal comfort (Fig. II-11).
3. Diurnal temperature fluctuation within the soil is limited and extends to a depth of about 10–30 centimeters.
4. Lag of time in the occurrence of average maximum and average minimum seasonal temperature also occurs. The maximum monthly average lag increases as the depth increases. Thus, this maximum may appear in the deep level when the monthly minimum appears above the soil. Therefore, the desired level for placement of the house can be determined by noting the depth at which the *minimum* yearly temperature of the soil coincides with the maximum temperature aboveground, or the other way around (Table II-3).

Although seasonal heat gain and loss in the soil depends greatly on air-temperature changes, seasonal rain causes changes in this pattern. Givoni indicates that the higher the water content of the soil, the greater its thermal conductivity.[3]

Table II-3. Average of Maximum and Minimum Annual Temperature and Annual Amplitude of Soil Temperature, as well as the Time Lag at Different Depths

DEPTH BELOW GROUND SURFACE	MAX AVG. TEMP.	MIN. AVG. TEMP.	AVERAGE ANNUAL TEMP.	ANNUAL TEMP. AMPLITUDE	TIME LAG
AIR	°C.	°C.	°C.	°C.	DAYS
2 IN.	30.0	10.5	20.8	9.8	0
FT.	35.2	11.1	24.1	12.1	0
1	33.9	13.9	23.9	10.0	5
2	32.9	15.0	23.8	9.0	15
3	31.9	16.0	23.6	8.0	22
4	30.8	16.6	23.6	7.1	32
5	30.7	17.6	23.6	6.6	39
6	28.7	18.5	23.5	5.1	47
8	27.1	19.5	23.4	3.8	62
10	26.3	20.3	23.4	3.0	74

Source: B. J. Fluker, "Soil Temperatures," *Soil Science*, Vol. 86, 1958, p. 43.

Around a building, the soil's temperature is related to its density; and this density also means there is less thermal activity than aboveground. Thus, basically the construction of a house within a subterranean environment will be associated with moderate thermal behavior, relative to aboveground conditions.

Therefore subterranean living is a *climatic comfort improvement* compared with aboveground living. This improvement is achieved by:

1. The reduction of the effect of solar energy radiation.
2. The use of the continuous "heat" of the soil itself as a moderating influence. Thus, the heat efficiency of the subterranean house increases as the depth of construction increases, since the massiveness of the soil increases.

In theory, the earth can store the summer heat and release it in the cold winter through the walls of the subterranean building so that less heating will be required for comfort. On the other hand, the earth will store the winter cold temperature and release it in the warm summer throughout the subterranean house so that further cooling may not be necessary. A few questions should be raised here if this system is to be applied in arid-zone regions:

1. How long does the summer heat take to travel between the earth's surface and the wall enveloping the subterranean house? If this heat supply wave (under the given climate and given soil) reaches the wall only when it is needed (at the beginning of the winter), how long can this heat supply last? Does it last until the end of the winter? What would the temperature fluctuation near the outer wall (and consequently within the house) be throughout the winter? Can we modify the time and the temperature of the wave by changing the soil type and the depth of the building? Finally, what would be the process for the summertime?

2. If this system is applied, do the outer walls have to be insulated? How much wall space should be insulated? What type of insulation?

In practice, during the summer, higher temperatures move down into the earth and reach a peak after the peak summer temperature of the air. In the wintertime, the temperature decreases from the upper soil down into the earth. The heated soil in the summer at a depth of 8 to 10 meters will store heat in the walls of the subterranean house at around 10°C (50°F), and this temperature will affect the house temperature. If the house is below a 3-meter depth, the house will remain at this approximate temperature level all year long. Thus, the "heat" pattern may well keep the building cool enough in the summer and require only a little additional heat in the winter for comfort. If the house is insulated, this pattern would change.

The question is: How much insulation should be installed? Of no less importance, should the outer wall be insulated or only part of it, and which parts? According to Bligh, Skipp and Meixel, the heat flux in uninsulated earth-covered buildings is maximum along the top and decreases toward the lower parts of the building. Thus, they suggest placing the "insulation vertically down the outer wall or extending the roof insulation horizontally out from the wall to cut the heat flux paths"[4] (Fig. II-12).

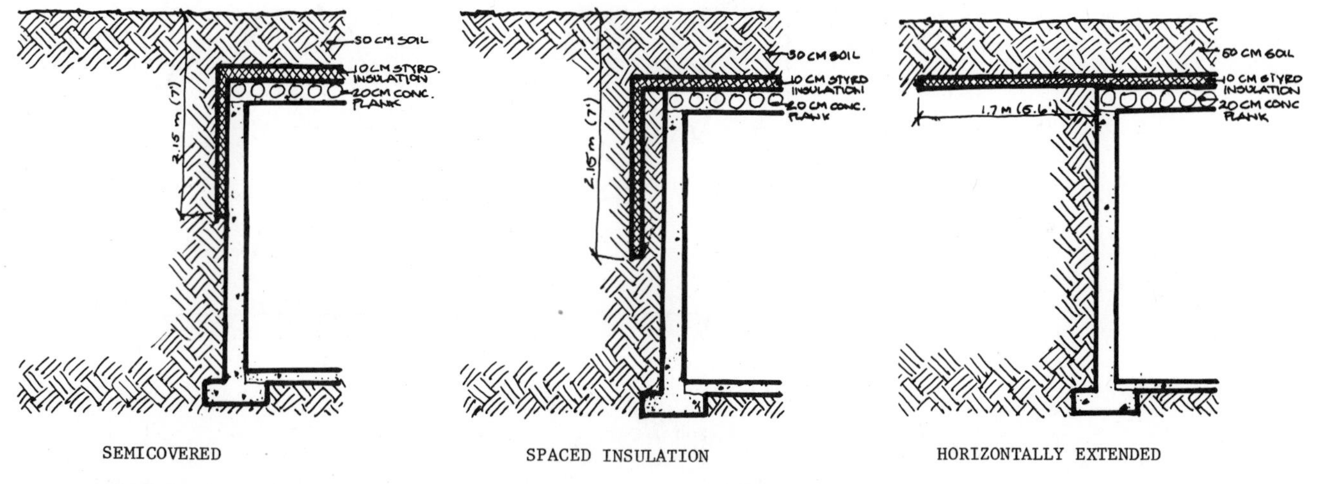

A. THREE INSULATION ALTERNATIVES SUGGESTED BY THE MINNESOTA UNDERGROUND SPACE CENTER THROUGH ITS RESEARCH

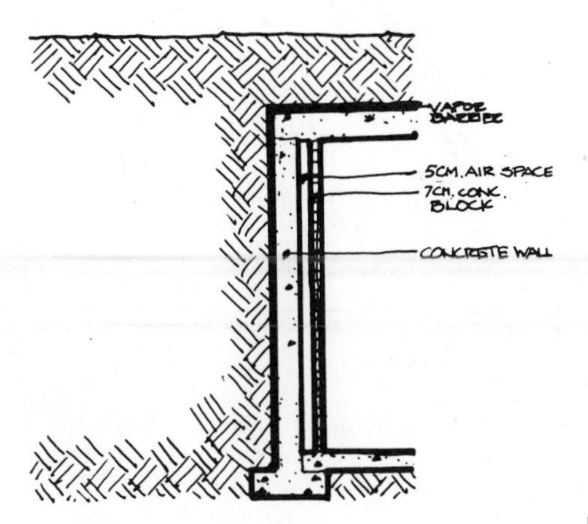

B. ALTERNATIVE FOR DOUBLE WALLS, USED IN MODERN HOUSES OF JERUSALEM, ISRAEL. MEDITERRANEAN CLIMATE: DRY AND WARM IN THE SUMMER, COLD AND RAINY IN THE WINTER. ELEVATION: MORE THAN 800 METERS ABOVE SEA LEVEL

Fig. II-12. Wall insulation alternatives.

This insulation should be installed only after the soil temperature fluctuation throughout the year has been studied to determine what parts of the wall need to be insulated, and what type and how much insulation are to be used. We could expect that the soil below the floor of the subterranean house has the least temperature fluctuation throughout the season. The case may be different if the floor is level with the outside ground (on slope areas). On the other hand, the soil above the roof will have the most temperature fluctuation. The side soils would differ, depending on their proximity to the surface. Figures II-13 and II-14 depict annual weekly means of extremes of soil temperature and heat gain and loss in soil at various depths in a representative location.

Therefore, the subterranean house will not lose heat through its roof or through the sides—if it is constructed deeply enough in the ground, if it is surrounded by a large mass of soil, and if required insulation is installed.

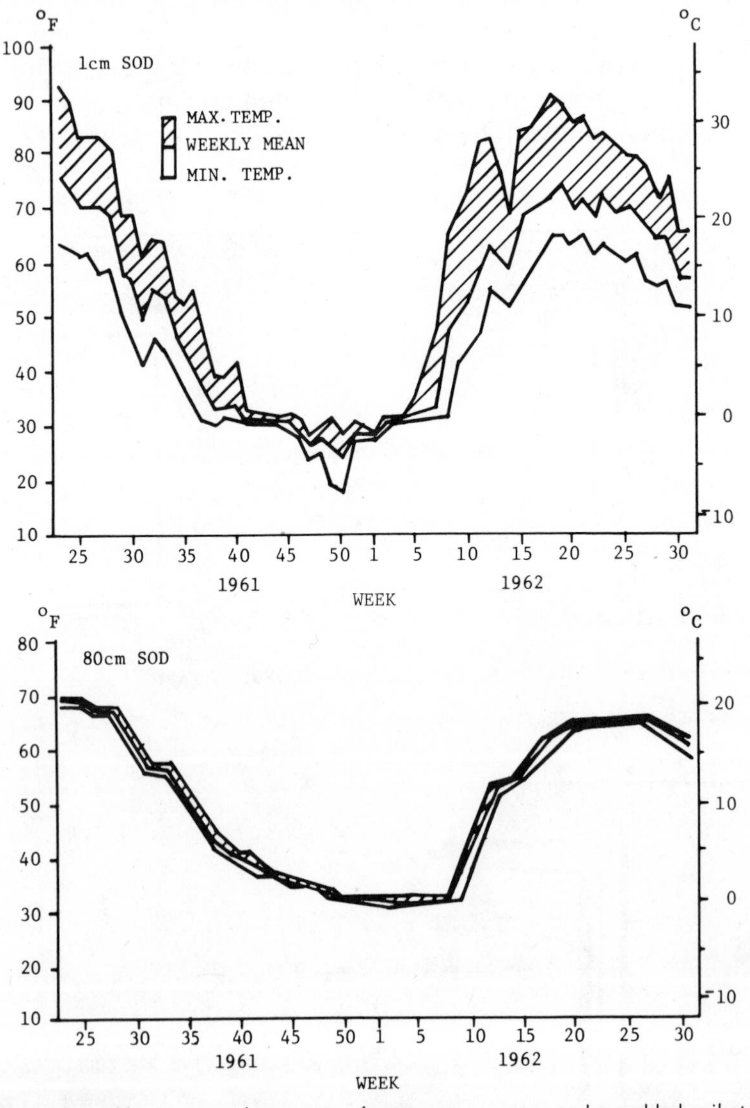

Fig. II-13. Annual weekly means and extremes of temperature waves under sodded soil at depths of 1 cm and 80 cm. Note the rapid fluctuation of the temperature at the shallow depths. Minnesota. Source: Maxwell, "Temperature Measurements and the Calculated Heat Flux in the Soil," p. 74.

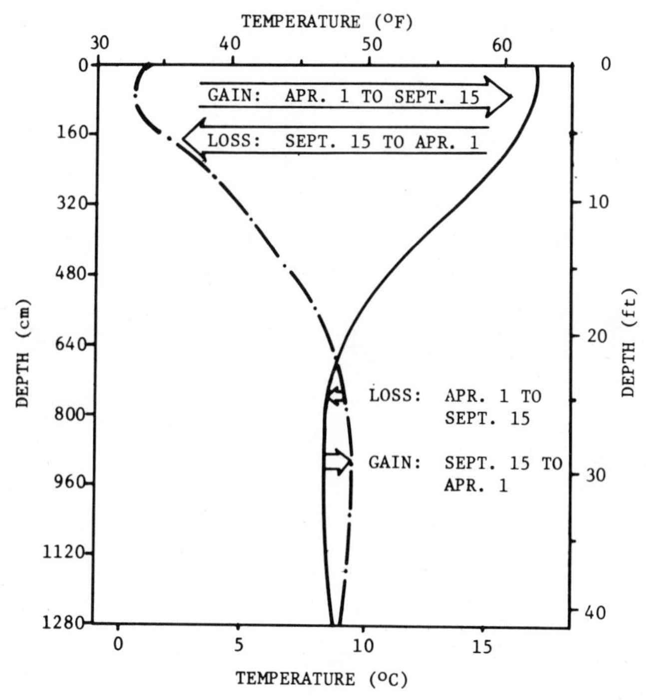

Fig. II-14. Heat gain and loss of the soil at different depths, St. Paul, Minnesota. Source: Maxwell, "Temperature Measurements and the Calculated Heat Flux in the Soil," p. 46.

However, the aboveground house may lose heat in excess of 50 percent of the total heat loss through its roof.[5]

We would be remiss if we did not indicate that another way of diffusing the heat pattern through the ground is by covering the soil with high grass or trees, thus filtering the direct solar radiation and shadowing all the soil (Fig. II-15). This arrangement will also reduce dust movement. One final way is to construct a conventional house above the subterranean house to cover the lower unit. In conclusion, we can definitely state that soil character and behavior are the most important and complicated factors to be studied during the design of the underground house.

SOIL AND HEAT GAIN AND LOSS WITHIN A STRUCTURE

Heat loss and gain of the structure itself is a factor of its volume and floor surface. Heat flows to the structure when a temperature gradient exists between the earth and the space of the house. Thus, the ratio of volume to surface is an important factor to be considered in the design of the house. The smaller the surface area the less the heat loss is. It may be good, however, to leave a portion of the wall uninsulated and backfilled in order to allow for heat transfer.[6] Soil sample testing for wet and dry densities (per cubic foot), compression strengths (per square foot), and moisture contents (%) are of great importance for saving investment.[7] Although in our subterranean house we consider the mass of the soil as an effective insulator, it should be mentioned that its thermal conductivity is approximately 25 times greater than that of modern insulation materials.[8]

In conventional buildings, heat loss or gain occurs through cracks in the walls and the walls themselves, the roof, and through many other openings

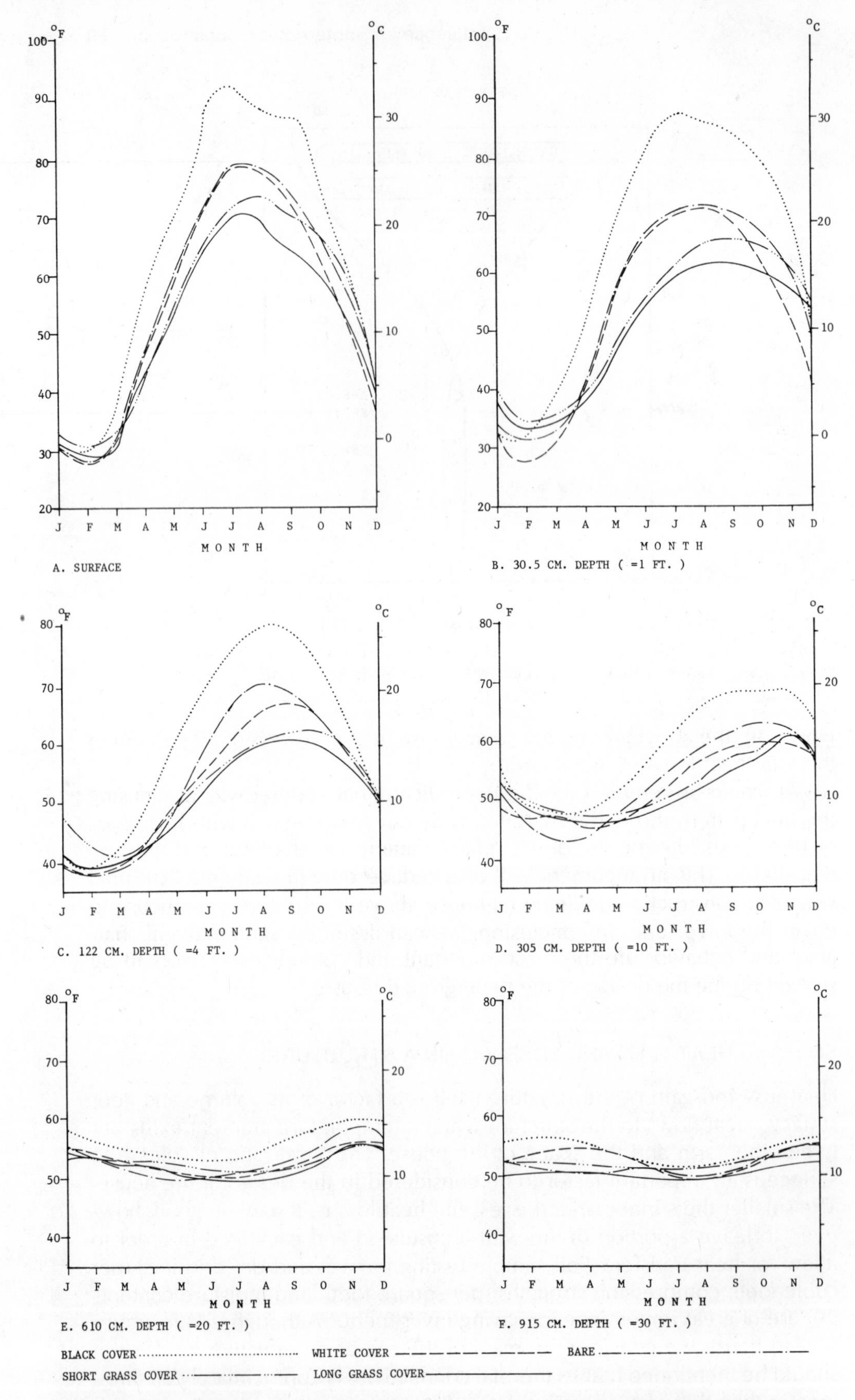

Fig. II-15. Monthly soil temperature changes under five coverage conditions and at different depths, Washington, D.C. Source: Adapted from T. Kusuda, "The Effect of Ground Cover on Earth Temperature," in *Alternatives in Energy Conservation: The Use of Earth Covered Buildings*, pp. 297–301.

in closed windows and doors. In the subterranean house, we would like to minimize or eliminate the influence of the outdoor climate on the indoor one throughout the winter and the summer. Two possibilities exist for this.

First, we could take advantage of the thermal properties of the soil mass enveloping the house. If the house depth is 8 meters or more, the stable temperature of the soil at around 48°F or 9°C will influence the house. If the house is not that deep, the time lag in temperature change will be of some advantage. Recall, too, that the heat gain in the soil throughout the summer will affect the building in the winter, and heat loss in winter will cool the building in the summer; and that the lag in heat loss or gain in the structure can be modified by influencing soil temperature—for example, with shadowing. As a second possibility we can insulate the walls from the soil enveloping it (when the house is not deep in the soil), and in this way minimize any fluctuation in the soil temperature which might influence the walls of the house and, consequently, its indoor space. It is clear that we do not want heat gain in the summer through the walls. In the winter, the stored "heat" in the soil will be a positive influence on the walls and, therefore, on the indoor temperature. Insulation of the upper part of the soil (by plastic cover) may be needed to prevent soil frost.

If the roof is not deep enough in the ground, then it is necessary to make sure there is a sufficient depth of soil covering the roof to support a ground-cover of grass. Such an increase in the soil layer is necessary in order to minimize heat loss and gain in the building, but it will increase construction costs. Therefore, adding insulation to the roof may avoid loss or gain risk and its consequent costs.[9] In hot, dry climates where cooling is necessary throughout most of the year, one approach is to develop the subterranean house in combination with the supraterranean house, regardless of whether both parts are owned or used by the same persons. Adding the supraunit will minimize the soil temperature fluctuation factor and the heat loss or gain through the roof and into the subterranean house.

As was previously explained, the negative energy gain of the subterranean house compared to the positive gain in a conventional house results from the earth mass which envelops the building and eliminates thermal exchange with the outside. Resistance to temperature gain is also a factor of the strong resistance of the subterranean house to wind pressure. When the wind (especially a strong one) blows around a conventional house, it creates pressure which cannot be resisted because of cracks and holes in the walls (Fig. II-16). This process of intensifying heat exchange is particularly significant in arid zones because wind turbulences, especially local ones, are frequent. In the case of the subterranean house, the wind will flow above the house causing little or no heat exchange.

According to Bligh, Shipp and Meixel who performed an experimental run during a summer in Minnesota, the heat gain of an aboveground house is six to eight times more than that of an earth-sheltered design.[10] The subterranean house walls store the heat within the house, and drop in temperature at sunset is minimal because of the earth mass surrounding it. In the conventional house, the walls do not retain the day's heat—and the temperature drops sharply at evening—because the house has low heat capacity.

Moisture migration through the soil is a complex process and not yet well

A. CONVENTIONAL HOUSE: WIND MOVEMENT CREATES PRESSURE AND INCREASES THE WIND'S PENETRATION INTO THE HOUSE

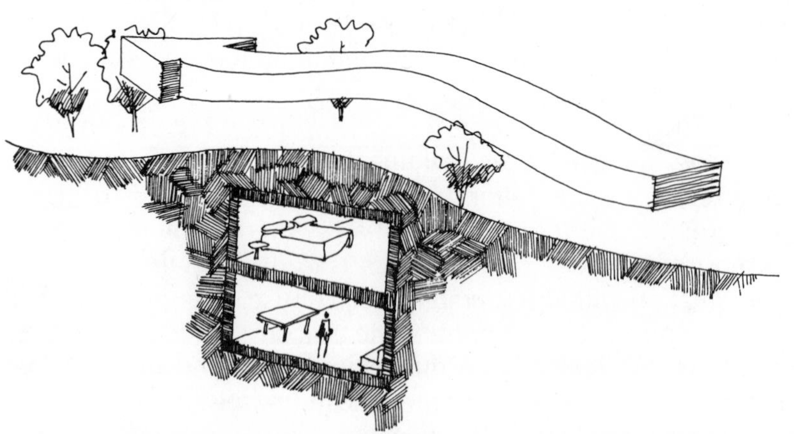

B. SUBTERRANEAN HOUSE: WIND MOVEMENT PASSES ABOVE THE GROUND AND CREATES NO PRESSURE OR PENETRATION

Fig. II-16. Wind effect on heat exchange in conventional and in subterranean houses.

understood. According to Bligh, in a snowy region, the soil is quite wet in the summer because of the melted snow and rain; consequently, the soil heat loss will be quite high which is good for cooling the subterranean house. In the winter when the surface freezes, the soil temperature, which drops slowly, pushes the moisture far from the building, thus drying the soil near the house. Then the heat loss is reduced, to the advantage of the house and the people who live in it[11] (Fig. II-17).

It has been found that conventional homes lose from 35,000 to 70,000 Btus per hour through cracks, doors, windows and other openings. It was also found that heat loss of earth-covered buildings on east, west and north walls and on the roof was dramatically low, 2 to 4 Btus per hour.[12]

The three factors which determine proper orientation of a building are sun, wind and view. Although orientation may result in a compromise among these three, a structure can be designed without such compromise (Fig. II-18). To save energy and heat exchange, most but not all of the openings should be grouped on one side of the building. In hot, dry climates (arid),

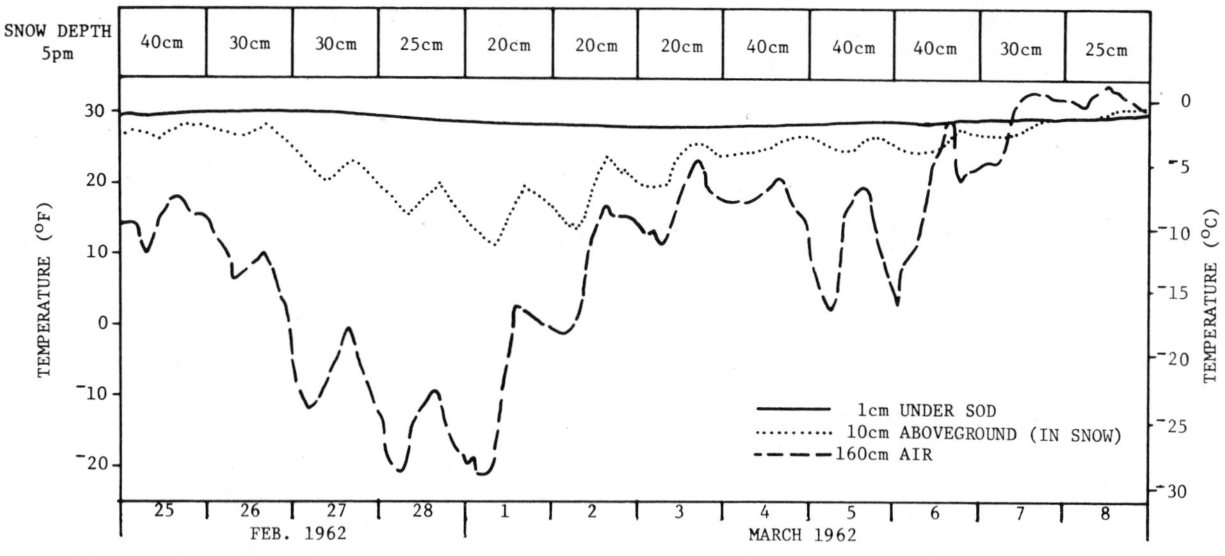

Fig. II-17. The insulation effect of snow during an arctic cold wave, Minnesota. Source: Maxwell, "Temperature Measurements and the Calculated Heat Flux in the Soil," p. 118.

the north side is preferable to avoid heat gain from the sun; while in cold dry climates (north Canada), the south-facing side is best to maximize heat gain. However, some minor openings should be on the opposite side of the major openings to allow ventilation. In any case, all openings must be controlled; when closed, they should permit little or no heat gain or loss.

SOIL TYPES AND THE PROBLEMS OF EXCAVATION

Soil excavation and treatment should be studied carefully, and soil experts should be consulted. Likewise, soil samples should be tested in a laboratory. A discussion of some soils and their ease of excavation follows[13] (Table II-4).

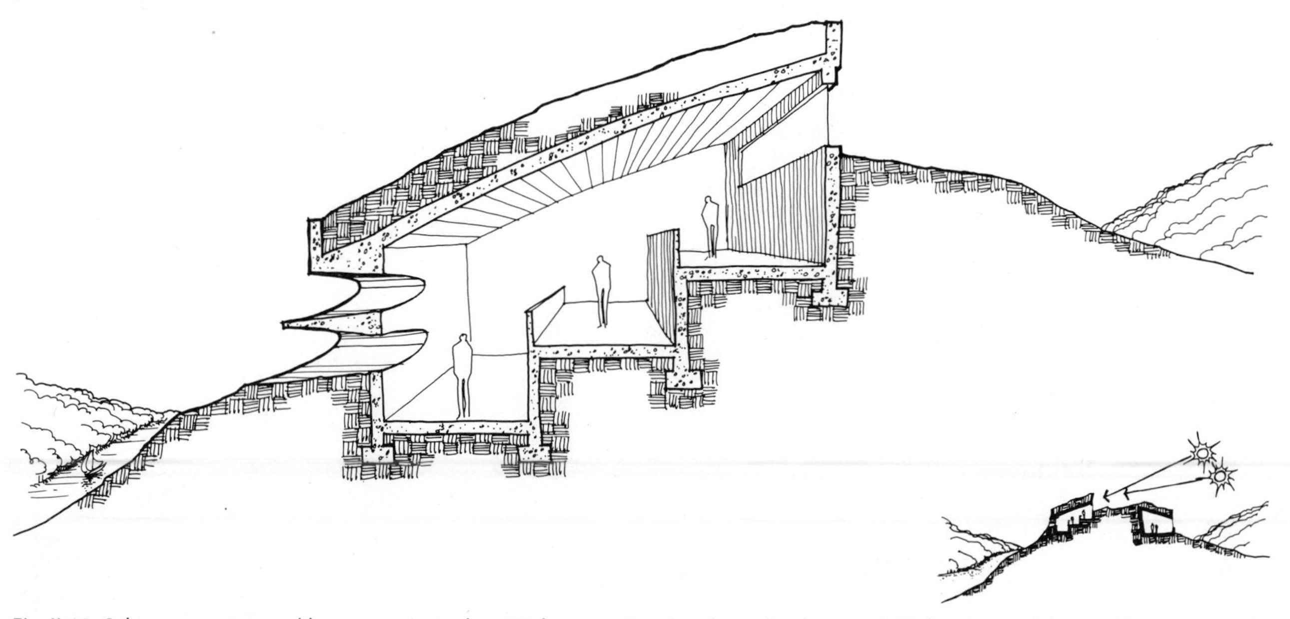

Fig. II-18. Subterranean terraced house concept when (1) the attractive view faces the slope, and (2) the view and the sunshine are opposite each other but both are desired. The ceiling angle coincides with the high angle of the sun in wintertime to permit maximum penetration.

Table II-4. Soil Conservation

	1	2	3	4		5	6	7	8	9	10
				GENERAL IDENTIFICATION							
	MAJOR DIVISIONS	SOIL GROUPS & TYPICAL NAMES	SUGGESTED GROUP SYMBOLS	DRY STRENGTH	OTHER PERTINENT EXAMS....	OBSERVATIONS AND TESTS RELATING TO MATERIAL PLACE	PRINCIPAL CLASSIFICATION TESTS (ON DISTURBED SAMPLES)	VALUE AS FOUN-DATION When Not Subject To Frost Action	POTENTIAL FROST ACTION	SHRINKAGE EXPANSION ELASTICITY	DRAINAGE CHARAC-TERISTICS
COARSE GRAINED SOILS	Gravel	Well Graded Gravel & Gravel-Sand Mixtures, Little or No Fines	GW	None	Gradation, Grain Shape	Dry Unit Weight or Void Ratio, Degree of Compaction, Cementation, Durability of Grains, Stratification & Drainage Characteristics, Ground Water Conditions, Traffic Tests, Large Scale Load Tests or California Bearing Tests	Mechanical Analysis	Excellent	None to Very Slight	Almost None	Excellent
	and	Well Graded Gravel-Sand-Clay Mixtures, Excellent Binder	GC	Medium to High	Gradation, Grain Shape Binder Exam. Wet & Dry		Mechanical Analysis, Liquid & Plastic Limits on Binder	Excellent	Medium	Very Slight	Practically Impervious
	Gravelly	Poorly Graded Gravel & Gravel-Sand Mixtures, Little or No Fines	GP	None	Gradation, Grain Shape		Mechanical Analysis	Good to Excellent	None to Very Slight	Almost None	Excellent
	Soils	Gravel with Fines, Very Silty Gravel, Clayey Gravel, Poorly Graded Gravel-Sand-Clay Mixtures	GF	Very Slight to High	Gradation, Grain Shape, Binder Exam. Wet & Dry		Mechanical Analysis, Liquid & Plastic Limits on Binder If Applicable	Good to Excellent	Slight to Medium	Almost None to Slight	Fair to Practically Impervious
	Sands	Well Graded Sands & Gravelly Sands, Little or No Fines	SW	None	Gradation, Grain Shape		Mechanical Analysis	Excellent to Good	None to Very Slight	Almost None	Excellent
	and	Well Graded Sand-Clay Mixtures, Excellent Binder	SC	Medium to High	Gradation, Grain Shape, Binder Exam. Wet & Dry		Mechanical Analysis, Liquid & Plastic Limits on Binder	Excellent to Good	Medium	Very Slight	Practically Impervious
	Sandy	Poorly Graded Sands, Little or No Fines	SP	None	Gradation, Grain Shape		Mechanical Analysis	Fair to Good	None to Very Slight	Almost None	Excellent
	Soils	Sand with Fines, Very Silty Sands, Clayey Sands, Poorly Graded Sand-Clay Mixtures	SF	Very Slight to High	Gradation, Grain Shape, Binder Exam. Wet & Dry		Mechanical Analysis, Liquid & Plastic Limits on Binder If Applicable	Fair to Good	Slight to High	Almost None to Medium	Fair to Practically Impervious
FINE GRAINED SOILS Containing Little or No Coarse Grained Material	Fine Grained Soils	Silts (Inorganic) & Very Fine Sands, Rock Flour, Silty or Clayey Fine Sands with Slight Plasticity	ML	Very Slight to Medium	Examination Wet (Shaking Test & Plasticity)	Dry Unit Weight, Water Content & Void Ratio, Consistency—Undis-turbed & Remolded, Stratification, Root Holes, Fissures etc. Drainage & Ground Water Conditions, Traffic Tests, Large Scale Load Tests, California Bearing Tests or Compression Tests	Mechanical Analysis, Liquid & Plastic Limits If Applicable	Fair to Poor	Medium to Very High	Slight to Medium	Fair to Poor
	Having	Clays (Inorganic) of Low to Medium Plasticity, Sandy Clays, Silty Clays, Lean Clays	CL	Medium to High	Examination in Plastic Range		Liquid & Plastic Limits	Fair to Poor	Medium to High	Medium	Practically Impervious
	Low to Medium Compressibility	Organic Silts & Organic Silt-Clays of Low Plasticity	CL	Slight to Medium	Examination in Plastic Range, Odor		Liquid & Plastic Limits From Natural Condition & After Oven Drying	Poor	Medium to High	Medium to High	Poor
	Fine Grained Soils	Micaceous or Diatomaceous Fine Sandy & Silty Soils, Elastic Silts	MH	Very Slight to Medium	Examination Wet (Shaking Test & Plasticity)		Mechanical Analysis, Liquid & Plastic Limits If Applicable	Poor	Medium to Very High	High	Fair to Poor
	Having	Clays (Inorganic) of High Plasticity, Fat Clays	CH	High	Examination in Plastic Range		Liquid &Plastic Limits	Poor to Very Poor	Medium	High	Practically Impervious
	High Compressibility	Organic Clays of Medium to High Plasticity	OH	High	Examination in Plastic Range, Odor		Liquid & Plastic Limits From Natural Condition & After Oven Drying	Very Poor	Medium	High	Practically Impervious
Fibrous Organic Soils with Very High Compressibility		Peat and Other Highly Organic Swamp Soils	Pt	Readily Identified		Consistency, Texture & Natural Water Content		Extremely Poor	Slight	Very High	Fair to Poor

Sources: Corps of Engineers, U. S. Army, *The Unified Soil Classification System.* Technical Memorandum No. 3-357, Vols. I, II, III. Vicksburg, Mississippi: Waterways Experiment Station, 1953, Table 1, Table A1, Table B1; and R. H. Karol, *Soils and Soil Engineering,* London: Prentice-Hall International Inc., 1960, pp. 40–41.

Sand and Gravel. These soils are easy to excavate and remove. However, because they are loosely packed, they require support if the house is to be built on a slope. Sand is easily moved by light winds and, therefore, necessitates occasional cleaning in the house. Sand is heavy when the water table is high and will require heavy machines for removal, thus increasing excavation costs. Because sand and gravel are not solid masses, they will increase pressure on the walls of the house.

Silty or Noncohesive Soils. Silty or fine-grained soils are noncohesive, especially when very dry; they are easy to excavate with scrapers. When the water table is intruded, soil excavation becomes more complicated. As is true with sand, fine-grained soils have the tendency to flow into the excavated space, so stablizing support is required. These soils also create pressure on the building. However, subsurface water has a negative effect on the excavation and the building, and a site with subsurface water should not be considered.

Clays. It is hard to excavate clay when it is dry; but when it is moist, it can be handled effectively by scrapers. Clay excavation is quite difficult when the clay has high moisture content or is situated in an area with a high water table. Excavation may also become costly when stabilization techniques are required. Soft clay exerts high lateral pressures, both when excavation is taking place and after construction of the structure.

Igneous Rock. These are the basic masses which formed the earth, including granite, diorite or basalt. They usually form a solid mass without any stratifications; these are very hard rocks and require blasting and energy for excavation and removal, a costly process. Excavation below the water table is not difficult unless the rock is full of faults and joints which cause water penetration. However, because it provides a vertical wall without any support, well-planned excavation may not require interior construction of walls for the subterranean structure.

Sedimentary Rocks. These are primarily the rocks resulting from sedimentation of mostly organic materials deposited within the sea, emerging later above the water surface by tectonic movement of the earth. They always form different layers of various thicknesses which distinguish them from other rock types; they are subject to water intrusion as well. Although their original layers were horizontal, they may have changed to become vertical, curved or slanted due to geological dynamics. As such, they represent a wide variety of forms and characters which range from loosely cemented conglomerates to soft or hard sandstone and to very hard limestones. In addition, their degree of hardness varies; they may even combine alternate hard and soft layers. Hard layers formed horizontally can be ideal for forming halls and wide spaces for subterranean housing, offices, factories and spaces for public assembly halls. Quarries which are systematically excavated can form ideal spots for such structures and land use, minimizing the cost. Because of their variation in character, limestones may require various techniques for excavation; therefore, the cost involved may vary. Some limestone would require blasting for removal. Sandstone or limestone which has a high porosity and is below the water table would require various means of water control and efficient waterproofing for the structure. Also the faults, joints or contact lines of the stratification are subject to increased water movement and must be handled accordingly. The most troublesome types of rock are the shales. Since they contain clay, they usually expand when they are removed from pressure, therefore endangering excavation slopes.

Metamorphic Rocks. These are rocks which have gone through a recrystallization process under high-temperature conditions. The product depends on the nature of the original materials and their deformation and decomposition throughout this metamorphic process. These rocks require blasting for removal. They usually contain faults and are uplifted which may result in unstable slopes. The water table can create problems with such rocks.

In conclusion:

1. Soil is both an efficient thermal insulator and a storage medium.
2. The significance of the soil mass is in its function as thermal storage. Moreover, the slow thermal movement between outside temperature and the zone enveloping the building constitutes a lag which can offer the right thermal condition in the right season.
3. Seasonal thermal changes have more impact on the soil than the diurnal thermal changes.

Innovative Design of an Energy-free Cooled House for Hot, Dry Climates

The following describes a proposed design for the innovative house mentioned earlier—one that embodies certain ancient key principles. The concept introduced here is a passive system for cooling, especially desirable in a hot dry climate. The method is based primarily on lessons learned from ancient civilizations. It is this author's intent to apply historical principles, combined and improved in order to meet our modern building standards and cultural needs. The new cooling system is based on three systems which are to be integrated to form one system:

1. Subterranean Placement: the use of the soil mass as an insulator, as in the arid zone of Cappodocia, North Africa, and northern China.
2. Passive Ventilation: used throughout time in Mesopotamia, Persia and other parts of the Middle East.
3. Evaporative Cooling: exercised throughout history in different places of the world, especially in the Middle East.

The ultimate goal is a combination of the three systems in order to cool the arid-zone house without consuming conventional energy resources.

SUBTERRANEAN PLACEMENT

Before dealing with the design per se, we must further discuss the nature of the soil which will surround the underground building and its relation to the theory underlying the proposed ventilation and evaporative cooling systems.[14]

Soil. Soil is a major factor in the development of the underground house because it determines the successful integration of the ancient concepts of underground ventilation and evaporative cooling. The nature of the soil must be studied first to determine its suitability for subterranean construction.

The four most common types of arid-zone soils are:

1. Eolian, such as the sand in the various forms found inland or on coastal plains;
2. Loess;

3. Sedimentary, such as that found in alluvial fans, alluvial plains, or piedmont regions;
4. Playa (actually an old lake bed), which is primarily saline.

In selecting from among the types of soil listed above as sites for subterranean structures, the following limiting criteria should be weighed:

1. Water table;
2. Humidity (diurnal and seasonal) and the capillarity of the soil;
3. Temperature at different depths (seasonally and daily) and degree of heat conduction;
4. Faults;
5. Sensitivity to earthquakes;
6. Possible existence of karst, caves, mines, tunnels or other hollow spaces;
7. Under- and aboveground erosion patterns;
8. Soil behavior (expansion and contraction) as a result of changes in temperature and humidity;
9. Density (which can be determined by measuring the sonic velocity or the electrical resistance of the soil);[15]
10. Ecological condition of the soil, such as the existence of plant roots and animal burrows which may have an impact on the foundations of structures;
11. Potential threat of landslides or rock movement.

Using these criteria, we find that loess seems to be most suitable for both construction and agriculture. Granular soil, such as sand, or soil which is subject to frequent contraction and expansion with changes in temperature or humidity, such as alluvial deposits, is not suitable for underground structures.

In other words, the arid-zone planner of subterranean structures must select a site where the soil will (1) minimize heat transfer, (2) retain humidity, and (3) be landscaped and shadowed to minimize the influx of solar radiation. We need to investigate these three planning goals more extensively.

Heat Transfer. As we have stated, soil temperature varies according to its depth, the nature of the ground and with regard to seasonal and daily external temperature fluctuation (Fig. II-19). The primary cause of surface soil temperature variation is the change in the intensity of shortwave radiation. As Van Wijk tells us:

> Absorption of both short-wave and long-wave radiation practically takes place in a soil layer of a fraction of a millimeter thickness. Thus, a flat surface of bare soil can be considered as a plane heat source during a period of positive net radiation flux, and as a heat sink when the latter is negative.[16]

More specifically, the temperature of the earth's surface tends to follow the daily or seasonal air temperature above it.[17] In hot, dry arid zones and under appropriate conditions, the fact that the earth will act as a heat storing and transferring agent and carry energy from the sun by conduction to the

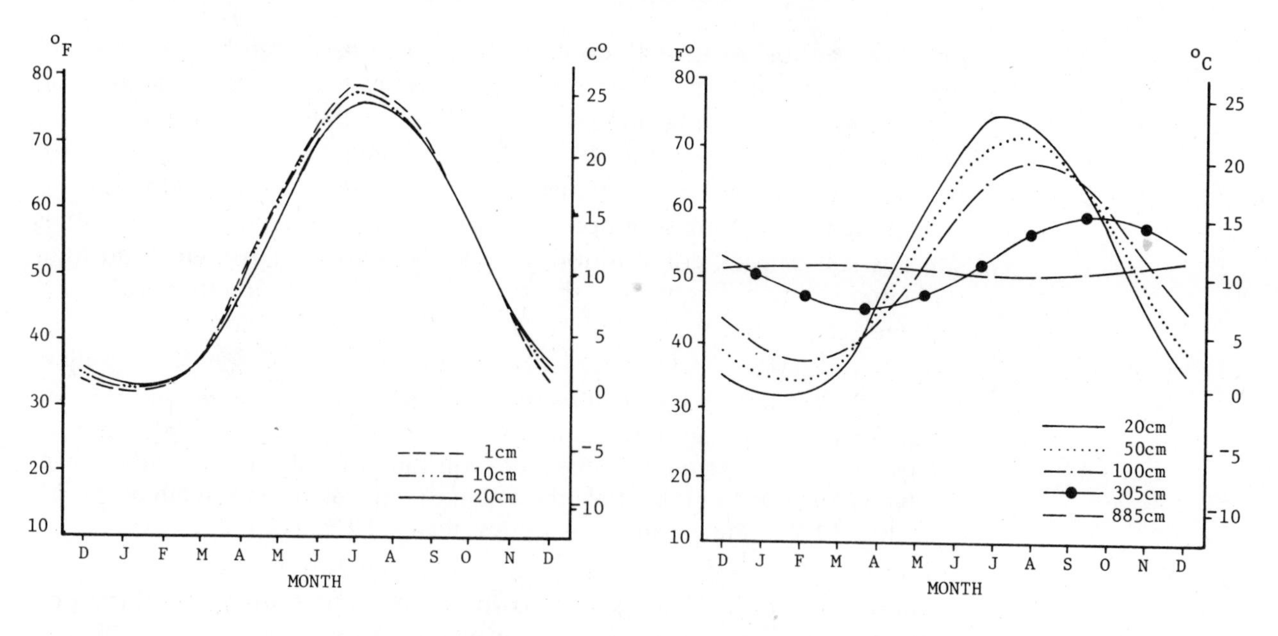

A. SHALLOW DEPTH, MONTHLY AVERAGE. ILLINOIS

B. DEEP, MONTHLY AVERAGE. ILLINOIS

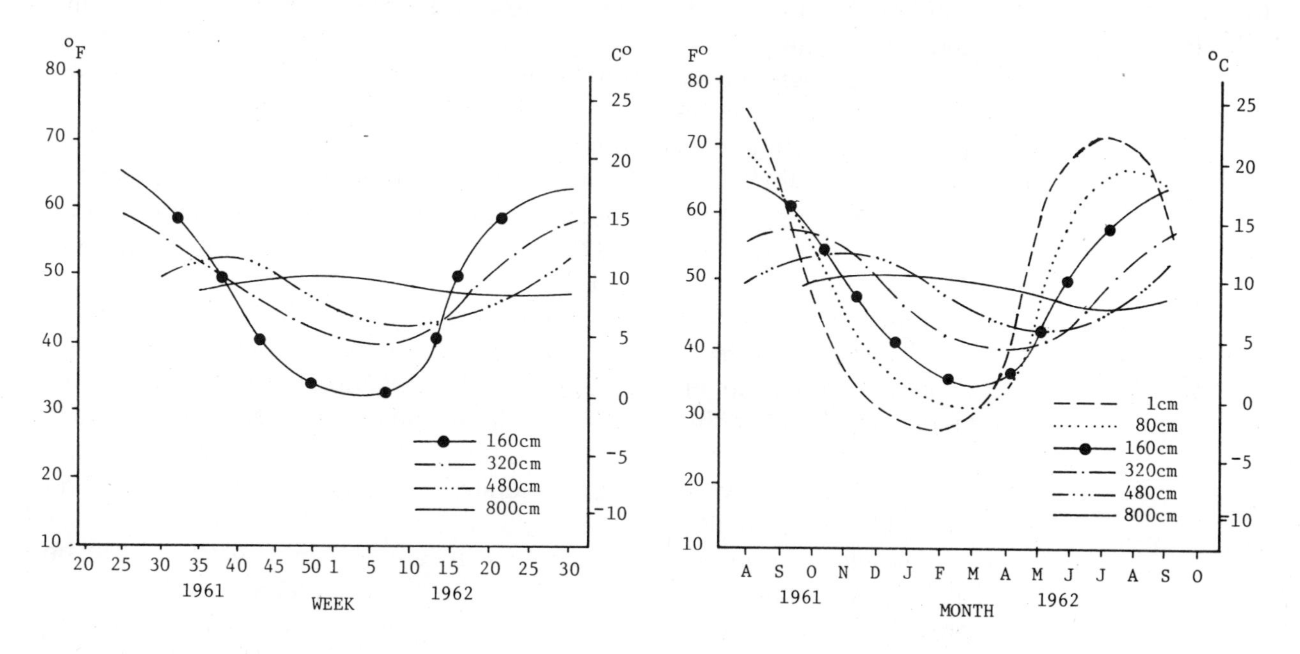

C. DEEP UNDER SOD,WEEKLY MEANS. MINNESOTA

D. DEEP UNDER SOD, MONTHLY MEANS. MINNESOTA

Fig. II-19. Weekly and monthly average progression of soil temperature at different depths. Both Illinois (A and B) and Minnesota (C and D) are cold and snowy in winter and warm and rainy in summer. Note that the closer to the surface, the more fluctuation in temperature. The graphs show more stability toward the depth of 8 meters. Sources: A and B are adapted from James E. Carson, "Analysis of Soil and Air Temperatures by Fourier Techniques," *Journal of Geophysical Research*, 68, no. 8 (1963), 2221. C and D are adapted from: Maxwell, "Temperature Measurements and the Calculated Heat Flux in the Soil," p. 75.

dwelling below has to be recognized as a phenomenon that is not wholly desirable.

However, one principle arguing for the concept of subterranean structures is that, beyond a depth of 40 centimeters (3 feet) or more, earth temperature is rarely affected by diurnal cycles of air temperature and solar radiation.[18] Thus, even a 40-centimeter layer of earth can be an excellent insulator, protecting a subterranean dwelling from extreme fluctuations in daily temperatures. "The annual fluctuation of earth temperature, however, extends to a depth of 30 to 40 feet."[19] We need to remember, then, that heat transfer cannot be totally discounted, at least on a seasonal basis.

Although the extent of temperature penetration depends upon variables such as soil or rock composition, the temperatures themselves, and their daily and seasonal fluctuation, the essential factor is the inverse proportion between depth and thermal penetration. As Spiegel has noted,

> The effect of surface temperature at a level 5 feet under the ground shows a substantial dampening of the amplitude, and at 10 feet below the surface, in most locations, the ground temperature will vary annually on the order of magnitude of plus or minus 6°F from the average temperature. At a depth of approximately 30 feet, the variation will be less than + 1°F.[20] (Fig. II-20)

Spiegel also notes that, "going down below 30 feet, the general temperatures will become warmer because of the warm earth core."[21] Thus, at some depth, the earth serves not only as an insulator, but actually as a heat supplier. Underground structures are especially suited to exploit this geothermal energy.

In an aboveground structure, heat flows readily from the external to the internal environment, because like a fluid, heat moves from one space to another if a differentiation exists between them. The actual flow between two spaces will depend on the quality and quantity of the insulation between them. In subterranean situations, the flow will still be from the warmer space to the cooler space, and this flow will stop when the temperature of the two spaces becomes equal.

Considering all of the above, we obviously must consider the question of heat transfer from the soil and earth to the underground dwelling. In fact, we must consider not only soil and earth temperature, but also the activities taking place within the structure itself; for among the dynamic variables which contribute to the thermal condition within the shelter's space are occupancy and activities, which generate heat, and ventilation, which will facilitate heat exchange.

Although the indoor temperature of an underground structure will be equivalent to that of the surrounding earth, this temperature will not always provide the precise conditions for human comfort. A few facts should be kept in mind: a comfortable interior temperature is generally considered to be 27°C or 68°F; and although a fairly constant temperature can be maintained in underground structures, their natural temperatures will often be below the level of human comfort. In general, consistent temperatures of 9°C or 40°F can be experienced at a depth of 10 meters.[22] In the winter the 18° difference between earth and ideal indoor temperatures can be ameliorated by human heat radiation, lighting, cooking, washing machines and dryers, refrigerators, televisions and radios, and other indoor energy sources, with the remainder provided by heating. In summer, the indoor temperature may not require any adjustment in nonarid zones.

However, in arid and semiarid regions, the indoor temperature may often exceed the level of human comfort, even the highest tolerable level of 85°F.[23]

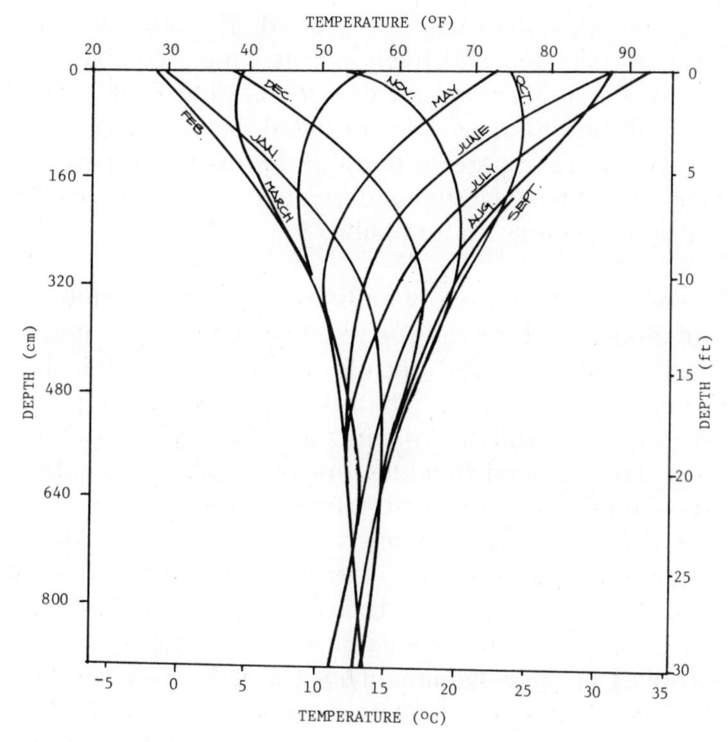

A. BLACKTOP SURFACE

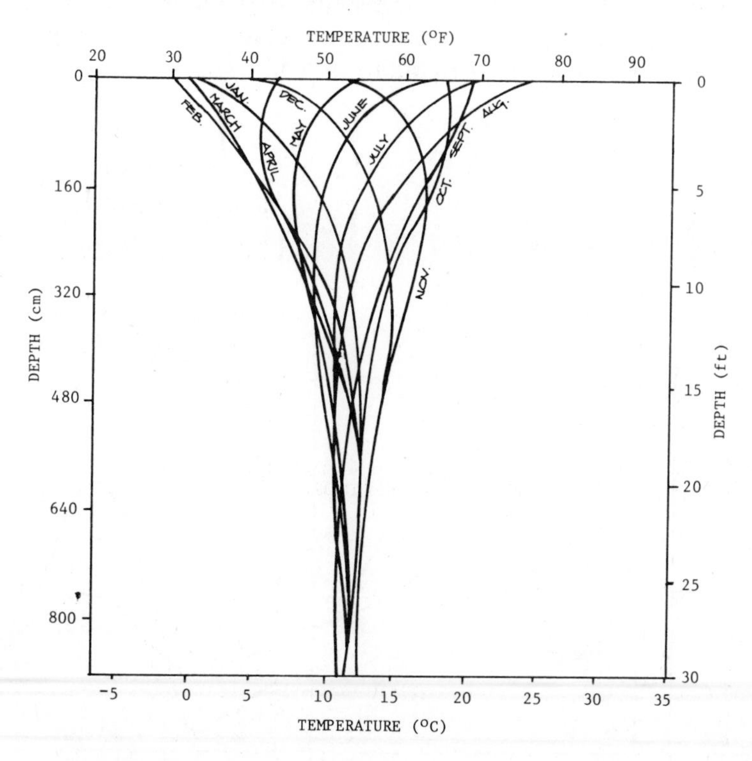

B. SHORT GRASS

Fig. II-20. The insulating effect of soil cover on the changes of the monthly average earth temperature at different depths, Washington, D.C. Source: Adapted from T. Kusuda, "The Effect of Ground Cover on Earth Temperature," pp. 302 and 303.

In this case, a cooling system will be required. The internal environment can, however, be regulated effectively by the simple combination of two systems, one for humidification and one for passive ventilation. Both of these systems will be discussed later in detail. Since, using conventional means, cooling and heating are responsible for nearly 20 percent of total energy consumption, this suggested underground system will result in a reasonable reduction in energy consumption.[24]

Humidity Retention. The impact of moisture and its fluctuation, especially in the form of precipitation on the surface, is second only in importance to the nature of the soil. However, because of the low precipitation in arid zones, this impact is less than in humid regions. Thus, underground structures are generally more suitable in arid zones than in humid ones; for the dangers of excessive ground humidity are practically nonexistent in arid areas, and special insulation may not, therefore, be required.

However, in arid zones, soil humidity can act as an effective buffer against extreme surface temperatures and is, therefore—in moderation—quite desirable. Evaporative cooling can thus be used to decrease the impact of hot air temperatures upon the interior environment of a subterranean structure, not only by cooling but also by humidifying. In situations where the natural environment does not provide adequate amounts of moisture to achieve this cooling effect, we can thus introduce a type of artificial humidification. This process will be described in detail later.

Landscape and Shadow. Soil temperature is not only affected by solar radiation but also by the amount of shade caused by trees, buildings and topographical forms, as well as by altitude, soil density and moisture level (Fig. II-21).

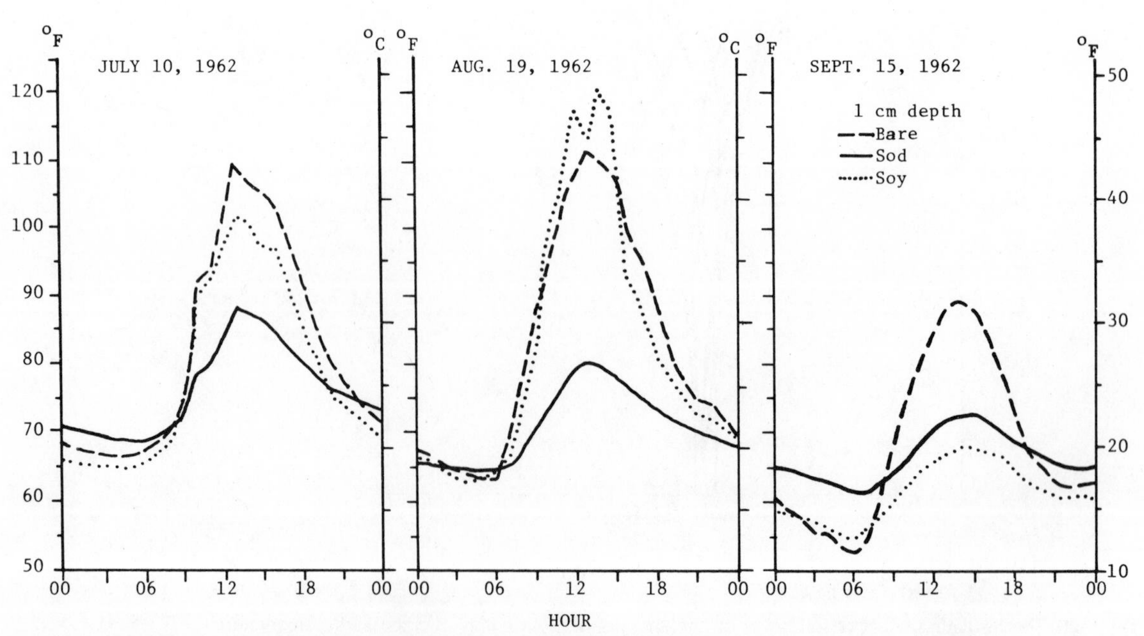

Fig. II-21. The effect of vegetative cover on the diurnal temperature wave of the soil at a depth of 1 cm, Minnesota. Source: Maxwell, "Temperature Measurements and the Calculated Heat Flux in the Soil," p. 110.

Dense vegetation will absorb a considerable amount of incoming radiation. Minimization of solar radiation's effect on the soil can be achieved not only by the presence of vegetation, however, but also by the orientation of the rows of crops or trees in order to cut off direct radiation and to increase shadow.[25] The presence of trees or vegetation does, in addition, reduce reflection, absorption of heat, albedo and the air temperature immediately aboveground. Because vegetation will probably be present to some degree, daily soil temperature will not be purely a function of solar radiation. Recognizing the variable role of vegetation and shadow is especially important in the case of semisubterranean structures; for to make them habitable, solar heat gain must be minimized by creating shadows and vegetative cover. The closer the shelter is to the surface, the higher the solar heat gain will be.[26]

An interesting extension of the role of vegetation and shadow outlined above can be observed in existing dense urban centers where little or no direct solar radiation ever reaches the ground. In these areas, the seasonal and daily stabilization line (beyond which no temperature fluctuations are recorded) moves very close to the earth's surface. Thus, compact urban construction aboveground—an advantage itself in arid zones—will make the subterranean sites below even more stable and indeed more suitable for construction.

Our review of the studies concerning soil (the results of which should be adjusted as we move from site to site) has led us to the following conclusions: (1) Soil temperature and moisture vary according to depth and soil composition; (2) in every potential soil there is a border line dividing the area affected by the diurnal and seasonal temperature fluctuations and that

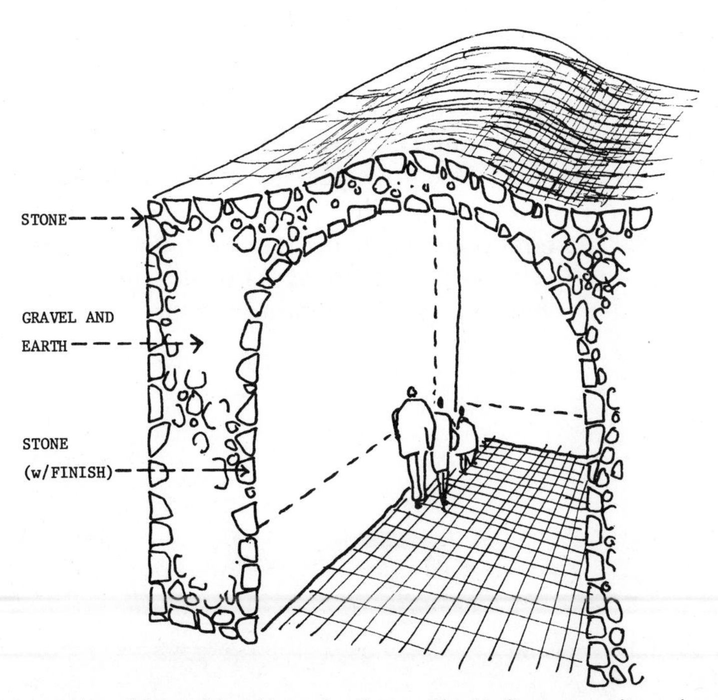

Fig. II-22. Cross section of the traditional Jerusalem house. The Mediterranean climate has a dry, warm summer and a cold, wet winter. The combination of this climate with thick walls of 1m or more (approximately 3 ft 3 in) produces a comfortable, moderate climate within the house during both winter and summer seasons.

featuring stable temperatures; (3) watering the soil cannot only change its texture but also increase its temperature and absorption of radiant solar energy. In evaluating the advantages of a particular site for underground construction, these lessons should be remembered.

We can see some of the issues discussed above exemplified in the traditional Jerusalem house (constructed until the 1940s) still in use and typical of many in the Middle East (Fig. II-22). Located in arid and semiarid zones, these houses were constructed with consideration given to their climate and environment, and the design reflects the experience of centuries.[27] Walls average 1 to 1½ meters in thickness and are composed of two shields made of stone with earth filling between them. The limited rain and humidity of the winter season is absorbed over a period of several months from the outer environment into the earthen core of the walls. By the beginning of the dry summer season, the inner walls are sufficiently humid to moderate the house's microclimate by decreasing the ambient temperature of the house. By the beginning of the winter, the situation has reversed. This system provides adequate environmental control during most of the dry, hot summer and the relatively cold, wet winter. In a very different climate, the Eskimos use a thick cover of snow or sod for creating a controlled environment (Fig. II-23).

PASSIVE VENTILATION

This is an important factor to be considered in the planning of the subterranean house, especially with regard to building codes and health needs. On the other hand, ventilation by itself has little impact on heat exchange in the arid-zone house. Ventilation, however, can and should be used in a passive way for air exchange. Differences in the temperature of air masses (between indoor and outdoor) and in relative humidity should produce air movement and, therefore, ventilation.

Ventilation in the subterranean house is as essential as it is in the aboveground one. In the latter, however, the house openings supply most of the ventilation needed through the doors, windows, walls or cracks in the house. In the subterranean house, the mass of soil enveloping the house limits the ventilation, which then occurs primarily through controlled openings: the ventilation chimney and the windows, if there are any.

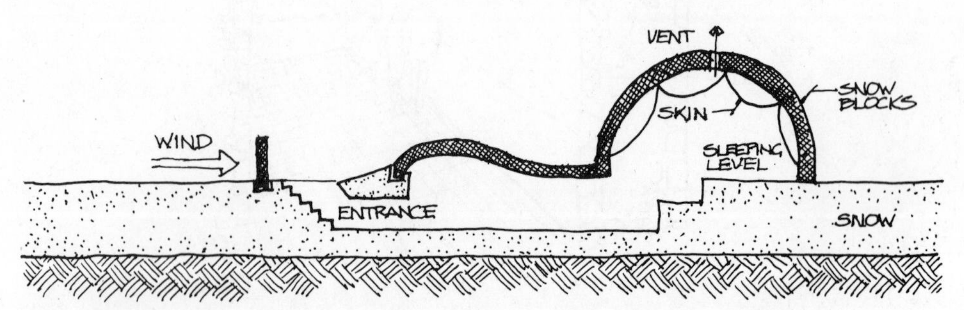

Fig. II-23. General cross section of the Eskimos' igloo, a semisubterranean shelter for the very cold winter. The summer shelter is similar in plan but made of stone or sod. The sleeping level is raised to be closer to warm air and to reduce drafts. Note the wind barrier near the entrance. Adapted from a number of sources.

Ventilation is required for health needs, to supply fresh air with oxygen, to take away odors (especially from kitchens and rest rooms) and to eliminate overheating produced by people and machines (from cooking, washing or large gatherings). In short, new air is needed for refreshing. If a fireplace is present in the house, it is necessary to plan adequate ventilation. However, it is most efficient to have a few forced-air systems in the house for ventilation and this can be done through the proposed ventilation chimney.

One other aspect which necessitates ventilation and exchange of air is the need to reduce relative humidity. High humidity within a subterranean house may result from cooking, sweating, crowded gatherings, washing, showering and bathing, indoor plants, etc. There is little, if any, possibility that high humidity will be present in the soil in arid zones, but it still may result from floods occurring in the area or from overwatering of surface plants adjacent to the house. However, in nonarid areas, high relative humidity may occur; if buildings are not waterproofed properly, the moisture might rust metal, cause dampness, form condensation, damage furniture, cause discomfort and perspiration and form mold on interior surfaces.

Thus, ventilation is necessary; and passive ventilation should be designed, although electrical types may be installed for emergency use. Since backup heating will be used sometimes, the failure of electric ventilation could create a very dangerous carbon monoxide situation. Thus, built-in passive ventilation is necessary as a safety precaution and is also suggested for the development of the cooling system. Henry Orlowski tells us that natural ventilation is most effective in high, narrow shafts called thermal chimneys. According to one authority, a ventilation rate of 1 to 2 c.f.m. (cubic feet per minute) per square foot is adequate.[28] The passive ventilation system in the subterranean house should be designed and constructed as an integral part of the house. The most common and effective design is one for a chimney (or roof-catch) high above the roof of the building and leading to its lowest interior level. To permit air circulation within the house spaces, it is necessary to have small openings at the upper level of the house and opposite the entry side. In a house to be built on the slope, the chimney may need to be horizontal rather than vertical (Fig. II-24). The study of the wind pattern of the site will help to determine which should be the intake side and which the exit.

In hot-dry or cold-dry climates, wind should be avoided because the dryness has a negative effects on humans, animals and vegetation. When it carries dust particles which hit facial and other skin, wind is especially undesirable.

However, in hot, dry climates, wind can be treated in special ways to make it a positive element by:

1. Combining the wind with surface water to reduce wind temperature;
2. Passing the prevailing wind through shadowed zones, especially trees, which will reduce the temperature.

Facilities of tunneled ventilation should also be made to meet the needs of a large gathering in the house. In dusty areas the shaft-chimney for air ventilation should be constructed high above the ground (4 to 5 meters) in

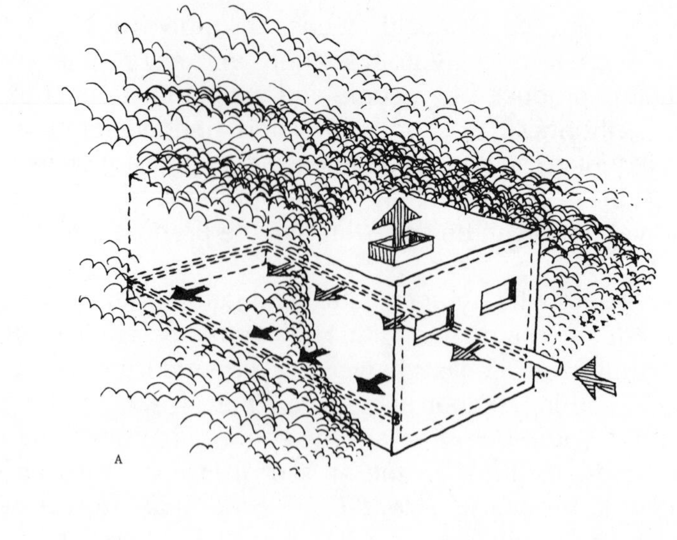

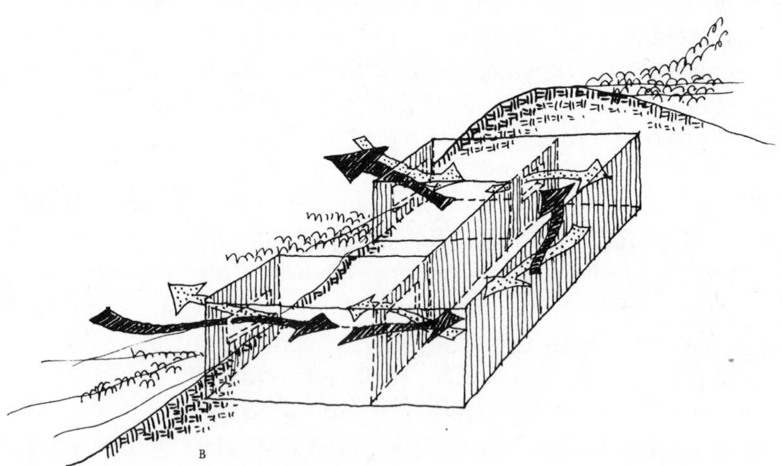

Fig. II-24. Two examples of a horizontal ventilation system for a house built into a slope: (A) single entrance with shaft for exit; (B) dual entrance-exit system.

order to avoid dust or other unnecessary items passing through, to catch upward moving air flow and to minimize vandalism from outside. The aboveground part of the chimney can be designed to join with the placement of the solar heat collector or with the garage, which may be placed aboveground. Openings for ventilation and for other purposes should themselves be designed in such a manner as to be safe for children and to avoid animal penetration and other dangers.

The success of any semisubterranean or subterranean structure such as the one introduced here is largely based on the effectiveness of the ventilation system designed for it. In an underground structure, ventilation will cause an exchange in temperature between the interior and the exterior space. This exchange will bring about either the gain or the loss of heat. Thus, if underground housing is not designed to maximize heat loss when the exterior space is cooler than the interior, or to minimize heat gain when the outside is warmer, then ventilation may defeat the purpose of the underground structure.

It should be clear now that ventilating the subterranean structure is vital for human activity within the shelter.

The American Society of Heating, Refrigeration, and Air-Conditioning Engineers (ASHRAE) requirement is a little more stringent than the previously mentioned authority:

> Ventilating with pure outdoor air is the most economical method for maintaining the necessary chemical quality of air in a shelter. The recommended minimum ventilation rate of 3 cubic feet per minute per person of fresh air will maintain a carbon dioxide concentration of about 0.50 percent and an oxygen concentration of approximately 20 percent, by volume, in a shelter occupied by sedentary people.[29]

This "air replacement rate of 3 c.f.m. per person," ASHRAE continues, "will not be, however, in itself sufficient to limit the resultant effective temperature to 85°F (29°C) under many conditions unless the supply temperature is less than about 45°F (7°C)." Furthermore, those norms will necessarily need to be higher in public gathering spaces such as schools or factories.

An automatic, self-regulated oxygen apparatus can adjust and provide the necessary air circulation in the shelter. Shelters housing many concentrated activities, such as manufacturing facilities or shopping centers, will be safer with such equipment; furthermore, such a system will enable the managers to regulate the thermal comfort levels without endangering life or impairing health.

Precisely controlled ventilation is more necessary for the semisubterranean structure than the subterranean; for, being partly above and partly below the surface, such a structure is more subject to the influence of extreme outside temperatures. Ventilation, however, causes heat exchange; therefore, when the outside temperature is undesirably high, the temperature of this imported air must be reduced. A ventilation system, when combined with an evaporative system to control humidity and temperature, can reduce the effect of such uncomfortably high temperatures.

Humidity control, we should point out, is essential in the semisubterranean arid-zone shelter if we are to reduce temperature to a comfortable level. Every soil, including that of the arid zone, contains some degree of moisture.[30] In the soil surrounding the arid-zone semisubterranean or subsurface structure, such humidity is absolutely desirable, since it lowers air temperature and thereby improves comfort. Since loess soil in arid zones retains humidity well because a dry and relatively hard crust forms on the surface to seal in moisture at deeper levels, it is preferable for arid-zone underground construction. In nonarid zones, however, humidity in the ground must be controlled when subterranean structures are built. Since soil composition is the primary factor determining the capability of the soil to retain humidity, this should be considered before selecting a site. Similarly, for any site under consideration, whether in a nonarid or an arid zone, it is necessary to measure the soil humidity throughout the year and consider the source of the moisture.

Moisture due to the wall humidity, together with that introduced through ventilation, will decrease the temperature of the air which flows into the subterranean section of a given structure. However, when the available moisture does not produce the habitable thermal environment desired, it will then be necessary to introduce an artificial humidification system, which will treat the air before it enters the subterranean space.

Not only do we have to be prepared to add moisture in arid zones, but we also must be ready to handle the effects of excessive moisture; for, although arid zones feature very low humidity and very low water tables (if any at all), deep subterranean structures may be subject to dampness and mildew because of the combination of soil moisture and inadequate ventilation. This dampness will occur especially when the shelter is not occupied. In such cases, "conventional methods such as mechanical dehumidifiers, silica gel, or calcium chloride can be used to control the stand-by environment."[31]

Having now established *how* a ventilation system must function for an underground house, we should turn to the system itself. To properly establish air circulation within the house, air may enter through a series of ver-

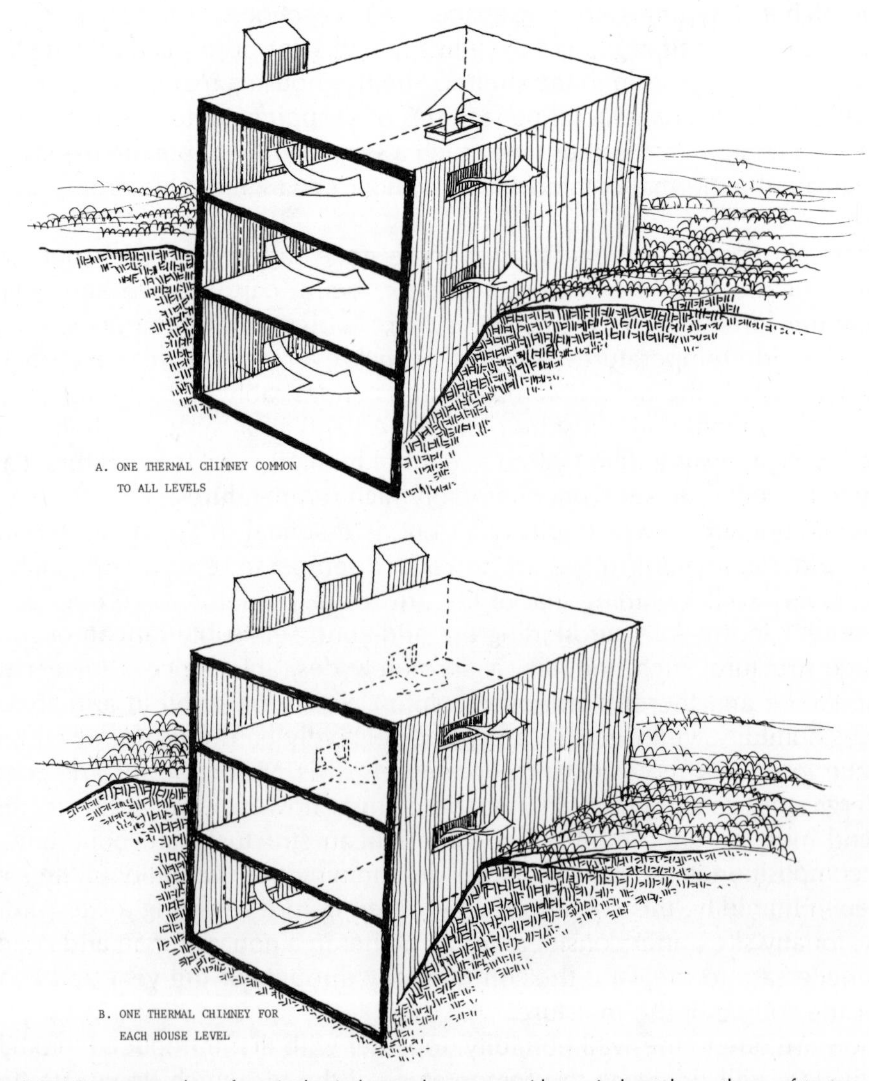

Fig. II-25. Two examples of vertical wind tunnel systems with a wind catch on the roof to trap and channel air and vertical (A) and horizontal (B) surfaces to expel it.

tical tunnels in the walls of the structure which begin at the roof and lead down to the subterranean dwelling space (Fig. II-25). Recall that we have suggested horizontal shafts in a house on a slope, and openings on exposed vertical surfaces which do not lead to shafts may be used. Different schemes of tunnels and openings can be combined as the site and the building's design dictate.

These air tunnels should have vents at each end which are maneuverable and thus can be opened or closed fully or partially for maximum control. The roof catches and the vertical surface openings can have two, three or four facets to catch air from or expel air to all possible directions (Fig. II-26). We may also design the system so that it is reversible as wind direction changes (Fig. II-27).

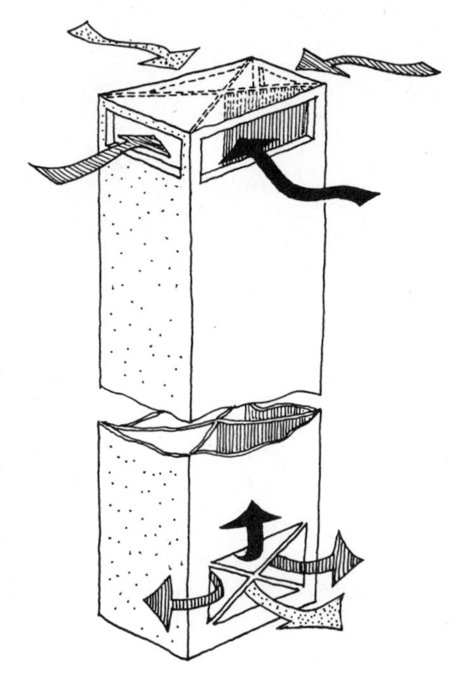

Fig. II-26. Four-faceted roof and vertical-surface air catch for possible diurnal changes of wind direction.

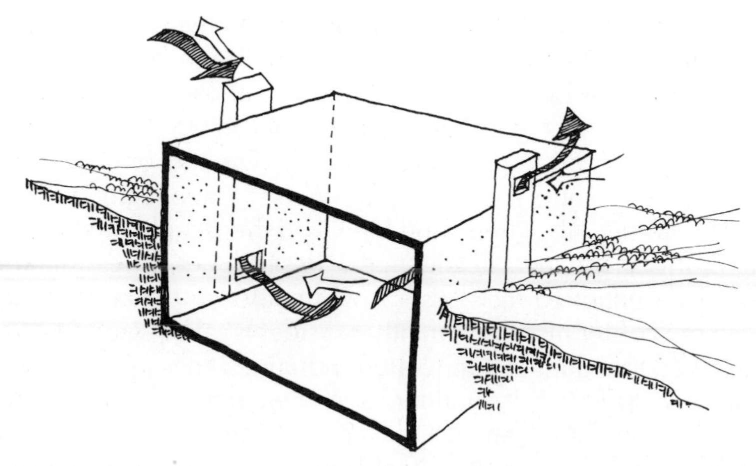

Fig. II-27. An air-circulation system which will reverse its flow as prevailing winds shift.

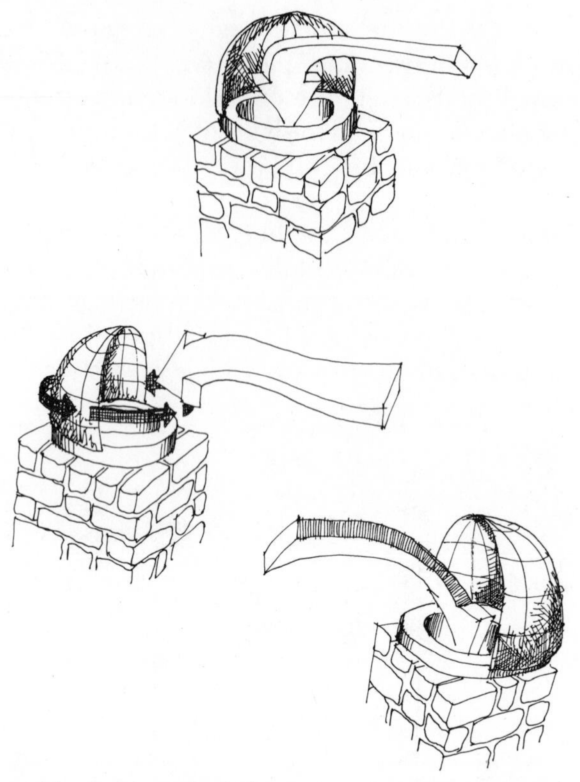

Fig. II-28. Rotating semicircular air catch which moves as the air moves. Based on a similar arrangement used in Australia for ventilating mines.

It might be very practical to use, with improvement, an Australian concept of an air catcher (Fig. II-28). To catch air from all possible directions of the shaft, a freely rotating, curved and half-circled lightweight form should be used. This catcher will rotate itself and force the wind to move forward into the shaft and to the house. The air catcher induces the air flow into the shaft. The air movement will begin from the catcher and not from the small window opening on the house because of its being much higher and subject to open wind flow. The shaft with the air catchers should be high, a few meters above the roofs. Its opening level should not be blocked by other buildings. High chimney openings also support proper ventilation and eliminate fumes and odors from the neighboring environs. We can state, in general, that the higher the shaft aboveground, the more effective the wind catcher, and the less dust will intrude into the house as well.

In interviews conducted by Sydney Baggs of the University of New South Wales, Australia, it was found that among the dwellers of Coober Pedy and White Cliffs the dugouts stayed cleaner than the aboveground houses because dust entered only when wind blew and the people forgot to close the opening of the shaft; such negligence also produces a heat build-up. The interviewees mentioned that ". . . it was cooler in the warm months and warmer in the colder months than other residents' aboveground dwellings." They observed that dugout temperature remained reasonably "constant" at approximately 20°C (68°F) summer or winter, day or night. Likewise, although a bathroom had been constructed "outside," it was hardly ever used. It was too hot in summer and too cold in winter.[32]

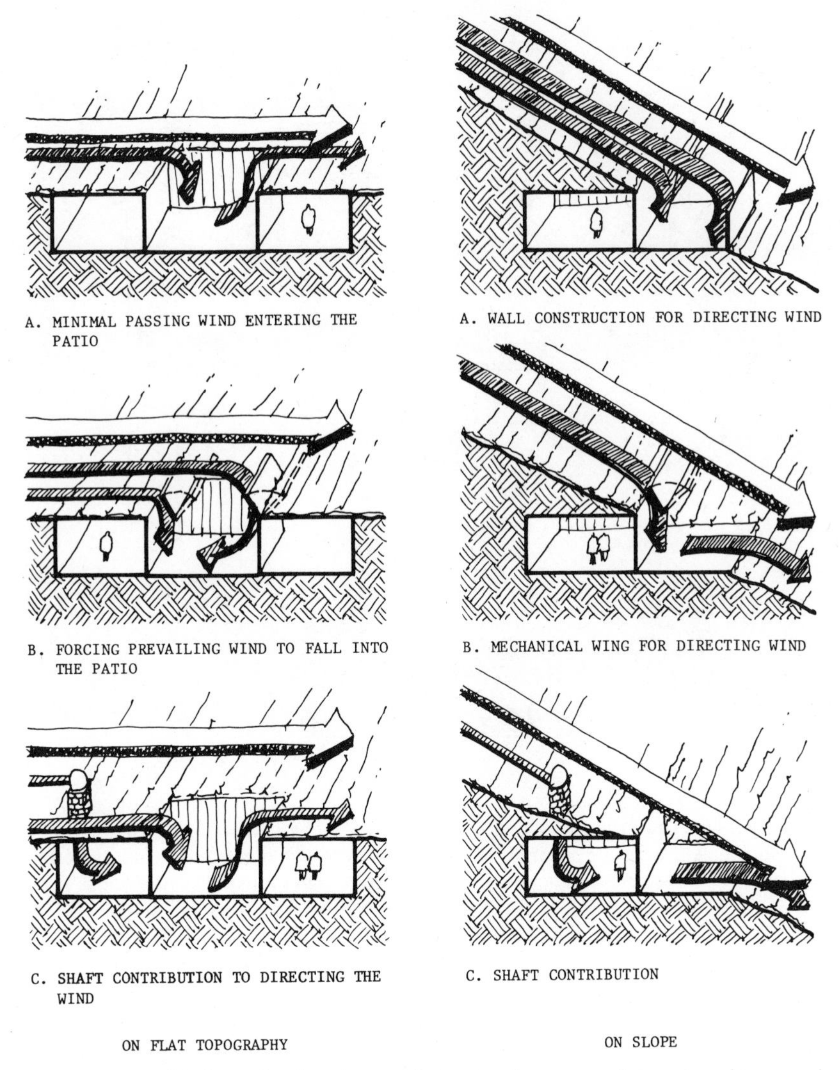

Fig. II-29. Various possible forms for directing ventilation in a house on a flat or sloped topography, with patio or semipatio.

Ventilation into the house can be obtained through other ways, too. In addition to the thermal chimney system discussed here, every design may require different and special elements (Fig. II-29). The patio house has a unique configuration in which the open patio creates air gradients when air moves above the house; yet when ventilation is desired, wind can be forced mechanically to enter the patio.

The aerodynamic pattern is subject to change throughout the day and the season. In warm, dry climates, the wind direction changes continuously throughout the day because of the air heating process. In the later part of the night, the area is mostly subject to inversion with its cold and stable air. Therefore, the ventilation system should be designed to be adjusted to changes in wind direction (Fig. II-30). Also, the openings of the shaft should be so constructed they could be shut completely when air movement is not desirable.

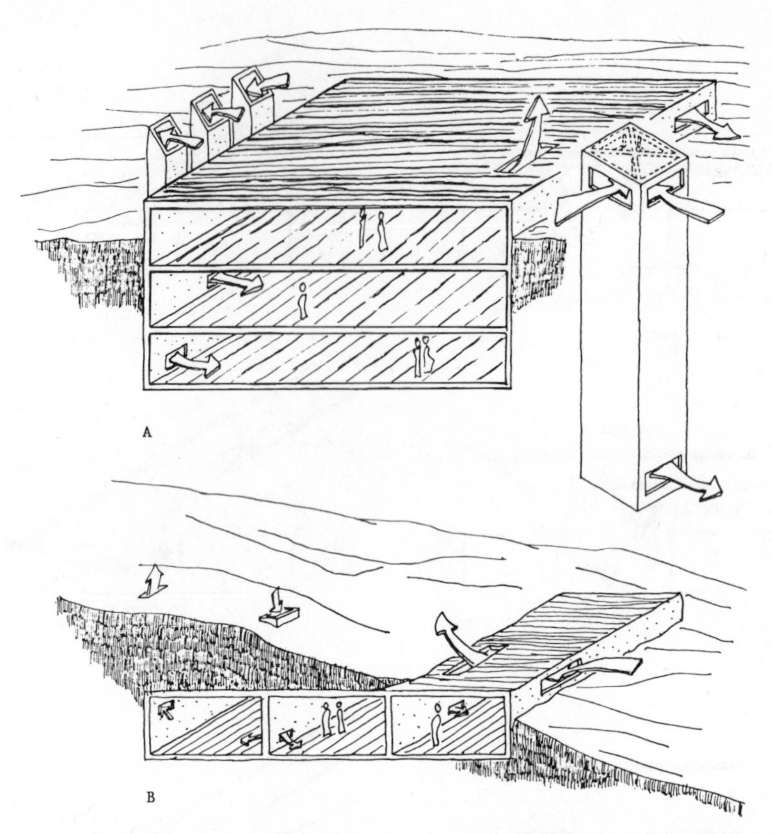

Fig. II-30. Various ventilation forms for the subterranean structure: (horizontal (A) and vertical (B)). The system should adjust to the changes in wind direction and meet the needs of horizontal or vertical structures. Nonresidential structures are usually built with more than one floor.

COOLING BY EVAPORATION

The design of the ventilation system cannot ensure air movement into the system and to the house. But passive ventilation can be developed if a differentiation in temperature, and thus in pressure, exists between the outside and the inside. Indeed, the rotating air catcher at the top of the chimney will support movement downward through the chimney, but such movement can be insufficient and may bring warm air into the house. An additional system is necessary if we need a cooling air movement within the house.

Heated air is subject to expansion; dynamic cold air is subject to contraction, stability and movement downward. Differences in temperature between one mass of air and another can establish air movement. Air mass is cooled and moves downward when humidity is increased. Water in a porous ceramic jar becomes cooler because of evaporation of droplets on the outside of the jar (Fig. II-31).

Natural ventilation can occur when there is a difference in pressure between outside (at the top of the chimney) and inside the house, and thus a difference in air temperature. It is possible that the height of the chimney from the ground will have an effect on the amount and the temperature of the air brought in from the outside. Thus, the higher the chimney, the greater the differentiation in the forces. The ratio of the dimensions of the chimney

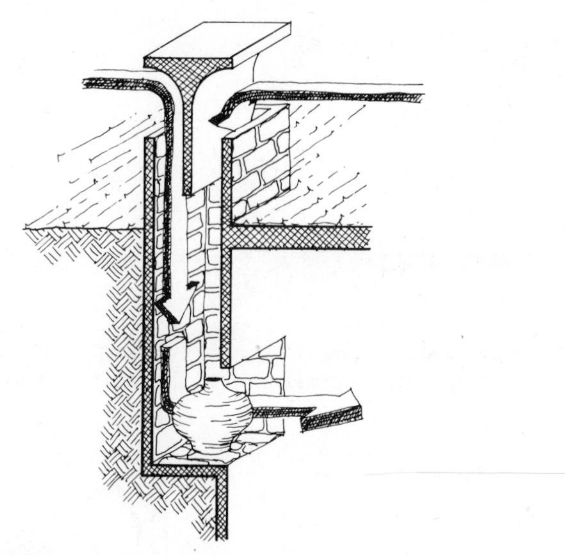

Fig. II-31. Traditional system of water cooling by evaporation in use in a hot, dry climate. In Iran, residents traditionally used indoor chimney openings for placing ceramic jars filled with water. The water moved through capillaries to the outer surfaces of the jars where it was evaporated by the dry air movement, cooling the water in the jars.

opening should not exceed 2 to 1. In addition, the chimney should be straight, extending a minimum of 1 meter above the roof line, it should have no air leakage which would reduce air draft, and, of course, it should be built of noncorrosive materials.[33]

At the subterranean level, as we have already noted, soil humidity will seep through the walls and help reduce the temperature. However, we still need to incorporate the flexible air-ventilation system described above in the design at these levels and use it when the need arises. As we also noted earlier, in a semisubterranean shelter, the higher the floor, the less humid the environment and the greater the influence of outside temperature. Thus, the higher levels will probably require artificial humidification in order to cool the air, create differentiation in its density and, therefore, force it to move downward. Any one of the following systems might be used to provide such results (Fig. II-32).

1. Wall Humidification: In this system, moisture is dispersed throughout the structure's soil-surrounded walls. The water would move by capillary action from outer to inner walls and then humidify the hot air which is entering the interior space of the structure. Watering in this system must be carefully regulated and supplied only when necessary. The watered soil close to the structure's walls should also be covered, paved or shaded to minimize evaporation. Water can be supplied externally from a water pipe extending around all the walls of the structure (Fig. II-32A).
2. Drip Humidification: In this system, regulated water drips from pipes built into the blocks, passes into the outer walls of the air tunnels of the blocks and moves by capillary action to the air passing through the tunnels (Fig. II-33). In this case (to distinguish it from wall humidification), the pipe is part of the prefabricated blocks (Fig. II-32B).

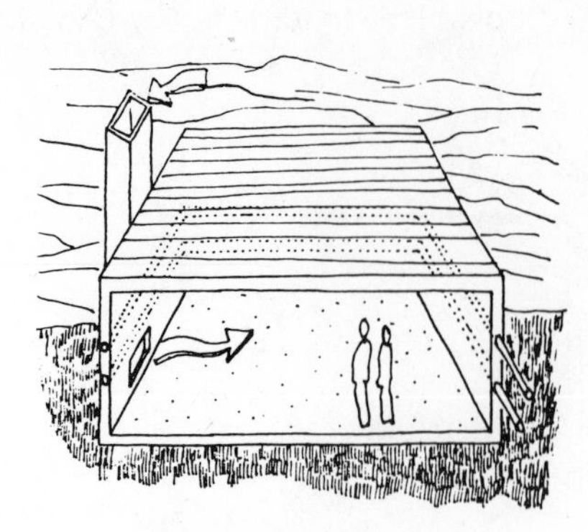

A. WATER DRIPPING PIPES BUILT CIRCUMFERENTIALLY INTO THE WALL TO PROVIDE CONTROLLED WATER FLOW

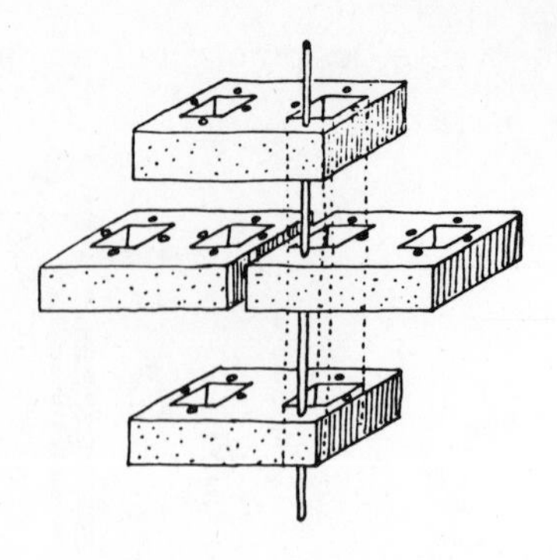

B. PIPES BUILT INTO THE STANDARD CEMENT BLOCK TO DRIP WATER AND USE THE BLOCK HOLES FOR VENTILATION

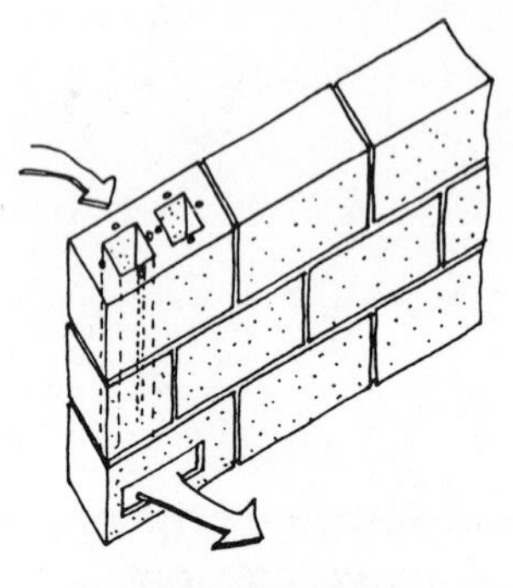

C. WALL HUMIDIFICATION BY AIR PASSING THROUGH WATER SURFACE OR BY DRIPPING PIPES ON CERAMIC AIR TUNNELS

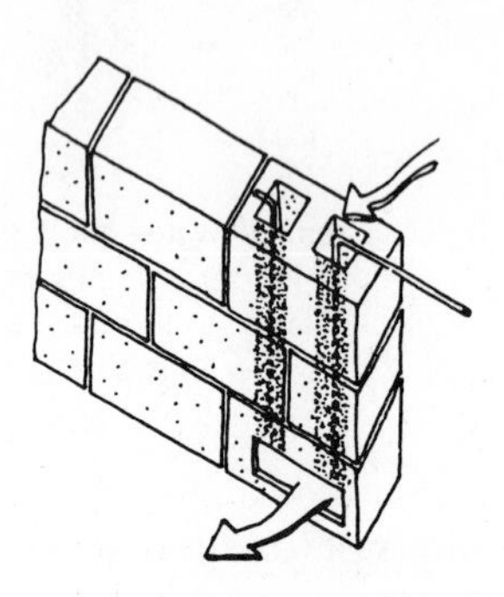

D. PROVISION OF MIST THROUGH PIPES WITH SPECIAL VALVE INTO THE VENTILATION BLOCK HOLES

Fig. II-32. Cooling by evaporation humidification. Four possible passive methods for air cooling by humidification. Humidified air becomes cooler and therefore heavier and is forced to move downward into the air tunnel and to the house.

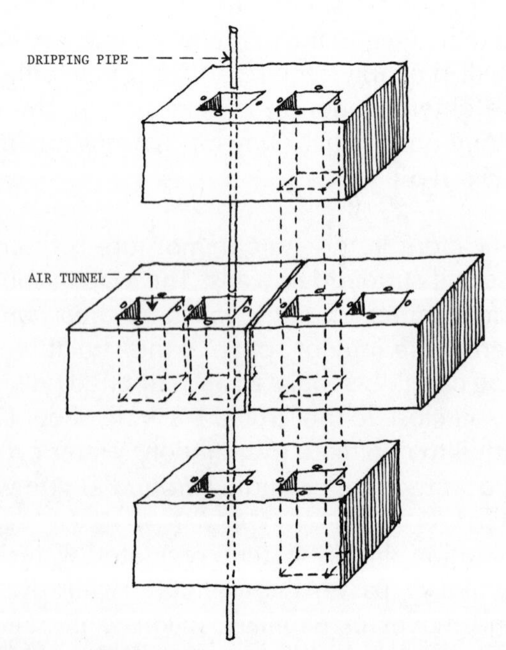

Fig. II-33. Dripping pipes built into the wall blocks near air tunnels. Humidification should be combined with the passive ventilation system in order to be effective.

3. Drip Humidification Through Tunnels: A prefabricated terra-cotta tunnel is built into the wall (Fig. II-34). This air tunnel, if constructed of relatively thin materials, will be most effective in accelerating the evaporative cooling of the hot air passing through the tunnel (Fig. II-32C).
4. Mist Humidification: In this system the hot outside air is mixed with water mist when entering the air tunnel or opening. The air temperature will fall as a result of evaporative cooling before the humidified air enters the room (Fig. II-32D). The density and temperature differentiation will force the air to move downward. This system seems to be the most effective and is easy to install.

Another cooling system design has a shallow water container within the tunnel as an integral part of it (Fig. II-35). Here too, the movement of warm air will cause water evaporation and, therefore, reduce air temperature before it enters the house.

Finally, we can use the system of intake humidification. In this system, water is dripped on a filter of shrubbery at the entrance to the air tunnel or opening. This process may bring about results similar to those obtained through mist humidification; but, overall, it will be less effective since it will require stronger air movement because of the impeding effect of the shrub filter. To strengthen air movement it will be necessary to have a separate air tunnel or opening for every floor rather than a combined system serving all or most of the floors.

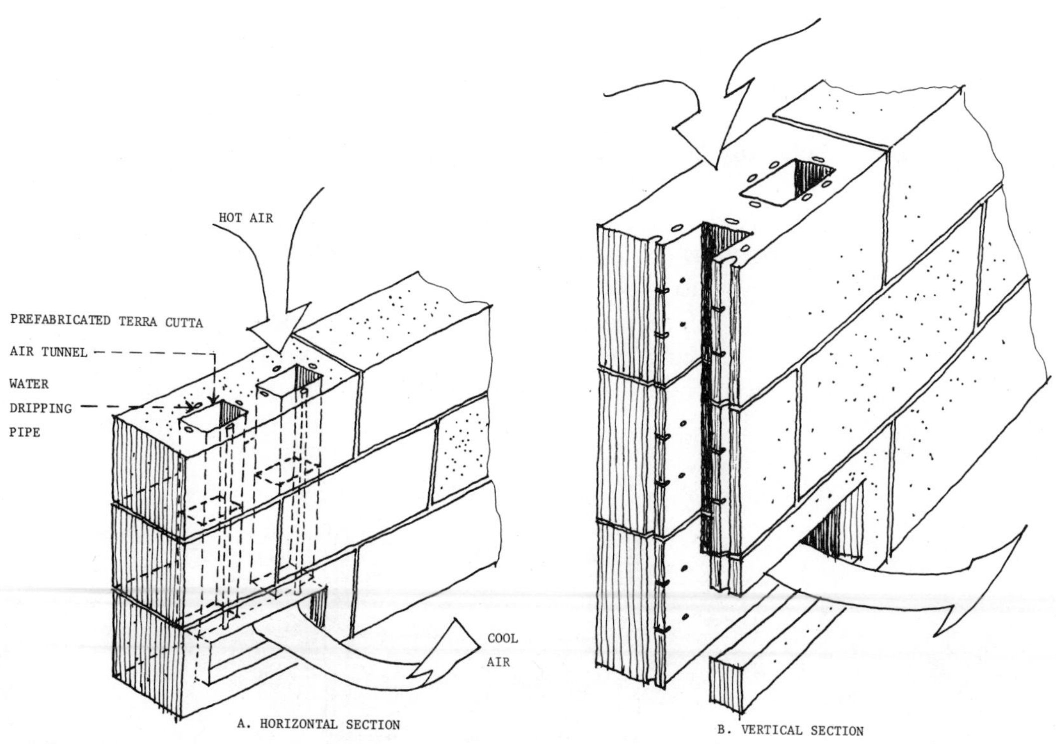

Fig. II-34. Cross sections of prefabricated air tunnels placed in dwelling walls with vertical dripping pipes, which will humidify the clay wall of the air tunnel and, through capillary movement, cool hot air passing through the tunnels.

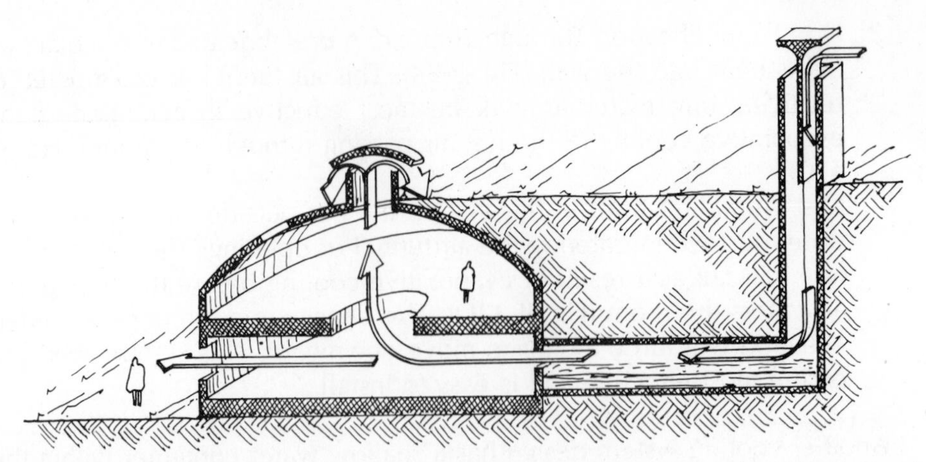

Fig. II-35. House design concept with passive evaporative cooling system for a hot, arid climate: air movement over water surface. Domed ceiling facilitates air exchange and ventilation. Such a form can be especially effective for building public gathering halls.

Humidifying the ground and walls should be terminated a few weeks before the winter season so that the soil can dry. It then will retain the still warm outside temperature's encroachment so that the underground space in the winter will be warmer than the outside temperature and, therefore, comfortable. However, all things considered, the mist humidification of the tunnel air is preferable to either wall intake or drip humidification because it is more easily regulated. Figure II-36 depicts such a humidification and cooling design's operation in a two-story semisubterranean and subterranean structure, featuring the ventilation system proposed earlier. Such a design may also, as Figure II-37 indicates, be used aboveground.

The ventilation and evaporative-cooling concept introduced above is a passive system employing wind force and air density differences. It works simply through the appropriate orientation of tunnels and the strategic location of interior vents. There is no need for fans or blowers since the incoming air, now cooler due to humidity-induced density, will move down toward the floor and then, as it warms, will rise within the subterranean space as it is replaced by further incoming air. The location of the inlet and the outlet openings is therefore crucial. Small outlet openings at high levels will be necessary to support this ventilation and cooling method (Fig. II-38).

Improvement of the evaporative cooling system also can be accomplished by (1) making the shaft longer, and (2) having the air of the shaft move through a water surface. This method was used in Mesopotamia and Persia.

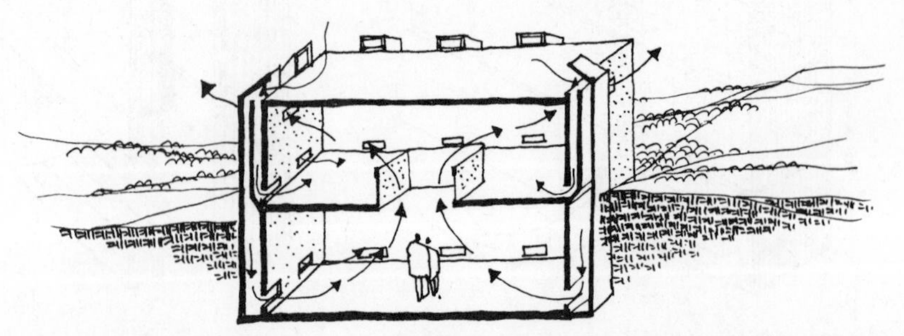

Fig. II-36. Air flow through a two-story semisubterranean house using the proposed passive ventilation and evaporative cooling system.

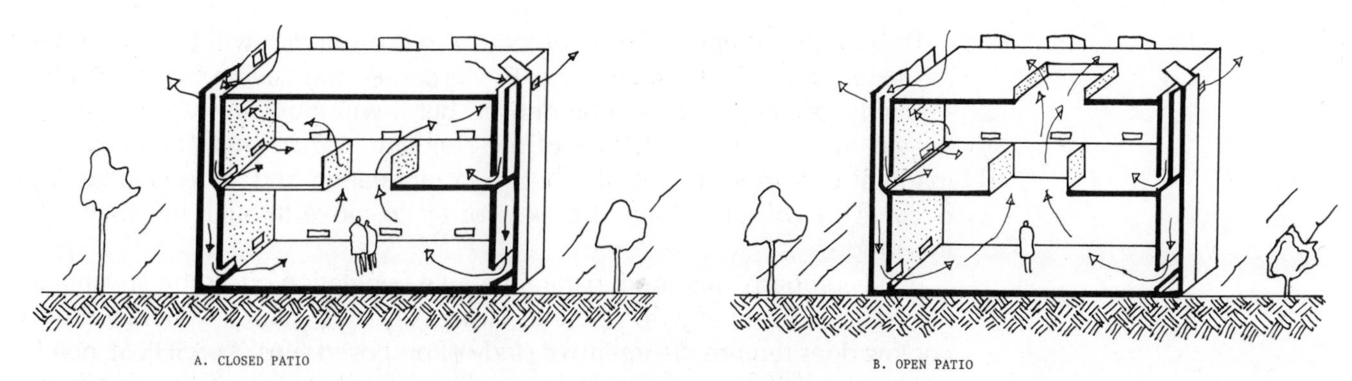

Fig. II-37. Use of the cooling systems of combined passive ventilation with evaporative cooling in an aboveground house is also possible, although it will be less effective because of the probable heat gain throughout the house walls.

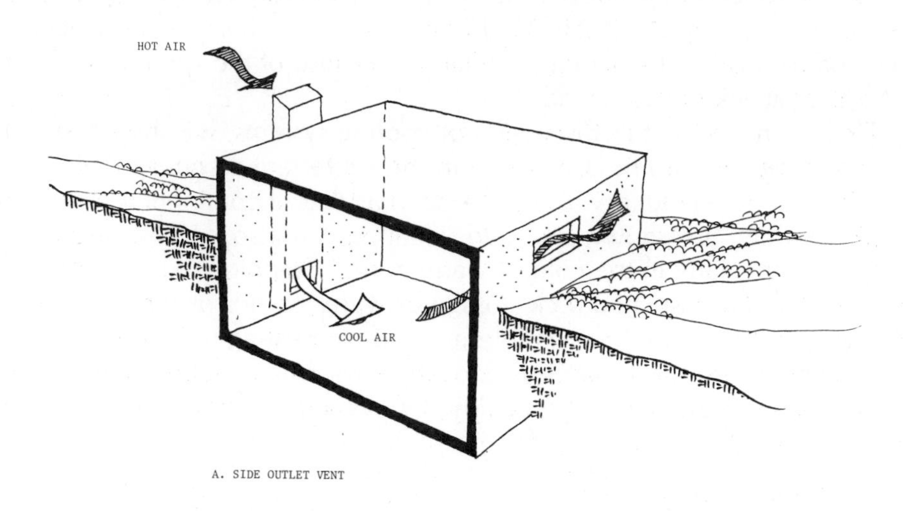

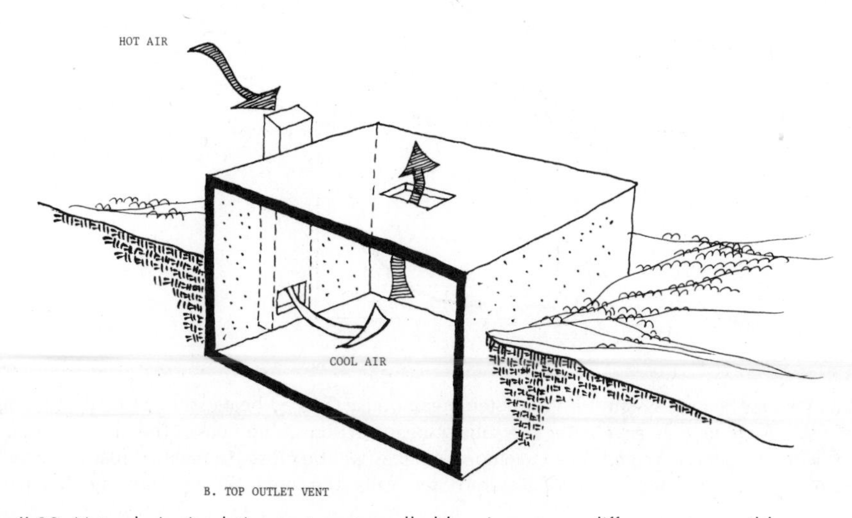

Fig. II-38. Natural air-circulation system propelled by air pressure differences caused by evaporation and channeled through passive system passageways.

The interior temperature of the house in the summer will be lower than the exterior ambient temperature. Being cooler and having higher relative humidity, the air inside will be denser, but it will move upward when it is heated by the varied activities of the residents. When small openings are made at the upper part of the house, a circulation and air exchange will occur. This will encourage the exterior air to move through the shaft into the inside.

If thus properly designed, natural passive ventilation offers the advantage of requiring no energy to operate. However, such natural ventilation and cooling does require the intensive study of proposed sites. As ASHRAE notes: "Natural ventilation is affected by inside-outside temperature or air density differentials, changes in wind velocity and direction, pressure distribution over building surfaces, nearby obstructions, and a network of flow restrictions." It must also be kept in mind that, "A ventilating rate that is satisfactory under a favorable set of conditions may be inadequate under other conditions."[34] As this caveat suggests, there are some disadvantages to a natural ventilation system because of the uncertainty caused by the possibility of changes. As ASHRAE admits, "there is presently no accurate method for evaluating natural ventilation because of the variable and complex post attack environment."[35]

Finally, in closing this discussion of cooling systems, we should mention that heating can also be a problem in the subterranean house. This would be the case in a cold, dry arid zone, such as that of continental plateaus or regions which are snowy and cold. Heating can also be passive in some regions, especially in the cold, dry arid zones. Trapping the solar heat and storing it within insulated rock storage or water barrels or directly blowing the air into the house are possible methods of heating (Fig. II-39).

In conclusion, the proposed innovative systems of subterranean placement, passive ventilation and cooling by evaporation have an effective re-

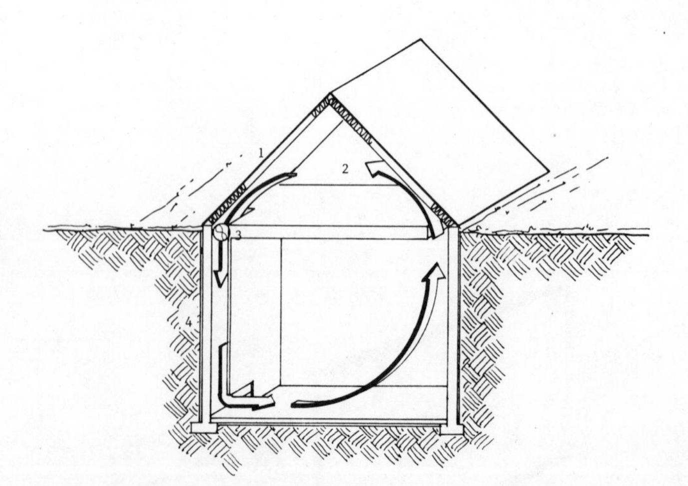

Fig. II-39. Winter heating system for the subterranean house. Forced heated air can be directed from the attic to 1.5 cm tunnels enveloping the subterranean section of the house. The fan will force the air only when the attic is heated. The envelope openings will be closed at night. 1. Black shingles, 2. heated air, 3. small fan, 4. insulated and waterproofed walls. The roof heating system: after B. Givoni.

sult when they are all combined (Fig. II-40). It should also be indicated here that the optimum benefits of this combination can be achieved in hot, dry climates where a cooling system is essential.

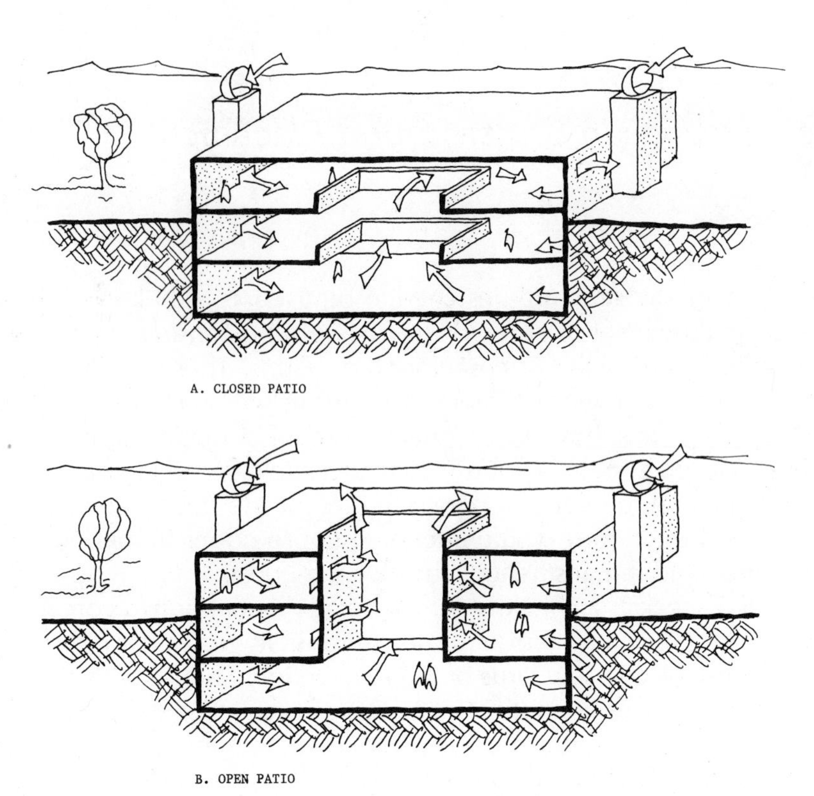

Fig. II-40. Combined application of the three innovative systems—subterranean placement, passive ventilation and evaporative cooling—in a semisubterranean house with closed or open patio.

Design Issues in Underground Placement

The quality of any structure designed to be used as a shelter for living can be evaluated by a set of criteria which should be consistent with the local building code and with our contemporary norms. These norms are formulated by societal, economic, moral or other values stemming from cultural standards. They are, however, subject to frequent modification throughout the life of a society and an individual; but their dynamic changes do not greatly influence the general principles set for the house's overall design. Instead, most often they would affect the house norms in the indoor space requirement and the aesthetic values for the house.

Table II-5 suggests a checklist for house quality under two conditions, the conventional house and the subterranean one. If we compare some design elements, norms and standards of the conventional house with those of the subterranean house, we notice many conditions which favor the subterranean house. Some of those conditions are determined by the reciprocal interaction between the environment and the house in each of the two cases, and also by the design and the construction standards required for the subterranean house.

An average American house usually includes certain standard elements which may be applicable to subterranean houses:

- Bedroom
- Living room
- Family room
- Recreation room
- Bath and shower rooms
- Laundry
- Storage area
- Garage
- Workshop space
- Closets
- Deck
- Space for garbage
- Sewage system
- Water system
- Telephone network
- Electricity network
- TV network
- Heating and cooling equipment.

Table II-5. Comparative Checklist of Norms and Standards of Conventional and Subterranean Houses

ELEMENT	CONVENTIONAL	SUBTERRANEAN
Building material	Varied	Must be strong
Design overall	Varied	High standard required
Noise protection	Varied but mostly plenty	Absent
Pollution	Varied	Absent
Energy consumption	High	Low
Safety:		
Fire	Varied	Highly protected
Earthquake	Varied but mostly risky	Secure
Storms	Varied but mostly risky	Secure
Light and air	Varied but mostly plenty	Can be limited and require special design considerations
Sunshine	Varied but mostly plenty	Can be limited and require special design considerations
Outdoor space	Limited	Ample
Privacy	Limited mostly	Good

To improve the standard components of the house, an open patio can be included, surrounded by the house on three sides and open to the north (in hot, dry arid zones) or to the south (in cold climate zones). This design will minimize solar radiation and reflection onto the patio, and maximize shadow (Fig. II-41); or in the case of the cold climate, the patio open to the south will obtain maximum sunshine (Fig. II-42). The open-air patio can function as a family center, and it can help provide maximum daylight for underground levels of the house.

The interior design of the subterranean house should be unique and original, at the same time responding to the special conditions present in the underground situation. We must:

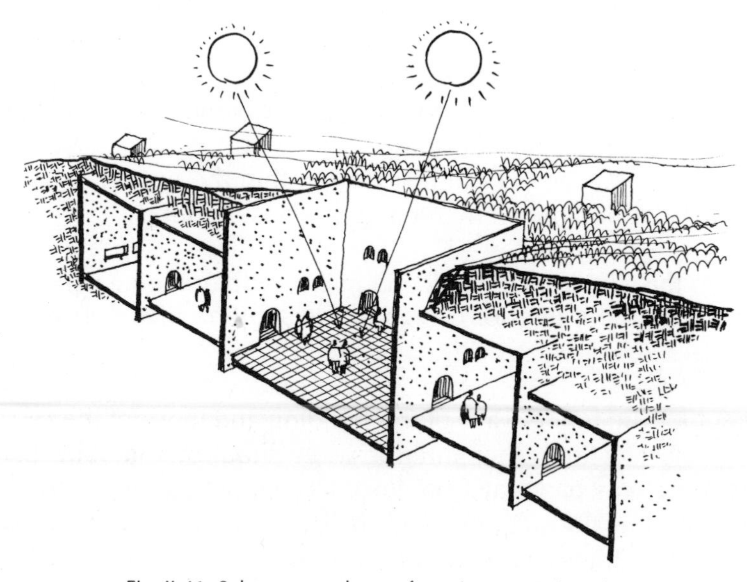

Fig. II-41. Subterranean house featuring open-air patio.

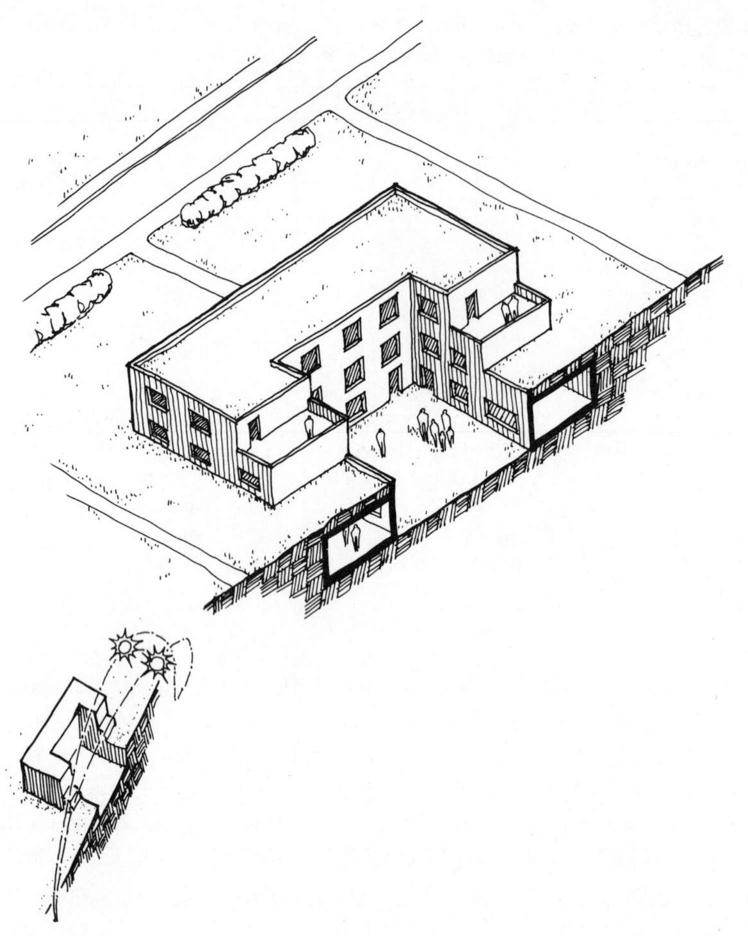

Fig. II-42. Combined subterranean and semisubterranean house. House orientation is important. In cold climate, for example, the house can be positioned to receive maximum direct sunshine.

1. Maximize light or sunshine penetration without disrupting the heat gain and loss system of the building. The house design concept, however, minimizes window surface exposure.
2. Introduce more than one entrance (or exit) to the house to meet code requirements for health and safety.
3. Secure efficient air circulation by passive ventilation into the house at the time and degree desired.
4. Ensure easy passage of the residents within the house.
5. Eliminate or minimize any feeling of claustrophobia.

It seems to this author that in accepting these determining conditions, the basic principles in the interior design of the subterranean house can be as follows:

1. Except for the frame walls of the building and those of the rest rooms, all or most of the other interior walls should be mobile partitions, extending only part way from floor to ceiling to allow ventilation and aid in temperature control and in light penetration. Some partitions may also be transparent curtains or drapes. By the formation of all the house as one unified space, the owners can form their house unit and

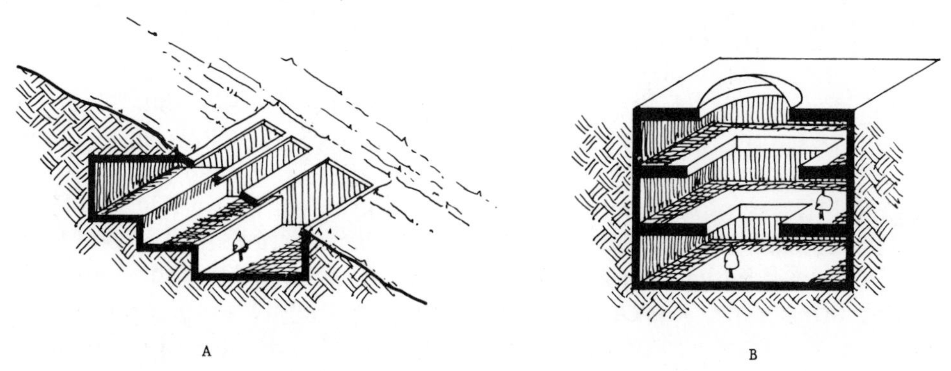

Fig. II-43. Two conceptual alternatives for an interior scheme of movable partitions in a subterranean house. Level divisions can substitute for conventional wall divisions in most cases. This form provides light, ventilation and sunshine penetration deep into the various parts of the house and eliminates claustrophobia.

reform it as desired and as the season, the natural light or the family size changes.

2. There should be no entirely closed, nonventilated unit in the house because of the possibility of dampness.
3. The interior space of the house unit can be made of different levels or semilevels in order to adjust to light penetration from different angles and maximize eye contact with outdoor views (Fig. II-43).
4. The design should be innovative, not conventional, in order to meet the challenge of its uniqueness.
5. Indoor plants should be introduced and integrated within for aesthetic, comfort and health needs.

It may be advisable to attach a greenhouse to the house itself, which could form a house wall, since greenhouses induce heat retention and improve the ambient microclimate of the house. This solution can be especially successful in cold, dry climates. Also, utility rooms for laundry and their machines and heating or air-conditioning appliances should be in a closed and separated unit in order to avoid noise transmission to the house.

TYPES OF DESIGN

There are several types of underground houses: (1) completely underground; (2) semiunderground, i.e., half underground and half aboveground; (3) an underground house with another unit aboveground, thus a combination; and (4) an underground house constructed on a slope—a major part of it below ground, but the structure is open to allow for light, sunshine (if it is desired) and a direct view to the outside.

Within this generalized classification, we can also have other varieties. For example, the completely underground house located on a flat area or nearly flat area could have an open atrium or patio with the other rooms surrounding the open patio from two, three or four sides. Here, the patio is a type of sunken courtyard. The interior division of the house would greatly depend on the overall configuration of the house itself.

The building code usually requires openings and windows in every living section of the house. Unless the code is modified, it is necessary to plan accordingly. However, the storage area, utility spaces, rest rooms, bathrooms and workshop can be in the back part of the building, while the living, dining, kitchen, bedroom and study room are in the front. Figure II-44 shows various houses: closed patio open to sky (Fig. II-44A), L-shaped patio house (Fig. II-44B), open-sided patio house (Fig. II-44C), square house (Fig. II-44D), semipatio house (Fig. II-44E), and linear house (Fig. II-44F).

It is possible to design most of the interior part of the house as one large space divided by light, mobile partitions if necessary. Within this large unit of one open space, we can include: livingroom, family room, dining room, kitchen and one bedroom. Divided and closed units are the bathroom, laundry room and storage space.

Regardless of what plan is used, being able to move throughout the house becomes very important and may create somewhat of a problem. In warm, arid zones, the patio can be used for movement between one section of the house and another (Fig. II-45). The linear house also creates a problem with circulation which may be solved by placing the circulation system (i.e.,

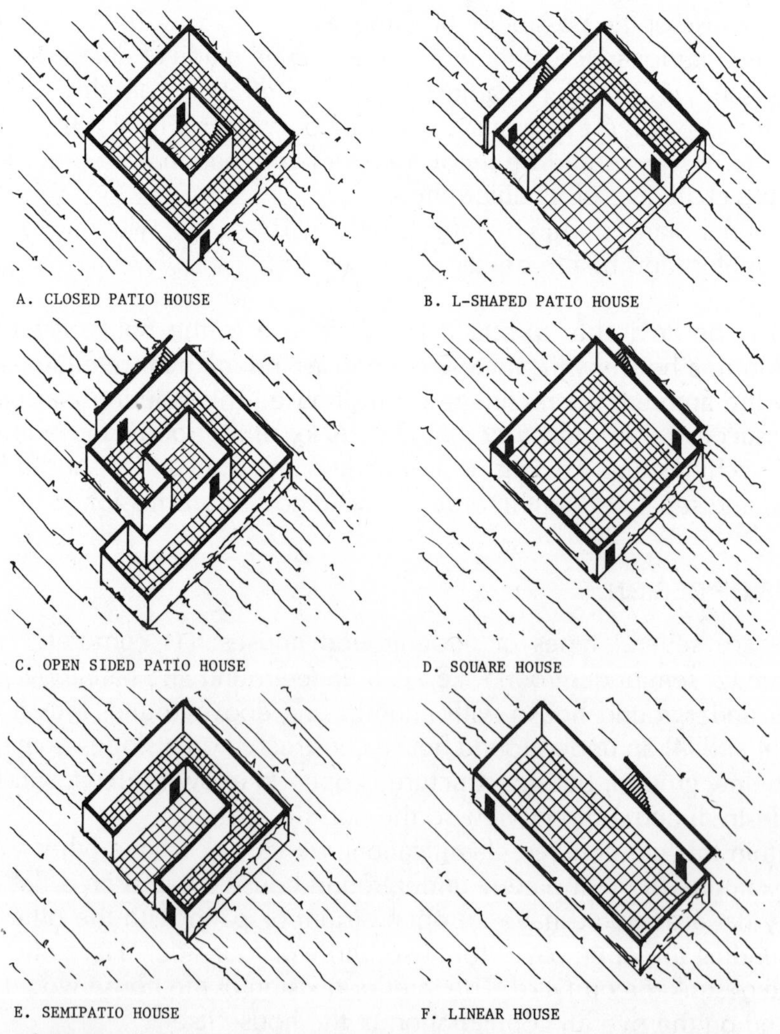

Fig. II-44. Subterranean house accessibility. To meet the building code requirements, the house should have an entrance and a back exit, natural light and natural ventilation in every habitable room, and convenient passage from room to room within the house.

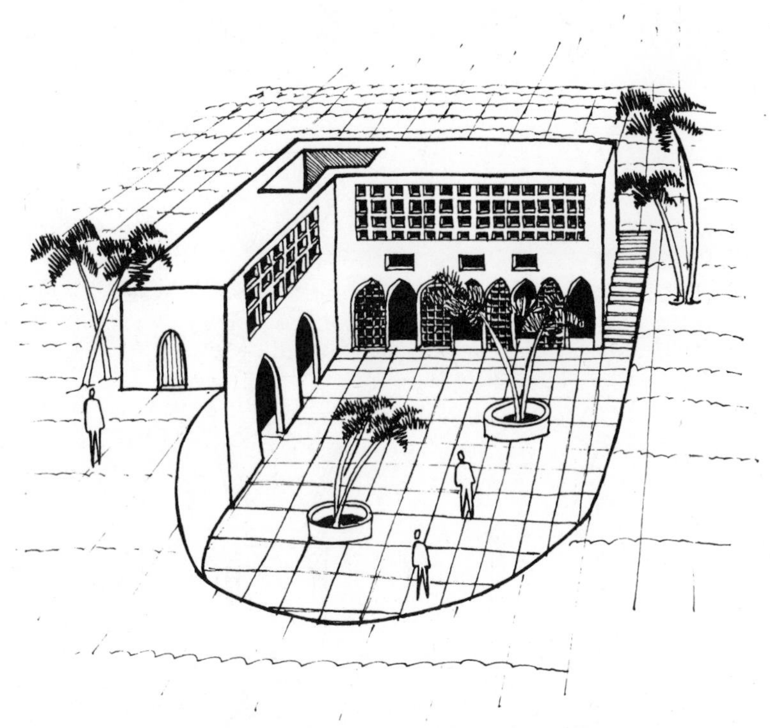

Fig. II-45. Combined subterranean and semisubterranean house suitable for hot, dry climate.

hall) along the front line of the house with glass forming one linear unit as a closed balcony.

In cold, arid climates the central patio can be overlaid with a transparent cover (double glass) to allow for traffic circulation within the house as mentioned above as well as light and sunshine penetration. This arrangement will be a tradeoff, of course, with possible loss or gain of energy. The double glass cover will, however, reduce or minimize the disadvantages of this tradeoff.

In conventional house architecture, attention is usually paid to the exterior as well as to the interior form of the house. In the subterranean structure, however, the designer must focus his attention almost exclusively on the interior part of the structure and its entrance. Furthermore, the mutual relationship between exterior and interior space which exists in conventional designs is less important for underground structures. Thus, natural landscaping will be related primarily to the lot's open space. Engineering, however, will be, as we have already demonstrated, focused on the interrelationship between the nature of the ground and the structure's walls.

DESIGN AND THERMAL EFFICIENCY

Buildings can have the same floor size and still differ greatly in their surface area. The latter will determine to a great extent the heat loss or gain of the building.[36] Thus, the selected design configuration is one additional element which determines the extent of energy consumption for heating or cooling the house. Figure II-46 shows that buildings of identical floor and ceiling area and of identical height but having different shapes will have different surface areas and therefore differentiations in heat gain and loss. Clearly,

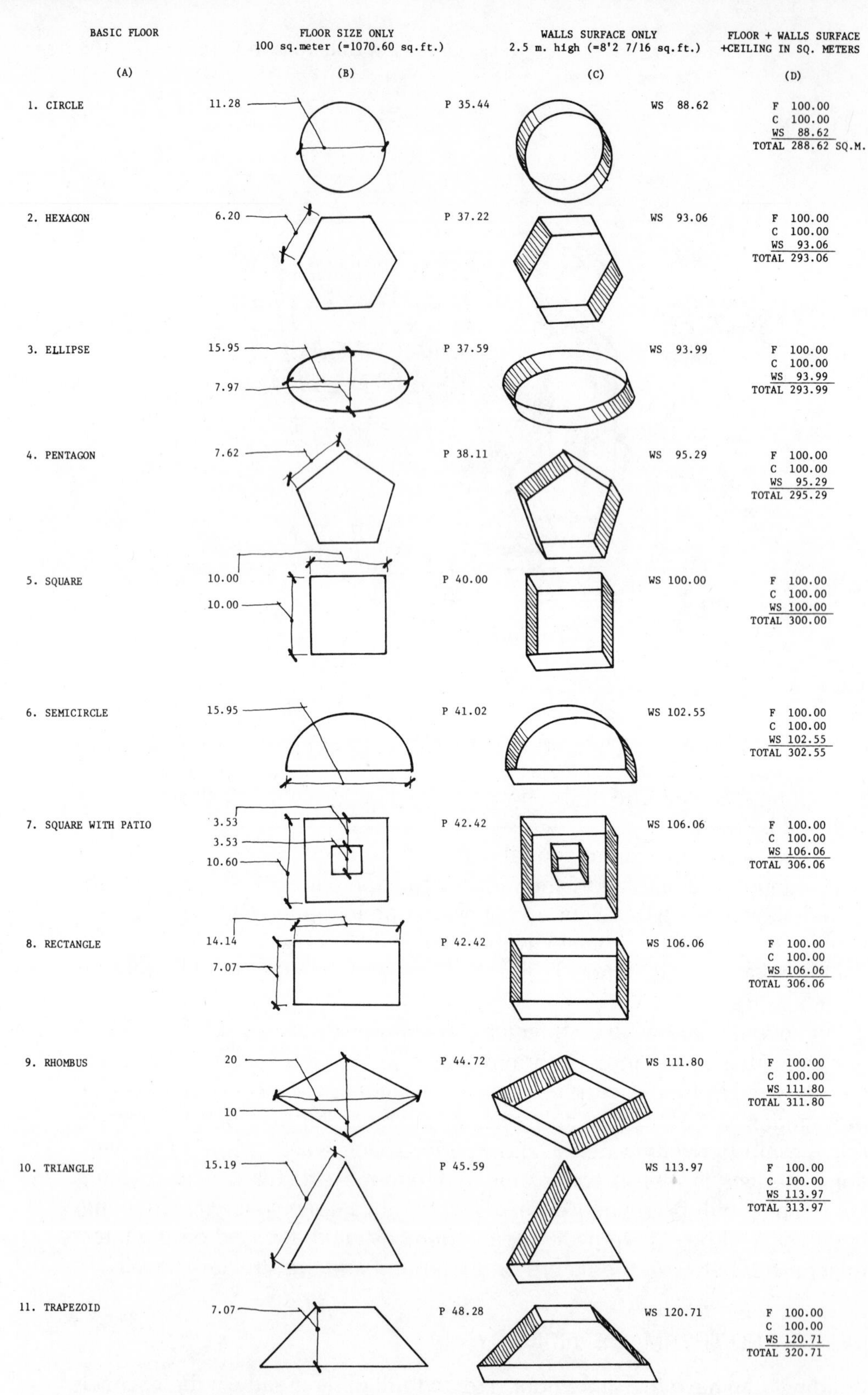

EACH FLOOR = 100 SQ. METERS (=1070.60 SQ. FT.). ALL FLOORS ARE EQUAL IN SIZE. ALL HEIGHTS = 2.5 METERS (8'2").
P = PERIMETER, F = FLOOR, C = CEILING, WS = WALL SURFACE

Fig. II-46. A variety of design forms with equal floor area as well as equal height to the ceiling. The total wall surface areas differ greatly. Building form and configuration have a prime impact on heat gain and loss and, therefore, on energy consumption. Surface wall area determines the amount of heat gain and heat loss which will differ from one building form to another. The minimum area wall is the circle; therefore, it is the most energy-saving design form for nonarid regions. The soil below the floor has the least temperature fluctuation, while the greatest fluctuation is in the soil above the roofs.

the circle form is the most efficient since it has the least total surface area. It also shows that the trapezoid has the largest surface area among all the eleven examples and, therefore, has the highest heat gain and loss. The forms closer to the circle are the hexagon, the ellipse, and the pentagon; while the triangle and the rhombus are close to the area of the trapezoid. It is particularly proper to apply this theory when energy is consumed for heating and cooling. In regions where the soil heat fluctuation coincides with the summer and winter needs, the larger surface size may still be acceptable.

Around the hot, dry arid-zone house, the soil is cooler than the house itself, especially in the summertime. Consequently, the more surface area is exposed to the earth, the cooler the house will become. The case will be the opposite in subterranean houses with cold climates (such as Canada) where heat loss is not desirable. Here the *smaller* the surface of the house (the more compact), the better it is.

Thus, the house design form of the circle and the forms close to it are desirable for regions with heat consumption needs, while the trapezoid and the forms close to it are desirable in regions where cooling is necessary.

In arid regions where the summer is warm and dry and the winter is cold and dry—such as southern Mesopotamia, parts of Iran and parts of Turkey, there is a need for a different adjustment. In such arid zones, the proposed house should be designed for year-round use, with interior spaces changing their functions with the seasons. In the semisubterranean house there are two sections distinguishable by level and function: (1) the semisubterranean and subsurface sections which are to be used primarily throughout the summer season when the temperature is high; and (2) the aboveground sections which are to be used primarily throughout the winter season when temperature is low. Sections of the house which are not used intensively during the summer (such as the garage and dining room) can be permanently located aboveground, while others in use most of the day—bedrooms, kitchen, living room and family room—should be underground, at least during the summer. The flat roof of the building, as depicted in Figure II-47, can also be designed for sleeping on hot summer nights. In fact, the roof may be the

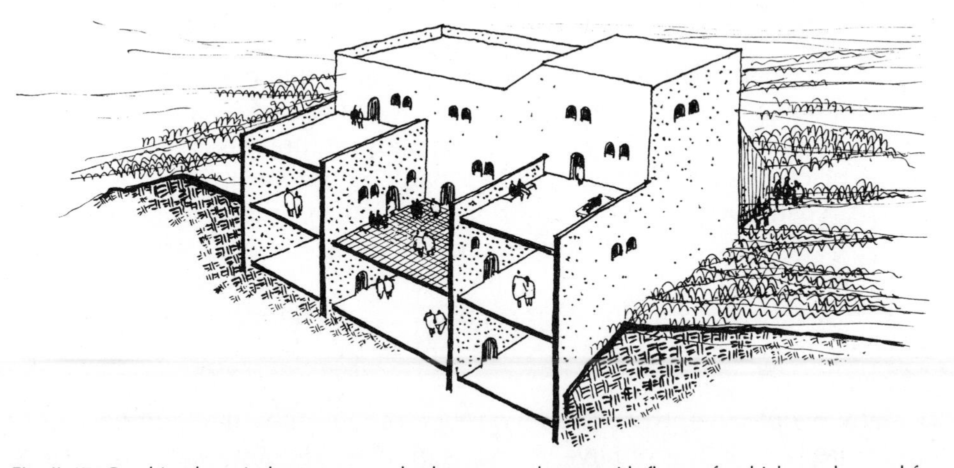

Fig. II-47. Combined semisubterranean and subterranean house with flat roofs which can be used for sleeping in the summer. The roofs can be at different levels and should be integrated with the covered sections of the house so that in a hot, dry arid zone the sleeping facilities will be open to the north and enclosed on the south, allowing for shading in the afternoon.

house's most important feature, for it is here that the designer's innovation and imagination should find their most expressive outlets.

In addition to studying the soil, the engineer and designer should consider the geography of a proposed site and the orientation of the proposed house vis-à-vis solar radiation and the prevailing winds. In hot, dry climates such as those in the Mediterranean region, residential sites should be chosen at a high elevation on slopes facing north, or south as necessary, in order to maximize or minimize shadow and take advantage of prevailing winds. Elevation, however, improves prevailing winds, increases humidity and reduces temperature.

PSYCHOLOGICAL BARRIERS AND DESIGN

We now should further consider psychological barriers that some people have against living underground. Such barriers are crucial to design considerations and may be categorized as follows:

1. Personal bias against living in an earth-covered building because of its connotations of living in caves, seclusion, primitiveness, poverty and backwardness, and associated feelings that it may be damp, unhealthy, unventilated, dirty and unsafe. Bias may also result from some previous experience of badly designed subterranean space or from an association with old basements or cellars, usually unhealthy for living space. Such bias can be overcome by education and explanations on the unique advantages of subterranean houses.
2. Claustrophobia among some people who are afraid of being trapped underground. Such people may always be afraid of being in a confined, closed form. This claustrophobic feeling can be eliminated by explanation and education, and most importantly, by good design. Such design must assure strong construction and, in this case, should include having one side of the building with direct eye contact to the outside. This can be achieved effectively by locating the house on a slope. The lack of windows associated with full quietness (lack of a view and sound) poses some psychological blocks for some people.

There are two other important psychological problems which are related to subterranean living: one is the problem of image and the other is the problem of life style.[37] However, being dirty or primitive is a socioeconomic or cultural phenomenon rather than a consequence of living underground. Living underground may be an option of middle-income people today; but it may be obligatory for low socioeconomic classes in some developing countries because of the availability of natural caves and the easy, low-cost construction in their particular location. Further, many indigenous low-income people realize the climatic efficacy of living underground.

Many people live, work or spend much of their time in confined spaces which are windowless or have windows on one side only, such as theaters, restaurants, offices, industrial and manufacturing spaces, shopping malls, museums, libraries, laboratories and many public gathering places. In any case, they are experiencing circumstances which are the same as those of subterranean structures, since even those which are built on flat areas can be designed to have one opening to the outside. Thus, the psychological effect is probably due to the idea of being underground rather than anything else. Problems with living underground may simply be a matter of getting acquainted with such new environments. To accomplish this requires understanding, usage, experience, choice and selected preferences.

A compact, isolated, underground living or working space deprived of natural light (or the sunlight necessary for health), sounds, time orientation and social interaction can certainly have a psychological impact on people if they are exposed to such conditions for a long time. Of course, these are abnormal conditions which people are not used to. Therefore, such people will have to become acclimated even if they are introduced to such conditions just for experimentation. Planners-developers and others who are introducing the innovative concept of subterranean living or working are not proposing such unvarying conditions of deprivation of natural light, sound, time or social interaction. In any case, there has been much misinterpretation of the subterranean and earth-covered shelters by uninformed people. When more informed people select underground living because of energy conservation and other good reasons, the negative or "neutral" images about subterranean housing should disappear.

Socially and culturally, every type of person needs different types of space. The planner-designer must consider those requirements and meet them. This process would certainly result in large variations in and experience with dwellings. For instance, the effect of contact with the outdoors and all its associations can be achieved by a picture window that looks out at a painted view that is recessed to create the illusion of the outside view. In public places, real outside movement can be simulated by having a "view" projected behind the picture window. This is still, however, a virgin area which needs to be researched. The research which has been done so far is only the beginning.

For example, according to Wunderlich, the acceptance of underground space for working is much greater than for living. On the other hand, this is the only study in this area, and more details need to be considered.[38] In any case, the absence of windows for direct contact with the outside environment plays a major role in this psychological reservation. However, often this lack of windows can be solved by the designer so that there is less psychological restraint. It should also be noted that the lack of windows protects the workers from seeing the outside climatic changes throughout the day, such as rain, snow or changes in temperature which may affect them negatively.[39] Swedish research on physiological conditions shows that there is no negative physiological effect on the health of people living underground if the climate is correctly adapted, while psychological reactions of importance do occur.[40]

Subterranean space can suit certain needs better than aboveground space because it is so unique. Consider, for example, the needs of education, musical and creative art activities, medical or research work, or certain manufacturing which requires a constant, stable microclimate (temperature and humidity) or vibrationless conditions.

To cope with the psychological problems, the designers should consider the following basic design recommendations:

1. The design should allow direct eye contact with the outdoor environment from living sections of the house.
2. The structure should be located on a slope and not on flat ground.
3. The design should develop an efficient ventilation system to eliminate entirely any odors associated with dampness or subterranean space.
4. There should be more than one convenient exit from the structure in order to strengthen the feeling of safety and security.
5. The design should avoid narrow, dark, unventilated or small-sized corridors and spaces within the structure which might cause some claustrophobia. Also, ceilings should not be low but rather a little higher than conventional ones.
6. There must be maximum natural light and/or sun penetration into the structure through windows, skylights or patios.
7. The design should introduce a free and flexible form within the house with minimal and mobile dividing partitions. This design will strengthen the feeling of spaciousness and reduce compactness. All the interior house space can still be introduced as one space of elevated small units surrounding a central, larger unit used as a living room.
8. The design must exhibit a free, clear traffic circulation pattern, connecting all parts of the structure with the exits.
9. Entrance to the structure should not lead downward but rather upward; or, at least, the interior should be level with the exterior.
10. Interior space design should introduce elements of nature, such as small plants, flowers and the like. If a central patio is open to the sky or a skylight is incorporated into the house design, the details of the patio design can provide a natural environment. However, in the case of windowless structures, it is desirable to introduce outdoor elements to the interior space of the house. It is possible to introduce visual simulation of the outdoor environment through pictures from a movie camera positioned outdoors, viewed through a "window" inside.

In Iran it is reported that 2 percent of the population is living underground. In Tunisia, the government built new homes aboveground for the cave dwellers, considering this as an upgrading of their living standards; but they have rejected the government's efforts, obviously for climatic reasons. The Turkish government has tried to remove the underground city and village dwellers of Cappadocia from their homes where Muslims have lived since the 14th century.[41] In the Middle East, contemporary living underground is associated with poverty and backward peasants, but it is doubtful whether this was the case when subterranean living was initiated. The climatic advantages, among others, were understood by those earlier people.

In conclusion, it remains to be seen if middle-income people will accept subterranean living (which is still quite an expensive enterprise for this economic group) as socially acceptable, even if it is an energy-saving venture.

STRUCTURAL NOTES, INSULATION AND WATERPROOFING

With so little fluctuation in soil temperature, the underground house certainly should be expected to have much less contraction and expansion than the aboveground building. The subterranean house must be a more massive, stronger structure than one aboveground. The best construction material would be concrete which minimizes the extensive loss associated with fire, in addition to providing low-cost maintenance for the house.[42]

The roof structure should be able to stand the vertical load of the earth above it. Vertical pressure may also come from beneath the floor, as when clay expands if watered. The horizontal load of the soil (the pressure of the soil on the wall) is another factor to be considered (Fig. II-48). We already know that for all pressure directions, flexible soil will create higher pressure than rocks, especially solid ones, and that soil saturated with water will increase the load and its pressure vertically or horizontally since saturation causes the soil to swell.

The structural aspects related to subterranean structures, according to Daniel True, are:

Loading: Load of ceiling and top surface soil added.

Support: The soil, when resisting compressive pressure, gives support to footings, floor slabs.

Buoyancy: The unit made of the enveloped space and the surrounding structure usually is buoyant in comparison with the surrounding soil.

Ground Motion: Motion caused by earthquakes and volcanoes causes ground movement and can result in the reduction of some structural support or differentiation in pressure.

Groundwater Percolation: The building must resist groundwater entering the structure, especially below the water table.

Deteriorative Agents: Water, especially the saline type, is corrosive to the building. Also, bacteria, fungi and their associated chemical residues can cause the deterioration of the materials of the structure.[43]

Insulation is another important element in subterranean housing construction. In an underground structure the soil itself acts as a good insulator

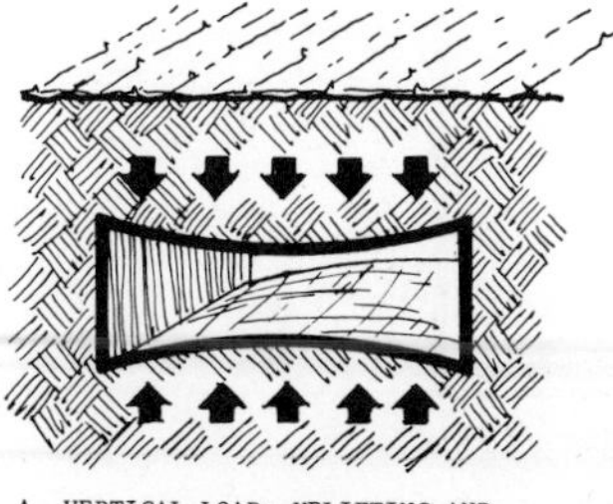

A. VERTICAL LOAD: UPLIFTING AND DOWN SETTLING PRESSURE

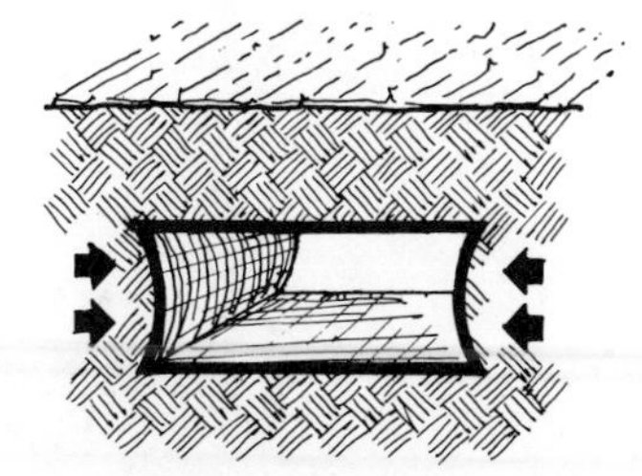

B. HORIZONTAL LOAD: SQUEEZING PRESSURE

Fig. II-48. Forms of load on the subterranean building. Temperature fluctuation, humidity changes or trees cause load increases. Soil saturation with water (water swells clay, increasing the load and the pressure) influences roots of plants; movement of trees by wind forces are transmitted to the soil.

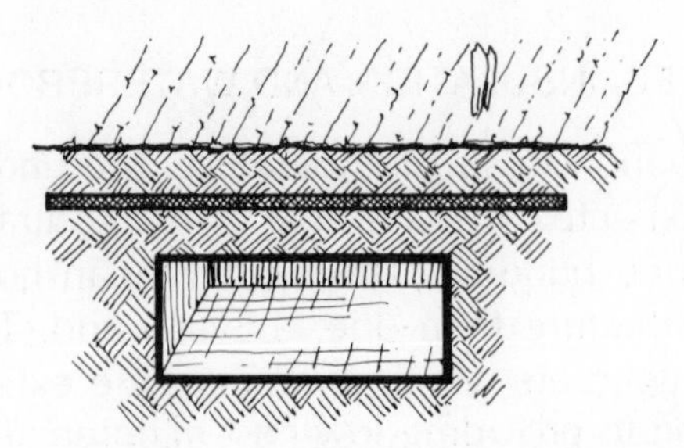

Fig. II-49. Insulation of the earth above the roof can reduce the effect of the outside temperature on the building envelope.

against the daily thermal changes. On the other hand, seasonal thermal change may influence the underground house to some extent. For these influences, different building materials contribute to insulating effectiveness (Fig. II-49. See also Fig. II-12). In the arid subterranean house, insulation of the walls may not be required, or may be only partially required so that some thermal effect from the soil is still transmitted to the building itself. In a cold, rainy climate, insulation can be achieved by building double walls of which the interior one is made of thin blocks with some space between the two walls. The modern houses of Jerusalem have had this double-wall system with the outer made of stone and concrete. Fig. II-50 shows use of the Jerusalem wall system in combination with earth covering and subterranean placement.

Cracking due to the miscalculation of the structural materials' properties, improper design, seasonal soil thermal changes or soil moisture changes which expand and contract a structure are serious potential problems in the subterranean house. In nonarid zones, cracks in the structure of subterranean houses can be a source of serious leaks.

The construction of a subterranean house in a humid area brings with it the problems of water, wetness or dampness, and mold. For the designer, the main challenge is to avoid any humidity on the outer walls of the house

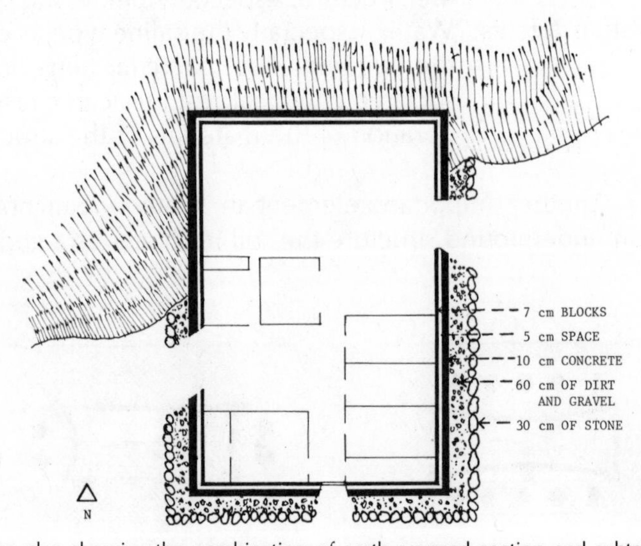

Fig. II-50. House plan showing the combination of earth-covered section and subterranean section built into the cliff, along with the use of the Jerusalem method of double walls for efficient insulation. This particular combination of earth-covered and subterranean forms allows a few openings for light and sunshine and is not expensive since it does not require much blasting in rocky slopes. Interior dimensions 15 x 10 x 2.5m.

and to envelop the house with lasting and efficient waterproofing. The first step in meeting this challenge is to make sure the house base is above any potential water table, not to mention, of course, far away from flood plains or areas subject to flood. We shall discuss waterproofing later.

In humid areas most of the rain seeps into the soil, because the rain falls relatively slowly and continues for a long time. The large quantities of rain which penetrate into local soil minimize the runoff. The ratio between the runoff and the penetrated portion differs from one place to another, depending on the duration and the nature of the rain, the character of the soil and the degree of the vegetative cover. However, water penetration into the ground is high between the earth's surface and the water table. This high degree of penetration means high humidity within the soil for long durations. As a result of these factors, the problems of outer wall drainage are worth major attention and investment.

In contrast, rain in arid zones is torrential and lasts only a short time which can cause large drainage problems associated with movements of large quantities of soil. Therefore, drainage treatment should be on a large scale within a city and its region, as well as on a small scale for each house. Within the city, treatment should primarily involve the creation of large drainage systems to handle runoff, the major threat as discussed in section III.

When considering drainage on the small scale, the first step is preventive: proper site selection. The optimal location of housing is on a slope rather than a flat area. Still, on a slope there is a need for special consideration in the design of the immediate environs of each house so that the runoff is directed to plants and not to the house.

Moisture in arid zones can generally not be considered a serious problem because of the very low, sometimes nonexistent, water table and the extreme dryness which causes quick evaporation. In spite of this, waterproofing should be undertaken. Although the water table may be relatively far away from the structure, it may still create pressure on the soil and consequently on the house itself. Thus, in addition to the building material considerations, we have to select the most appropriate wall form to withstand load pressure (Fig. II-51). Recall that the circle form is the most resistant to load since the pressure is distributed equally on all sides. In an arid land, very low night temperatures and dew accumulations may cause frost to form around the house and to influence adjacent soil behavior. Capillary action may draw water into the wall, creating moisture and dampness, especially along the northern wall since it receives very little solar radiation.

Possible runoff can be diverted away from the house by constructing a gravel trench with drain tile or perforated plastic pipe around the lower part of the house. This will not eliminate the need for effective waterproofing. In addition, an effective and workable drainage system for the roof and the patio should be constructed. Closed, paved patios can create further problems since they are subject to flooding which could cause great damage to the house.

Under certain circumstances groundwater can be the most important factor in excavation, influencing its cost and method. Therefore, in humid or moderate climatic regions, the water table and drainage control should be given prime attention. In any location, the interaction between water and

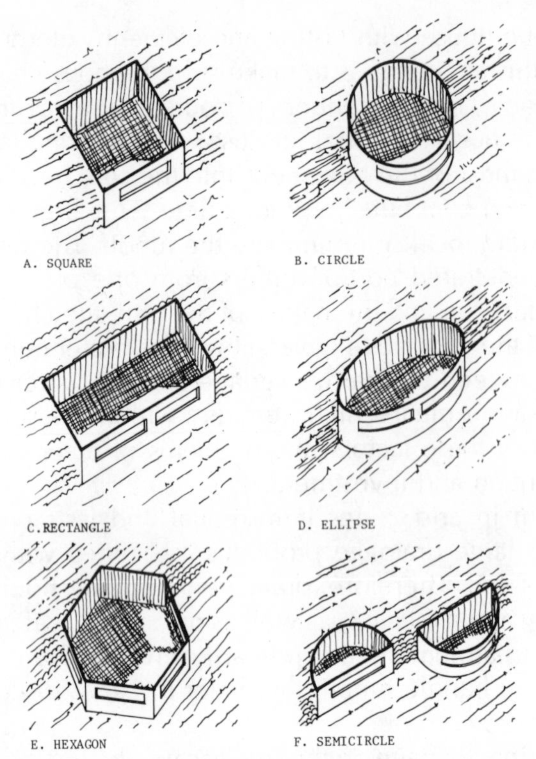

Fig II-51. Various wall forms for subterranean structures are related to load pressure, to light and to opening(s) toward the sun. The circle form is the most resistant to load pressure. Curved horizontal windows can receive maximum light or sunshine throughout the day.

soil must be well studied since there are some soils which expand with moisture, depending on their permeability, or become more difficult to excavate, as previously discussed.

Water can be absorbed in the soil in small amounts, establishing a little moisture which is removable. In larger quantities, it moves toward an outlet by gravity or by pressure. In these circumstances there must be consideration of a means for water movement and removal in order to drain the soil. Pumping tests should be made in order to find the degree of free water that can be anticipated in the finer soil grain. Thus, soil investigation requires knowledge of the grain size and the moisture and water state.[44]

Second to the psychological issue, waterproofing is the most important item in the design and construction of the subterranean house, and it is the major technical problem. This problem arises primarily in humid, temperate or moderate climates where some rain occurs and where the portion of rain which seeps into the ground is high compared with the runoff. On the other hand, subterranean houses built in a compact city may have less of such water penetration and dampness because of the compactness and the fact that most of the city surfaces (roofs, roads, etc.) are sealed against water penetration. In this case, there is a minimal amount of water which may seep into the ground; most of it continues on as runoff.

There are at least three effective ways of waterproofing:

1. Membranes coated with tar, wax or asphalt on the exterior walls, under the floor and on roof surfaces. They should consist of layers of felt made of asbestos and wood fiber or of cotton and fiberglass fabric.

It is also possible for membranes to be made of rubber or polyvinylchloride sheets.

2. Hydrolythic coatings, bituminous asphalt and plastics on walls and floors or cement mixed with ferrous particles used inside buildings.
3. Concrete admixtures, liquid paste or powder admixtures used to render concrete impermeable, for use primarily on floors.

Nonporous membranes should stop all types of water movement from one side to the other. However, membranes are subject to weather-related deterioration or cracking; fiberboard backing would help with the latter problem.

The waterproofing envelope of the house must be very efficient and worth the investment since it is expensive to make repairs at a later time. On the other hand, because of its efficiency and the massive cover of soil, the subterranean house must "breathe" and have a good ventilation system to avoid condensation. If it is adequate, waterproofing will protect the structure from dampness, from organic materials included in the soil and from corrosion, which causes the deterioration of the building materials. Thus, the whole building should be waterproofed, regardless of whether or not it is insulated. Again, as in nearly all aspects of underground construction, consideration must be given to differences in soil, its permeability and the water table.[45]

In cold climates, such as the northeastern and northern United States and Canada the foundation of the conventional house must be deep beneath the surface of the earth to avoid frost on the floor, a requirement which increases the cost of construction. Consequently, most of those houses use these foundations as basements. On the other hand, in the southwestern United States where the climate is arid or semiarid, such a requirement for a deep foundation is not necessary and, consequently, basements are not common in this region. However, climatically and practically the basements are needed more in the warm climate of the Southwest (where they don't exist) than for the cold and snowy climate of the North (where they do exist).

PATIO

The patio, open to the sky, located in the center of the house and surrounded by other rooms, is an ancient traditional setting used for many thousands of years, especially in the early civilizations of Mesopotamia, the Indus Valley and others located within warm, dry climates. The Greeks modified this patio house to meet the Mediterranean climate, and the Romans modified the patio by covering it, to become the atrium.

In each case, the patio house provides indoor privacy, easy circulation of traffic within the house, convenient social and working activities, as well as climatic advantages in the patio itself; light, ventilation, sun, etc. Within the traditional compact city of the hot, dry climate, where houses are connected and grouped together in clusters and only the one side of the house facing toward the road is free from obstruction, the patio becomes the only side free for any openings (such as doors and windows) to provide light, ventilation and other access within the whole house.

In contrast, a closed patio, usually used in a flat site, can limit ventilation

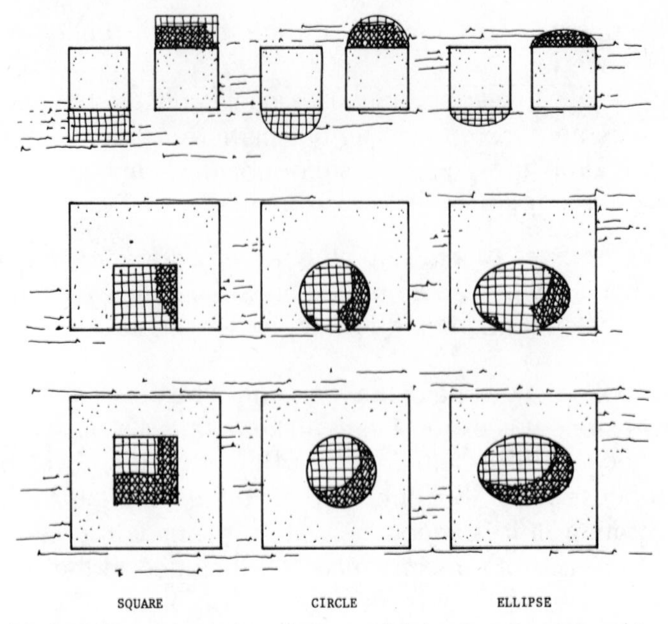

Fig. II-52. Various patio form designs and their orientations to sunshine.

and sunshine, although such a patio may support air movement within its space. Such a patio may create drainage problems and prevent view. The design of the patio form is very important (Fig. II-52). A patio with proper orientation and with one side open enables use of the sun's radiation when it is desirable. The size, form and depth of the patio determine the amount of sunshine and light and the aerodynamics.

Sun penetration into the rooms surrounding the patio is deeper in the winter than in the summer because of the differentiation in sun angle between the two seasons. Although this angle is fixed, the zenith, of course, is different from one region to another due to the latitude, an item to be calculated by the designer. Avoidance or reduction of direct sunshine penetration can be achieved by colonnading or shadowing. However, in humid regions such penetration of direct sunshine is most desirable in the winter. This fact was fully understood by designers of ancient Greece when they designed their patio houses for the Mediterranean climate. On the other hand, the house designers of ancient central and south Mesopotamia considered the avoidance of sunshine important because of the dry and mostly warm climate year round (Fig. II-53).

If sunshine penetration into the rooms is desirable, the patio should be wide and not too deep. On the other hand, small, deep patios minimize sunshine penetration. The first case is, of course, usually recommended since the patio can be shadowed by a movable cover. A wide patio also helps the aerodynamics. To determine the desired dimensions and the orientation of the patio would require the detailed study of the sun's movement.

Patio shadowing can be natural, with trees, or artificial, using various prefabricated materials, either light (mobile) or heavy (fixed) (Fig. II-54). However, a distinction should be made between shadowing and ventilation; shadowing should *not* block ventilation when the latter is essential.

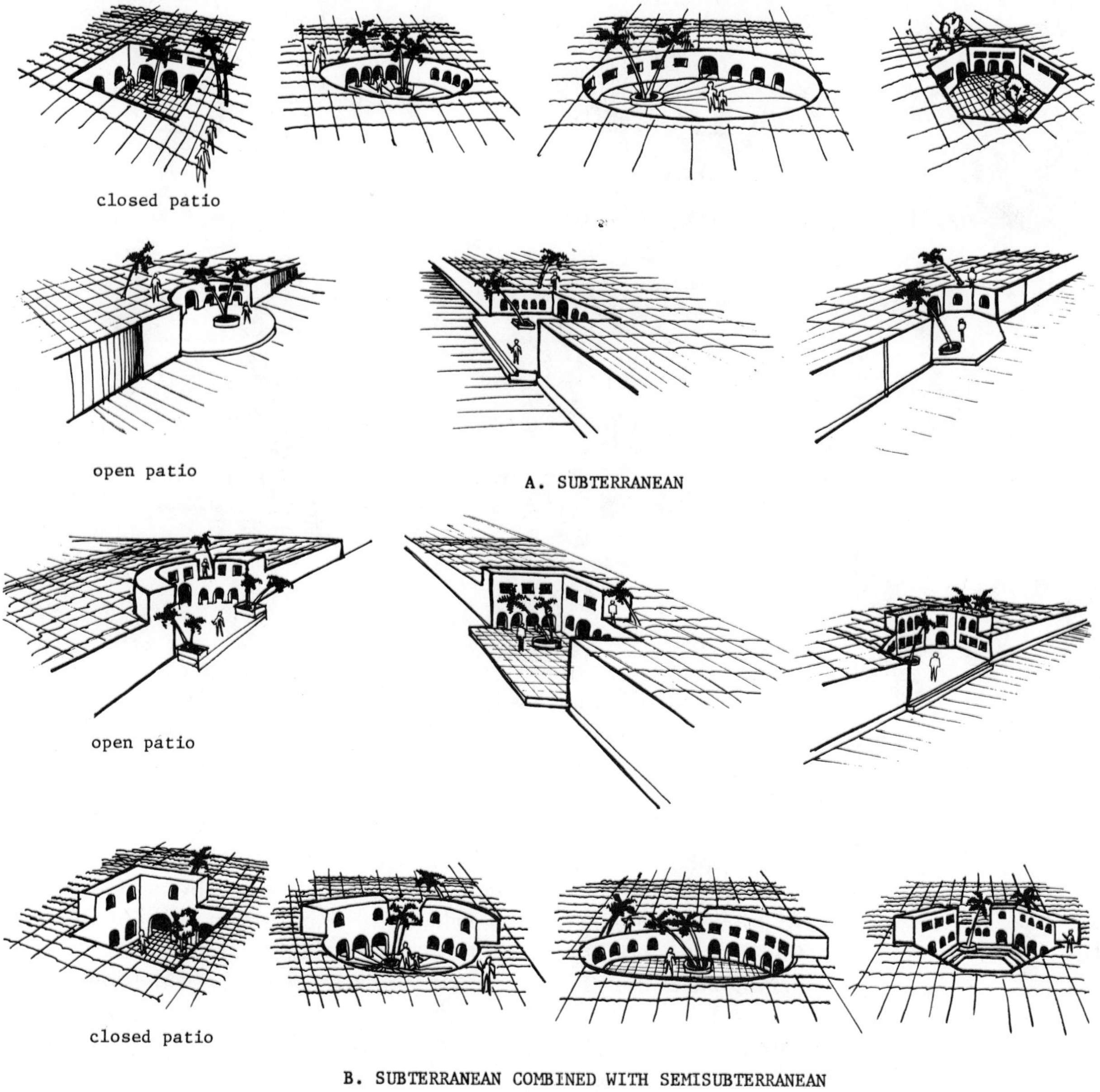

closed patio

open patio

A. SUBTERRANEAN

open patio

closed patio

B. SUBTERRANEAN COMBINED WITH SEMISUBTERRANEAN

Fig. II-53. Alternative patio design forms for combined subterranean and semisubterranean dwelling units.

NATURAL LIGHT, SUNSHINE AND WINDOWS

Some daily functions require natural light and sunshine—outdoor recreation, transportation and agriculture; others do not demand conditions of natural light and sunshine throughout the day—office work, commercial retail activities, entertainment, warehouses and storage; and still other functions require partial exposure—education, housing, industry and manufacturing.[46] Whatever the activity, the psychologically negative image of subterranean living, associated mainly with darkness or absence of natural light, imposes mandatory requirements on the designer to make the structure attractive, especially to those who have this negative image. Those

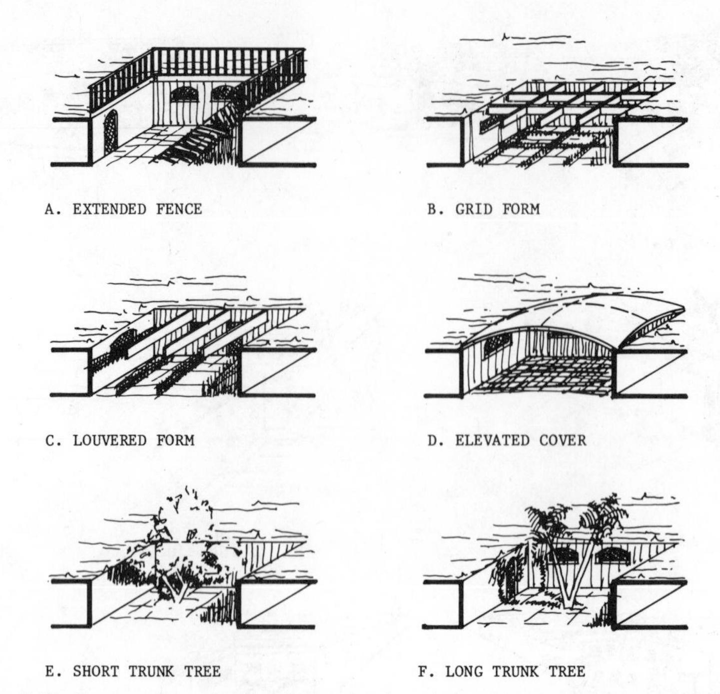

Fig. II-54. Cross section of patio shadowing for a house in a warm climate on a flat site.

basic requirements include plenty of sunshine and penetration of natural light deep into the house, as well as a direct view to the outside.

Natural lighting within the subterranean house can be made direct or indirect. If the site is on a slope, then orientation of the openings of the house (windows, door, shaft) become crucial, especially for windows. Window positioning in relation to light and sunshine can have a large variety of forms (Fig. II-55). Indirect light can also be attained by a special shaft with a prism which brings natural light to the inner parts of the house through continuous reflections of mirrors positioned in interrelation with each other (similar to that in the submarine). Another indirect way is by outside mirrors positioned to reflect light to the house through the windows. Use of these outdoor mirrors, especially movable ones facing house openings at a right angle, is of great importance. If they are movable, they can be adjusted to the sun's position throughout the day and the seasons. For greater convenience, the movement can be automatically adjusted.

Window openings can also be useful in lighting with special treatment. A diagonal window pedestal with a wide inner frame can spread the light. Also, high windows help to receive and spread the light throughout the room. To obtain more penetration of the light, it is also necessary to raise the roof threshold. Part of the covered roof or the upper walls of the building can be made of special thick glass to further promote light penetration.

Proper coloring of the indoor space and furniture is important, too. Light colors reflect penetrated light and spread it over the space, while dark colors absorb light and reduce reflection. The same can be said for the patio which can also support reflection of the light into the house. Indoor walls facing the openings of the house and designed at a right angle in relation to the outdoor reflection are also to be considered in the design.

In arid zones, outdoor light is intensive and establishes strong albedo in

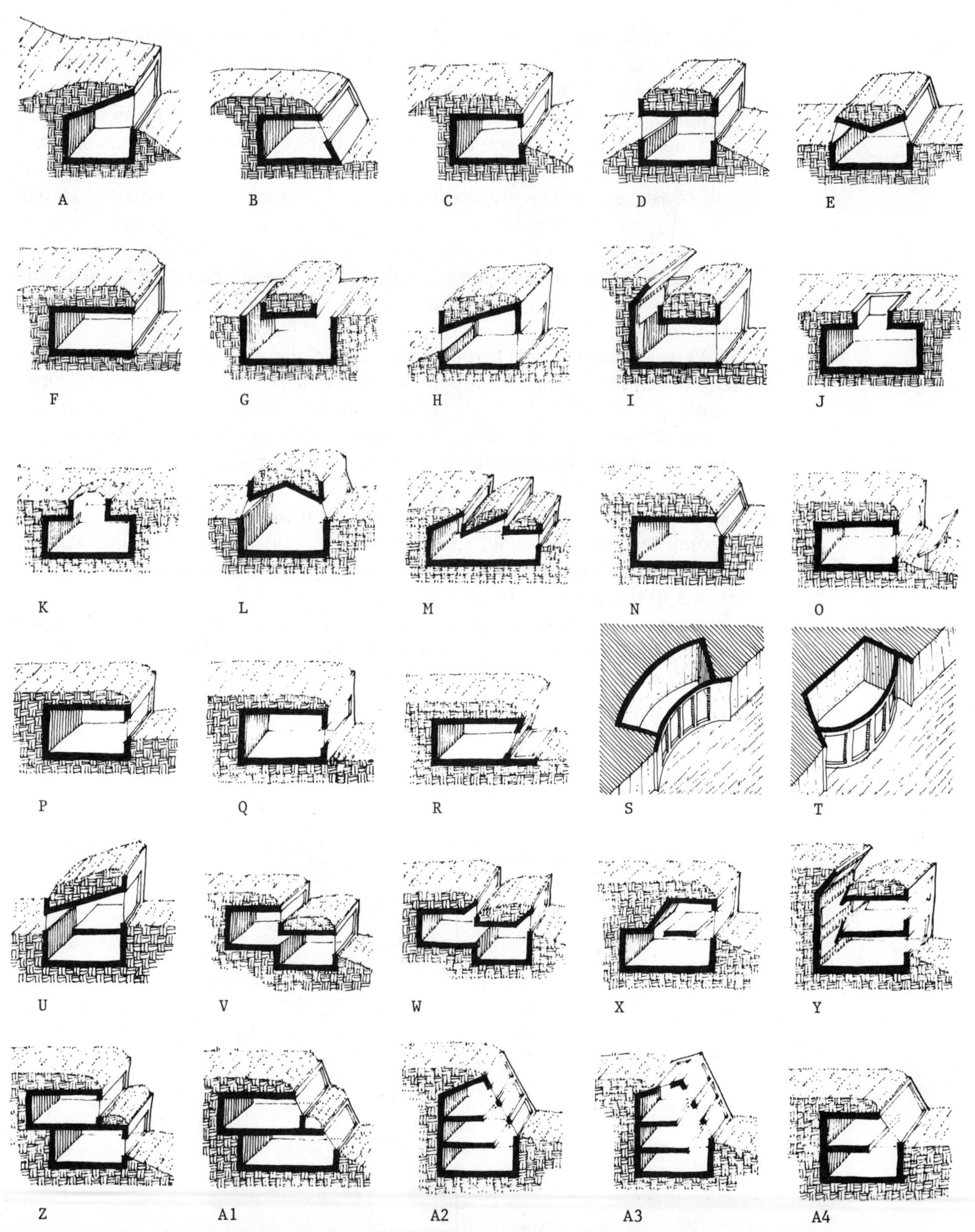

Fig. II-55. Window positioning as related to light, sunshine and view. Window form, size, positioning, orientation and height determine the amount of light and sunshine throughout the day and the quality of the view. In extreme climates the subterranean design attempts to overcome the dilemma of the need to minimize heat gain and heat loss (by minimizing the openings to the outside) and the need to maximize light penetration (which requires an increase of the openings). Each site, as well as each region, has its own uniqueness; the designers have to find the synthesis of these two requirements.

areas of light-colored stone such as chalky limestone. It is important for the house designer to prevent albedo penetration to indoor space and reduce it in the immediate environs of the house by paving, plantation or vegetation.

Although lighting in underground buildings will require the expenditure of some energy, it is more feasible than it may be first seen to use natural light through the use of light wells or light shafts. In general then, lighting will consume a relatively small portion of energy when compared to that normally used for heating or other industrial uses. If energy consumption must be reduced to the absolute minimum in an arid zone, solar radiation is so intense throughout the year that solar energy can supply all of the house's energy needs (as well as most of the needs of underground manufacturing facilities).

We now can see that through the proper planning of underground shelters, the designer can provide natural direct light to the shelter. If the site's topography is flat, he can use patios as noted above. If it is sloped, he can design accordingly. Hillside underground development permits the penetration of substantial light to the innermost parts of the house, especially if the slopes face south. Even on slopes facing east or west, a good deal of light penetrates for half a day. Sunshine may be desired in winter time and not in summer time. To avoid sun penetration, trees with long trunks can be planted to allow the winter sun's penetration and still provide direct eye access to the outdoor (Fig. II-56).

In hot, arid zones the designer should try to avoid the penetration of sun into the house to avoid more heat again. But he should also attempt to bring maximum natural light into the house throughout all the days of the year.

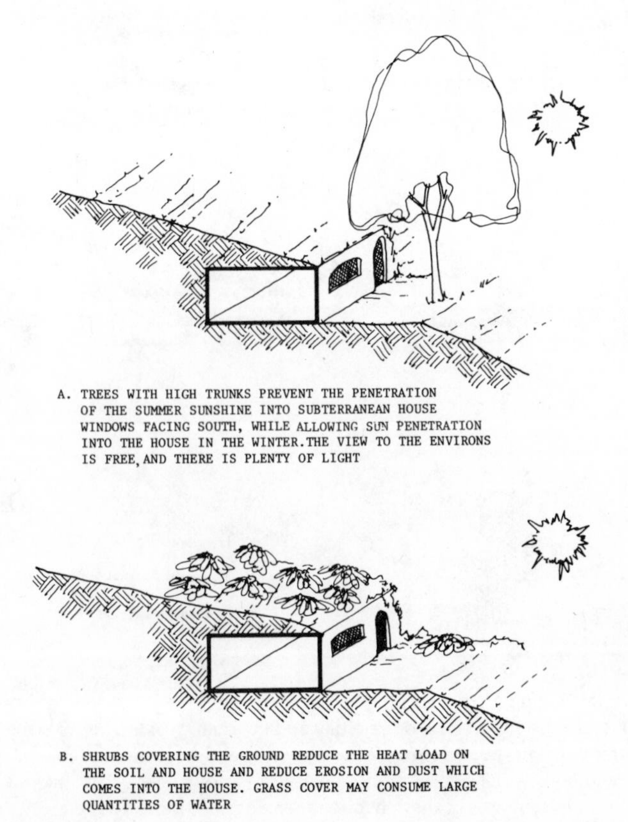

Fig. II-56. In a subterranean house, the vegetation can be used as an effective tool if selected properly.

These two policies may create a conflict for the designer. The historical lessons of ancient cities in hot arid climates is to avoid exposure to the south and to make intensive and extensive use of the north, where ample natural light can be obtained but not direct sunshine (Fig. II-57). In Abadan in southern Iran, all balconies of the city face north. The same orientation was used for the patio house of the Mesopotamian city. It is also possible in such a hot, arid climate to use an eastern exposure for obtaining light; sunshine penetration would come early, before most of the heat of the day. Other possibilities include using exposure to the northeast or to the northwest.

In cold, arid climates this conflict hardly exists, but exposure to the south is the most desirable. The Indians in Pueblo Bonito, New Mexico, used this principle very effectively by exposing all their settlement houses to the south in a terraced, half-circle form to obtain maximum sunshine. It is possible to moderate the exposure, if necessary, by orienting the buildings toward the southeast side (moderate) or the southwest side (less moderate because of the afternoon peak thermal load of the solar radiation).

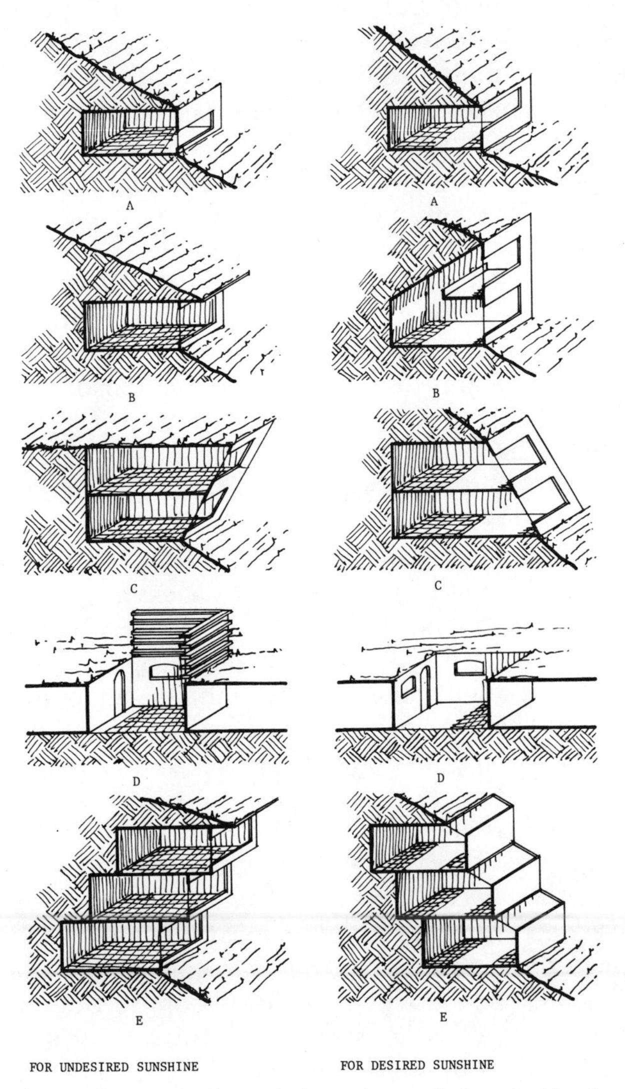

Fig. II-57. Subterranean house design and sun radiation considerations.

As we can see, the use of the sun for subterranean houses is dependent on the average daily temperature of the specific area discussed. In arid regions where the climate is hot and dry, such as in Egypt, it is generally best to keep the sunshine from entering the house, especially in low elevations. On the other hand, if the house is located in much higher elevations, above 400 or 500 meters where the average temperature is lower than that of the lowland, sun radiation is desired in the house. In such a case, the sun's radiation can be tapped in two ways:

1. Passively, by trapping the solar heat through a glass wall built to be oriented toward the sun to obtain the maximum heat throughout the day. Since arid areas are nearly free of clouds, sun radiation is ample and its availability is secure throughout the year. This system requires much attention in the house design: orientation (especially the glass wall) in relation to the seasonal movement of the sun; glass walls or windows with no openings at all or controlled openings; interior space design in relation to sun movement; arrangement for avoiding sunlight when so desired at certain hours of the day or periods of the season. This method requires that only one side of the house be exposed; all other sides can be underground.
2. Passively, by trapping and *storing* the sun's heat to be used at night when temperatures become very low, especially after midnight. This is particularly important in arid zones where there is great amplitude between daytime and nightime temperatures. This method, however, can be used in hot, arid climates as well as in cold, arid climates. Solar heat can be stored in water tanks exposed to the sun during the day and covered in the evening. Or we may direct the solar heat by a small fan (from the attic in the normal house; in a special place in the underground house) to a special rock storage area. The heat which is absorbed by the stone is released at night by another fan and is directed through the house.

In both systems, the main cost is for investment; maintenance is minimal.

It is important to emphasize to the public that underground homes can still have windows, open to natural landscapes and facilitating natural ventilation. Use of natural light to penetrate the subsurface is valuable because artificial light introduces one range of color appearance, while natural sunlight can cover a wide range and create a better appearance in the space. Window design is an essential consideration to support our previous analysis, and it is one of the major elements in the design of the subterranean house. Recall, too, that we have already mentioned building code requirements for windows which would apply equally to above- and underground placement.

This problem is less complex with aboveground buildings on flat sites where openings can be provided to almost all sides. However, in the subterranean house, the second level of habitable rooms below the surface should be designed with more than one exit (or entrance) built in the room.

The window constitutes a small but important portion of the physical components of the underground structure. People living or working in a subterranean structure are influenced by the characteristics of the space and its contents, as well as by their perceptions of those contents. Windows affect these perceptions; they provide fresh air, a sense of outside climate

Table II-6. Pros and Cons of Windowed and Windowless Components

	WINDOWED	WINDOWLESS
1. View	*Visual contact with outdoor environs	*No direct eye contact with the outdoors. Possible claustrophobia.
	*Relative safety and view control around the house	*Security from burglary.
	*Exposure to dust, noise and pollution	*Limited or no impact of the outdoor environs.
	*Limited concentration. Diffusion	*Supports creativity
	*Limited privacy	*Complete privacy
	*Introduce the sense of the dynamics of life movement & of changes: people, trees, air, birds, etc.	*Break contact with outside world *Bring focus to the interior & increase concentration with no interruption.
2. Natural light	*Healthy	*Artificial or indirect light
	*Natural motion	*No glare
	*Sense of cycle	*
3. Sunshine	*Healthy	*Deficiency
	*Warm	*
	*Sense of life cycle	*Loss of time feeling
4. Ventilation	*Natural and fresh air as a health necessity.	*Fresh air limited. Requires special design and artificial support.
	*Sense of the weather	*
Other	*Maintain the feeling of time and the natural cycle of day.	*Potentially increase personal, intellectual, creative production: art, music, writing, learning.
	*Minimize feelings of isolation & monotony.	*Cut noise.
	*Enhance the feelings of wide and open space when the viewer is in a confined environment.	

changes, a view (moving or static), sunshine and the psychological image of wider space (Table II-6). At the same time, the window in the subterranean house has some disadvantages; heat loss in the winter and heat gain in the summer contradict the ultimate purpose for being underground. Thus, a conflict of interests exists.

These conflicts can be resolved with proper planning of the underground structure. The orientation of the structure and its windows can play an important part in alleviating some of the problems. Windows and other openings facing north, east or west will have much greater loss of heat than windows facing south. However, any openings or walls to the outside not covered by soil will lose or gain heat quicker than those parts covered by soil. Therefore, any vulnerable wall can be built of double thickness to minimize the loss (as has been done in Jerusalem). In addition, double glass windows can be installed.

Light coming directly through large windows may cause uncomfortable glare to the eyes and fading of furniture, fabric and ornaments. Failure of the designer to consider such problems may result in residents' closing the curtain and, thus, avoiding the passive heat gain from the sunshine. Bleaching, fading and glare can be reduced by slim blinds.[47]

In many of our modern buildings, i.e., theaters, cinemas, restaurants, night clubs, libraries, department stores, museums, art galleries and factories, we have eliminated many of the advantages of the window. Air is provided by artificial ventilation in a controlled environment with closed windows, no sense of the real outside climate, an obstructed or eliminated view, and no sunshine. In some conventional structures, such as shopping centers, industrial parks or even offices, windows seem unnecessary or do not exist. In fact, a structure such as a school may benefit from being underground with fewer windows. Among the benefits are: no window breakage, no vandalism to windows, no excessive glare, no heat gain or loss, no wasted wall spaces, no noise from outside and an indoor focus of attention. In studies to which we have already referred comparing windowless and conventional schools, no significant differences were found with reference to personality tests, school health records, grades, function of intellect and academic achievement. Thus, windowless schools do not alter scholastic performance.[48] However, in other structures, such as offices, industries and homes, a view of the surroundings may be quite important. Although problems with light can be resolved by artificial means, there is still the problem of health which is related to light; the sun does have a physiological influence. In addition, natural light provides variation throughout the day (by its movement) which artificial light does not provide. Since most people prefer natural to artificial light,[49] we have to plan around this preference.

To summarize, the component of sunshine in lighting is related to health, psychology, warmth and excessive glare. In cold climates especially, the sunshine is expected to be very important to many people. But in any climate, windows may make the difference between the acceptance or rejection of underground living.

From this standpoint, the patio subterranean house is the best; every room can have direct natural light penetration, and the structure can still have a centered form and circulation of movement. The patio is an especially suitable design for dry, hot and temperate climates, but it is not an adequate one for very cold climates.

Our design must balance the positive functions of a window—penetration of light and sun, radiant heat, ventilation and direct eye contact with the outdoors—with the probably undesirable loss/gain of heat/cold through such a window. Planning windows that are horizontal, narrow and long is one answer to our design problem. Horizontal windows permit a higher penetration of the sun throughout most of the day than do vertical windows (Fig. II-58). Making the windows narrow minimizes heat gain or loss. The height of the window is also essential. Windows in the lower or middle section of the wall will limit the light and sun penetration, especially to the lower part of the room, while higher windows will support penetration into most of the room. Sections of the window can be fixed, perhaps made of thick blocks of glass; while others should be able to be opened. Horizontal windows curved toward the outside (Fig. II-59) and oriented to catch all the

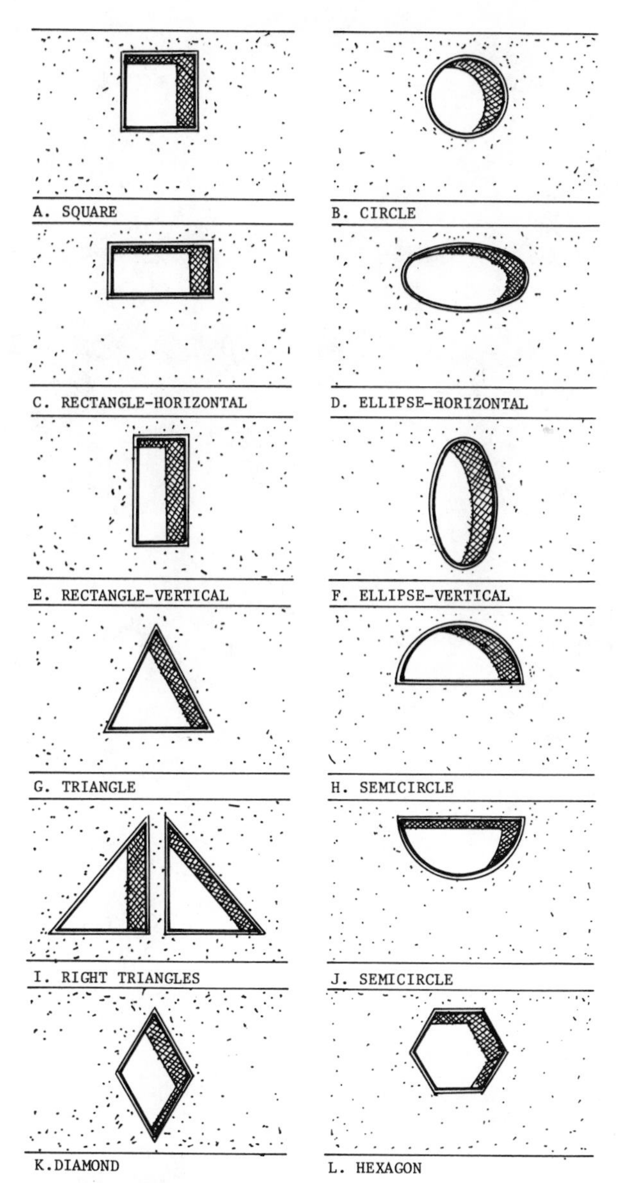

Fig. II-58. Alternative forms for windows. Windows for subterranean houses should be designed to maximize light and minimize surface area to avoid heat gain or heat loss. They should also maximize or minimize sun radiation penetration as desired. All the forms are presented with equal surfaces. A horizontal-rectangular window (C) is most effective for light and sunshine penetration throughout the day. It also retains a wide view to the outside.

daily sun movement can be quite effective in providing sun and light penetration throughout most of the day, in spite of their narrow form. All of these suggestions should be considered, but another important point is that windows should provide an outdoor view when a person is sitting indoors; yet it should be high enough for privacy.

Windows facing the sun's movement can be horizontal or vertical. It seems to this author that the horizontal one is preferable since it will spread the light throughout the space of the room. Windows can also take elliptical shape, which will maximize light entrance to the room. In any case, here too, window orientation is vital and requires a thorough study of the sun's movement throughout the seasons

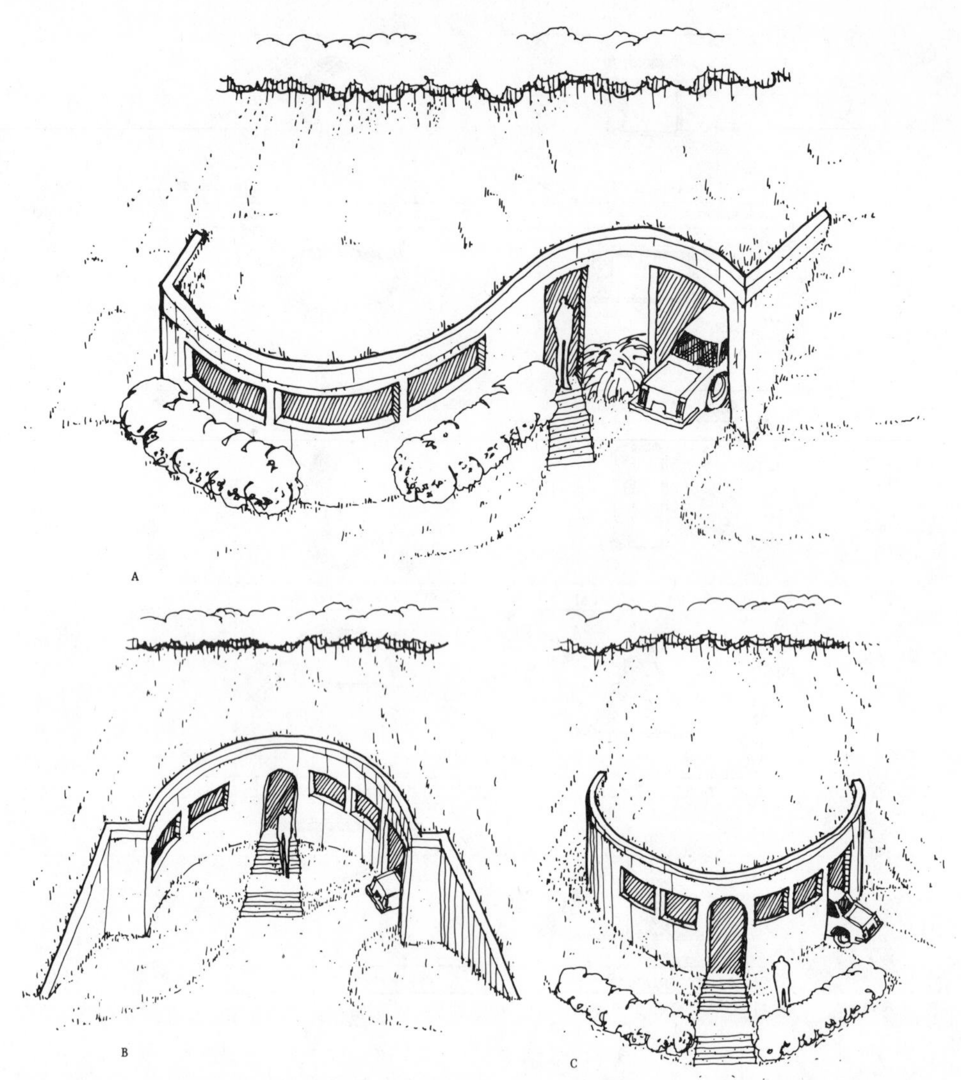

Fig. II-59. Subterranean house adjusted to topographical forms, light, sunshine and view. The uncovered wall can be insulated with stone and two spaced walls. Each form shapes the window form and, therefore, the amount of light and sunshine to be received.

There is very little, if any, research on windowless houses; however, they seem to be unacceptable to many people. Therefore, we want to make it clear that we are not suggesting windowless subterranean housing, nor are we suggesting only an artificial replacement for the window, although today's technology makes this possible. On the other hand, there is a reasonable amount of research on windowless nonresidential units such as schools, factories, offices or public buildings.

Conventional schools are built with large windows for access to light and ventilation. The fuel crisis, however, brought a revision of this approach. Wide windows admit extensive heat, bring noise, invite vandalism, expose the classroom to the outside and interfere with concentration. Classrooms without windows provide more wallspace for exhibitions or bulletin boards, and they are free from outside noise.

The case of windowless factories is quite different from that of schools. Many factories require a controlled temperature for special processing, free from noise or dust. Employees, however, seem to have negative responses to windowless factories.

OUTDOOR SPACES

Although there usually is ample outdoor space around a complex or single subterranean house, this space should be well planned and maintained. In a sense, with the house hidden, the outdoor space (the lot) is the only visible form or configuration conveying the house's image. For the inhabitants, it is the only immediate open space available. Therefore, the design of this lot should be thoughtfully conceived. However, there are many uses to be considered for open space of the complex or the single house:

- Playground for children
- Family recreation
- Clothes drying
- Mail
- Landscaping
- Garbage/trash storage before removal
- Parking space and driveways
- Gardening or other outdoor activities
- Access paths to the units
- Space for ventilation chimney and heating chimney
- Sewage ventilation pipes
- Space for solar energy collectors, etc.
- Space for control of the incoming infrastructure network (TV cables, water, electricity, gas pipes, sewage) and their meters
- Fences for protection and private lot alignment

Of course, these uses will vary from one site to another and between the complex of houses and the single house. It is the designer's role to integrate those features within the open space environment of the house. If an open patio exists, integration becomes essential, aesthetically and functionally. Many of the above features should be designed and erected to avoid vandalism. In addition, the design may be difficult when houses are located on slopes.

The distance the house is set back from its lot line should be predetermined. Labs suggests that this setback be determined by the solar rights, so that the lowest solar radiation of the winter (north of the equator) will not be hidden from neighboring lots by the house construction.[50]

Outdoor landscaping where trees are used should be planned so they don't influence the subterranean house structure, especially in humid regions where plants have an extensive root system. Plants can cause soil expansion and should be kept far enough from the structure so that such expansion (and any watering) will not affect it. Grass or plants can also influence the heat gain and loss of the soil.

In arid zones where water is scarce and dryness is intense, dust can be a problem if the outdoor space, especially space closest to the house openings, is not well treated. It is suggested that native plants be used which require little or no water—cacti or tamarisks, for example. Also, all the space left should be paved to avoid dust movement into the house or reflection of intense heat. The variation in stone colors and forms usually found in arid zones helps in the design of beautiful paved areas.

The combination of semisubterranean with subterranean structures can ease the design of much of the outdoor elements discussed in this section by integrating them with the aboveground area. It also is possible to integrate most of the elements with an aboveground structure. However, these elements cannot be left to be treated after the house construction; rather they should be integrated in the early design stage, inside and out.

Outdoor landscaping of the subterranean house environs is essential for reasons of air purification; reduction of soil erosion; wind control; dust reduction; beautification of the earth-covered roofs and structure; privacy; noise reduction; safety (especially when house location is on a slope); temperature control (by reflection, absorption, shadowing, or by forms and configurations); screening an undesirable view; giving a sense of identity; direction and delineation of space.

One landscaping consideration is the chimney, which is an important element in the design; since it stands alone above the ground, it should be integrated with the landscaping of the environs. Indeed, the landscaping of the house environs is essential, even more so than in a regular house. First, it should be a landmark for identifying the house which is totally subterranean. Second, it is a structural necessity for the house plan. If the ground is covered by plants, the soil will absorb less heat and, consequently, be cooler. Thus, the temperature fluctuation found within the soil will be less. Third, it is a practical need in relation to penetration or blockage of solar radiation through the house openings. Fourth, landscaping contributes aesthetically to the total environment.

In contrast to our previously stated concern, increasing the moisture within the soil by landscaping can contribute to the cooling of the subterranean house throughout the summer. But watering must be done at night. Watering during the day will increase the soil temperature in the deep levels. One side benefit of having grass and other shrub roots occupy some space within the soil above the roof is that it lightens the weight of the soil cover somewhat. In any case, a proper drainage system should be secured for the roof soil and for the area adjacent to it.

Because the house is underground, lawn and grass areas are larger for subterranean houses than for conventional ones. This increased size will require more landscaping and perhaps more watering. Also, shrubs may be better than fencing in the aboveground part of the building to keep people from falling into excavated areas. In some cultures, the tracing of the lot boundary is necessary to avoid trespassing and to indicate that privacy must be respected. This is another reason for the use of shrubs as fences.

Aesthetic values to be introduced in the subterranean house are limited by nature of its location. In a slope-located house there would be one wall with windows and entrances exposed to the outside. This facade must be planned aesthetically to be attractive. The subterranean house with a patio open to the sky is another form which permits aesthetic considerations, although these may be mostly facing inward.

The possibility of a view of the environs from the house is an important determining factor in: (1) the site selection of the house, (2) the house orientation, and (3) the design of the side openings, especially the windows, glass walls and the door, perhaps partly transparent. Consideration of these three factors will have a direct impact on the view, the amount of light entering the house, the ventilation of the house and the total value of the house itself. Although it would be ideal to achieve all the advantages, in reality, the positioning of the house is a compromise between these factors and others.

Artificial landscaping of the immediate environs can improve the view to a large extent, while a larger view of the natural topographical variations can be achieved by proper location on an upper slope or at the top of a hill, sites which provide fresh air with less dust than on the lowland. On the other hand, accessibility to a site located on the lower slope is easier and requires less investment. To this author, the total advantages of the latter are great and should be considered the first priority.

Indoor plants have to be selected with special care. Because indoor space may not be exposed to direct sunshine or light, special indoor plants which require little light or sunshine have to be selected to meet the limitations of the house. Interior vegetation in the subterranean house is necessary not only because of aesthetic needs but also for functional aspects. The ambient house climate inside is different from that outdoors because of the impact of the soil mass as an insulator and storer of heat and humidity. In an extremely dry and warm climate, indoor plants can reasonably moderate the climate through evaporation which depends on the plant's structure, design and size. Plants also purify the air, recycling it through humidification, oxygenation and filtration. Oxygenation is accomplished through the process of photosynthesis during which the plant absorbs carbon dioxide and releases oxygen during the day. An ultraviolet lamp may be required. At night the process is reversed. It is estimated that 30 to 40 square meters of vegetation are needed to supply one person's oxygen needs for one year.[51] The plant humidifies by evapo-transpiration, the degree of humidification depending on the amount of water stored in the plant, the degree of dryness in the air, and the number, the size, the structure and the design of the surface of the leaves. The air can be cleaned by dilution of the polluted air by the oxygen produced by the plants. At the same time, the fragrance (odor) of the plant can "capture" the odor in the air.[52] However, in spite of these advantages, plants cannot provide all oxygen and humidity needs.

ROOFS

The geometry of the structure and proper design of the roof determine significantly the load-carrying capacity of the roof of the structure[53] (Fig. II-60). Thus, the type of geometry and design of the structure should be determined by the type of the soil within which the structure will be built. Whatever the soil's characteristics, safe building design can be achieved.[54]

Soil to be put on the roof should be examined for these aspects:

1. Its response to erosion (by water) and deflation (by wind).
2. Its ability to hold and support the growth of grass.

Fig. II-60. Various forms of subterranean roofs and their resistance to soil load. Flat forms resist loads at their midspan less than curved forms, which tend to deflect more of the load.

3. Its response to temperature fluctuation and humidity and especially to frost; its potential expansion and contraction and swelling with absorbed water, i.e., clay, for example.

Because of its load, the roof is subject to bending; therefore, structural support should be given to the roof by vertical walls or columns.

Roofs and ceilings can be constructed in a dome shape which is safe and strong and supports drainage toward the periphery. In any case, roofs must be sloped in order to increase the ratio of the runoff and to drain efficiently. To supplement roof drainage, a thick layer of rocks or gravel, and French pipes (perforated pipes) can be put over the roof. Runoff should be diverted downhill. Beneath the topsoil there should be granular materials at least 4 inches thick. To avoid penetration of particles and grains amongst the gravel and in order to retain the efficiency of the gravel's drainage, it may be necessary to put a filtering sheet, such as fiberglas, above the gravel. Above the sheet would be soil. Water draining through can also be led to perforated pipes.

There can be a variety of roof-ceiling forms which could be selected both for their structural advantages to support the heavy load and also for their aesthetical values. We can divide the forms of the roof into three major types: the flat, the vault and the dome. From the structural aspect, the flat or straight type roof resists load pressure least, while the curved forms resist load most. On the other hand, the curved forms are aesthetically attractive but more expensive to construct. They also use more space than the flat and would require more energy for heating or cooling.

ENTRANCES

The entrance to the house should be designed to be pleasing. Light or sunshine, overall size, color, vegetation, protective overhang, identification, originality in the design, owner self-expression, convenience and the angle of entrance to avoid heat gain and loss—all these establish a pleasing environment for the owners and visitors. Since almost all of the house is buried underground, the main emphasis in exterior design can be on the house entrance.

To minimize the bias and the claustrophobia usually associated with subterranean living, it is essential perceptionally to have a careful design and planning of the house entrance. Consider then the following:

1. It would be preferable that people not descend into the house from the entrance. Rather, people should enter at the same level as the first floor or ascend to it. When the house is located on a slope, this is quite easy.
2. The entrance must have enough light, preferably natural. To support this, the entrance color should be light and pleasing.
3. The overall design of the entrance and the door should be thoughtful and attractive. Natural plants may add to the attractiveness.

Another important entrance is that to the garage and its location in relation to the house design. Two possible areas may be allotted for the private car: on top of the house aboveground or in a recessed space adjacent to the house if the house is on a slope (Figs. II-61 & II-62). Within a subterranean environment, it is not desirable to have the car visible; it would be better to have an in-house garage. Another factor making an in-house garage preferable is that climatic changes and dust add to the deterioration of the car.

NOISE

Subterranean houses have the advantage of being separated from much of the outside noise from highways, airports and other kinds of outdoor activity. Soil is a poor element for sound transmission. Being underground, the structure has minimal vibration and can support the housing of industries which require precision; such industries have been constructed in Kansas City. In addition, the absence of outside noise enables music schools and concert halls to operate in an ideal environment. Similarly, in classrooms,

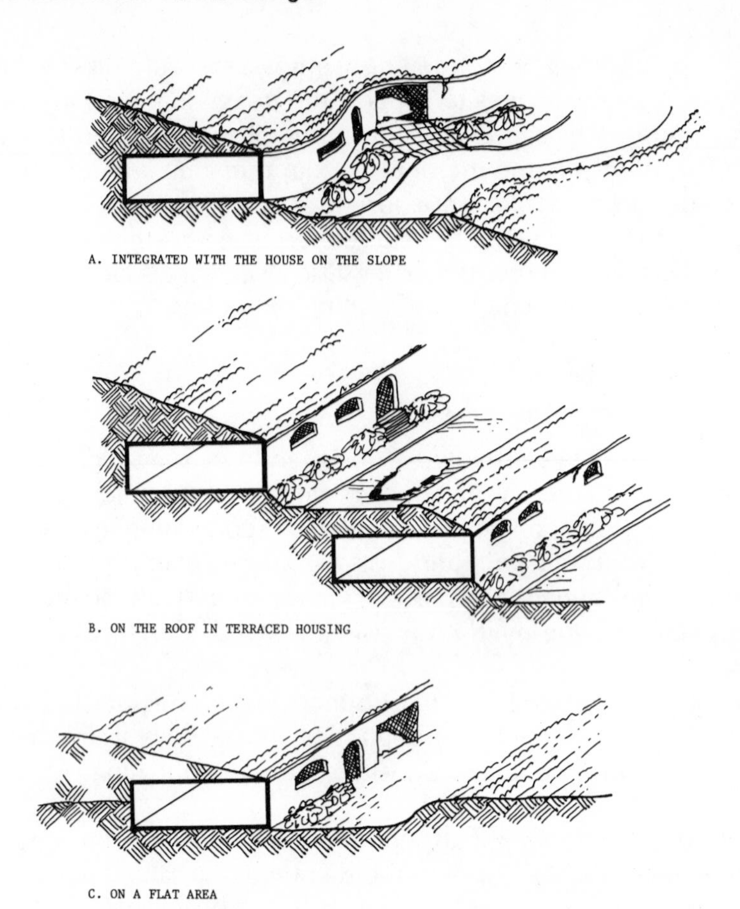

Fig. II-61. Alternative garage locations for subterranean terraced houses.

libraries or art studios which require a quiet environment, the elimination of noise may be beneficial.

The mass of earth surrounding the building contributes to the elimination of outside noise and vibration almost totally, while establishing the feeling of privacy and isolation. Indeed, it may be necessary to introduce some noise so that the isolation does not seem too extreme. Sources of noise inside the structure (furnaces, dehumidifiers, air conditioners, fans for air exchange or ventilation, washers, dryers) should be treated effectively. Because of the closed environment, such inside noises become much more noticeable and distracting than in the aboveground house.

Compared with the Western house which is partly oriented outward with many large windows, the subterranean house offers more privacy since it is more inwardly oriented with fewer or smaller windows to minimize heat exchange. An aspect of social importance is that noise generated by residents or their appliances (such as radio, television, utilities, etc.) is less transferable to the neighboring unit. At the same time, noise generated by the street (cars, children playing) will have less potential to enter the units of the residents.

Obviously, the subterranean house will have more aboveground open space than the conventional one on an equivalent lot size. This will encourage outdoor activity, especially for children. Socializing among age groups

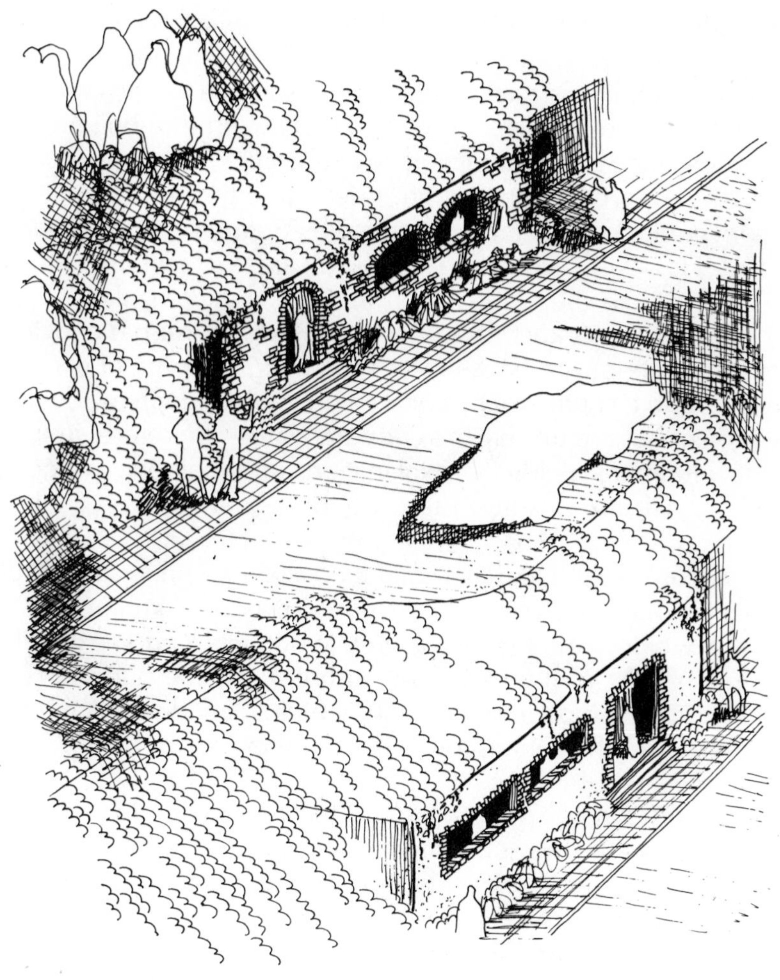

Fig. II-62. Terraced houses on a slope with pedestrian and car ways.

is expected to be intensive. However, subterranean houses will have the least outside noise.

APPLICABLE LEGISLATION AND BUILDING CODES

Not all countries of the world have legislation concerning subsurface construction. Some countries, such as the United States and Sweden, have just begun to deal with the necessity for such legislation. As laws are developed, they should be directed toward the following:

1. Ownership of the above- and the underground space and the owners' rights and limits;
2. Accessibility rights;
3. Planning and code requirements;
4. Construction rules, regulations and norms, and building permits;
5. Control and code enforcement;
6. Protection of public health and interest.

Building codes are constituted for public and private safety, health and property protection to ensure that the house is a safe shelter. The code should specify sound construction techniques, suitable for the area in which construction takes place. In addition, it should govern the building quality, structural safety and stability, function, fire resistance, durability, health and hygiene, accident prevention and life safety. It is the responsibility of the designer-developer of underground structures to prove the acceptability of these items. However, subterranean houses are innovations and do not always conform with the existing housing norms and regulations. Consequently, they may contradict those norms, in spite of the basic logic of the details of the subterranean house plan. Moreover, this type of plan may be introduced for the first time to the code enforcement people who may favor the new project but are tied by the code standards. Therefore, in this early stage of development, publicity, explanations and lobbying are necessary to persuade legislators to adjust the code to the new needs of subsurface construction.

Building codes respond to details of the house as an entity and in relation to its immediate surroundings. Labs proposes a set of guidelines, including discussion of window orientation, angles to sunlight and setback, height, and also a definition of the subterranean house for the purposes of code enforcement.[55] Some basic requirements of building codes which have special implications for the subterranean house follow.

Light. Most codes require that natural light be provided in every living space of the house. In the subterranean house, such requirements create difficulties. Since at least three walls of the subterranean house are covered by soil, the house forms which can allow natural light to enter every space unit are the linear one and the central patio house. The linear form will expose much of the house surface to the outside and defeat the original purpose of such placement. Also, circulation systems within the house have to be linear and can be made in the back part of the house (and require artificial light) or in the front part along the linear house unit. The situation is similar with the patio house. One solution may be to eliminate the code requirement for natural light from some of the house units: the bedroom, the storage area, the study room, the rest and shower rooms. Another solution could be to introduce different designs for the house (Fig. II-43, II-55, II-56).

Ventilation. The code requirement for natural ventilation in every unit of the house should also be modified. Mechanical ventilation by forced air is possible to every unit of the house and can be integrated into the house design. However, since forced ventilation is subject to failure, we again emphasize consideration of passive ventilation.

Emergency Accessibility. In the case of fire, movement is an important issue, especially at night between the bedrooms and the outside. Though standard codes require windows as access from the bedrooms, a good designer of subterranean housing can certainly meet the accessibility requirement by other means.

It should be mentioned here that code requirements, in the United States for example, are influenced by the fact that wood construction is so common. Houses constructed from concrete, as most of the subterranean houses are, should promote modification of the codes. In a concrete house with light metal doors the only burnable items are the furnishings. For further safety, a back exit can be constructed and connected to the rear circulation system away from the front entrance. Another alternative is a skylight with an escape ladder leading to the outside.[56] After all, in subterranean structures, the risk of fires spreading between two units is lower than in conventional houses. Furthermore, within the house unit itself, a fire is more easily controlled; it can usually be limited to a specific unit of the house.

Because firefighters are used to a traditional house design with various accessible sides (windows), underground houses may pose special problems for them. With an underground structure, however, the fire can be allowed to burn out without danger of spreading. Even so, the house should be easily accessible to fire fighters and their equipment.

As an underground house is cost effective in other respects, so it is in relation to construction for fire safety. In the conventional house, fireproof materials, such as nonflammable insulation, have to be installed, increasing the cost of construction. This is not required for the underground house, since it is built mostly of concrete.

On the other hand, it seems ironic that home insurance for subterranean houses may be a little higher than conventional home insurance. The subterranean house does not fall within any defined insurance categories because it has limited access and openings. Most subterranean houses are fireproof by virtue of their concrete construction.[57] Fire insurance companies obviously need more time to include the subterranean house in their categories.

The general standard building code in the United States is based on several definitions which are important to subterranean structures.* In the code, the basement has to have lower ceilings than standard rooms; while it is partially below ground level, the vertical height of the part within the ground is to be less than the vertical distance from the ground to the ceiling. The cellar, on the other hand, the space of a building which is wholly or partly below the ground, must have its vertical distance below ground either greater than or equal to that from the ground to the ceiling. The code also defines habitable spaces as those used for living, i.e., sleeping, eating or working; unhabitable spaces are bathrooms, toilet compartments, halls, and storage or utility spaces.

For patios, the code requires that the court should not be less than 3 feet in width; when windows are on opposite sides of the court, the minimum width should be 6 feet. If courts are bounded by three or more walls, the width should not be less than 10 feet.[58]

If the number of occupants is 10 people or less, only one entrance is required. In addition, the space requirement for each occupant is 300 square feet. Distance to the exit should not exceed 150 feet from any point, regardless of intervening rooms. However, exits must not require occupants to pass through kitchens, storerooms, restrooms, closets, etc. The exit should not be less than 3 feet wide and 5.7 feet high. It should open at least 90°.

*For example, the BOCA (Building Officials and Code Administrators) code.

Doors must open from the inside. Sleeping rooms must have an extra exit, either one operable window with a minimum height of 24 inches and a minimum width of 20 inches, or an exterior door to each sleeping room. The property must have a pedestrian or vehicular access from a public street. All habitable rooms must have natural light in an area of at least one tenth of the room or 10 square feet. There should be natural or mechanical ventilation for the habitable rooms. The site must be above flood level, with surface runoff drained from the building.

There are also other factors not related to subterranean housing which can bring changes in the building code: new building materials (fire-resistant materials, new waterproofing materials), new health hazards, energy conservation needs, new solar energy uses, and, finally, the adaptability of public officials to innovative methods of improvement.[59]

When a large number of subterranean houses are constructed and/or attached to each other, a few ownership questions will probably develop. Problems probably will arise involving open spaces, delivery through the infrastructure system, gas pipes, water supply, electrical power, monitoring points, sewage, walls, garbage collection, centralized heating or cooling systems, landscaping of the immediate environs, fences, noise generated by outdoor activities or neighbors, recreation and common spaces, visual obstructions, space expansion, maintenance, location of solar collectors, landscape obstruction, parking areas, lighting of passages and pedestrian paths at night, etc. It may be necessary, therefore, to develop some standard rules and regulations with input from subterranean homeowners and associations. Another type of question will relate to the deed to the house and to the land (because a portion of the land presumably will be owned by each family), election of representatives, monthly dues payment for common services, maintenance responsibility of the owners, and the use of common open space.[60]

In addition, apartments or houses built underground involve considerable sharing as opposed to the traditional privacy of the conventional, aboveground house. Families with several children may welcome the new adjustment, because of the energy savings and because of the more intensive interaction of their children with others. Planners/designers should, therefore, carefully plan for equipment and other needs of children within the common space of the units.

Many of the services in subterranean apartments will be communal to reduce expenses: garbage collection space, central heating or cooling, playgrounds and equipment for children, laundry facilities, tennis courts, swimming pools, announcement boards, storage areas, parking spaces for cars and for campers, TV and radio cable or antenna, sewage networks, water meters and electricity meters. Thus, the family must adapt both to the subterranean life style and also to the sharing aspects of the common housing-apartments. This change will require new perceptions, expectations and attitudes.

Conclusion

In this chapter we are primarily concerned with solving the problems of extreme climates, such as extreme cold (in the north) or extreme hot or cold and dry (in the arid region). Subterranean housing meets the needs of these climates more than that of other climatic regions. Interestingly, most of the earth-covered or subterranean houses built recently in the United States are either in snowy regions of the North, such as Minnesota, or in the arid regions of the Southwest. Similarly, the majority of subterranean houses used since ancient times have been developed in arid zones.

A question usually raised in connection with the subterranean house concerns the cost of construction and maintenance, especially when it is compared with the conventional aboveground house. A subterranean house's cost is basically composed of the cost of the land, grading or blasting, materials and labor. In general, we can state that land price is reduced greatly because of the possibility of simultaneous land use under- and aboveground.

In a sense, the proposed subterranean concept makes use of resources which have not been used before by multiplying the land resource space in a given site. Furthermore, such buildings make use of land such as slopes which are not otherwise usable conventionally. However, the cost estimate should take into consideration also the life cycle cost, utilities and energy cost, and the maintenance costs of a subterranean house as compared to a conventional house. In such a cost comparison and with the expected energy price increase, the subterranean house probably will involve lower costs than a conventional one.

If blasting is to take place, the site preparation cost will go up. However, the planning and design price should be lower because the subterranean house is simpler without windows. Also, maintenance costs will be lower since there is little of the subterranean house exposed to the outside.[61] Further reduction relates to the fact that most of the outside construction needed in a conventional house is not required in the subterranean house: gutters, siding, painting, windows, roofing and other finishing materials.

In general, we can state that energy required for the subterranean house is limited to heating, ventilation and lighting. Heating energy may be required to increase the temperature as much as 15°C, but this may not be necessary when the house is inhabited; in contrast, the aboveground house may require three to four times this amount of energy. Lighting can be a very small portion of energy cost if proper design is introduced, and ventilation will also be a small fraction if passive ventilation is present. In many of the subterranean homes constructed in the United States, the range of energy savings for heating reaches 50 to 80 percent. Furthermore, these savings do not reflect the differences in capital investment for heating equipment and its maintenance in the subterranean house versus that for the aboveground one.

In any consideration of the house cost, we should include, in addition to regular factors, two other aspects: the future continuous increase of oil prices (and with it energy cost) and inflation. Those two factors will continue to determine the increasing cost of houses. However, the latter (inflation) will influence the subterranean house as well, while the first (energy cost) will influence the subterranean house less. Thus, the gap in energy cost between the conventional house and an equivalent subterranean house will continue to grow (Table II-7). The cost of insurance should be lower in an underground house, because of its "relative immunity to hail, wind, and fire damage."[62]

From the owner's point of view, the short term is of most concern since the life cycle of ownership may range from 5 to 10 years in the United States, for example. The economic advantage must be strong enough to offset the disadvantages associated with uncertainty and marketability. It has already been proved in many places in the United States that the construction costs of the subterranean houses are about the same as the costs of the conventional residential units, and in some cases are lower by 10 to 15 percent.[63] Finally, according to McWilliams and Findley who studied the cost of the earth-covered building compared with the equivalent conventional one: as a long-range investment the earth-covered building is economically feasible.[64]

The concept of the subterranean house is new and unconventional. Yet it is a promising solution to many modern problems. Therefore, barriers such as the following will probably be resolved as more development of such houses takes place:

1. Psychological bias of the public and lack of awareness of the economic and health advantages of the subterranean house.
2. Lack of expertise of planners-designers, architects, developers and others for such a unique and innovative development.
3. Existing city codes, regulations and zoning which are prepared for conventional constructions and are based on historical experience, excluding this type of an innovative house and construction. New codes, regulations and zoning will have to be based on today's pioneering experiences in subterranean structures.
4. Very conservative financing systems of mortgage and loan institutions, based on conventional ideas of selling and buying real estate. The likelihood of their agreeing to finance a subterranean house would come with the increase of such development.[65]

A second look at the above-listed obstacles will show that all of them are artificial and, therefore, can be eliminated or diminished as time passes and as the new movement of earth-covered shelters and subterranean houses gains strength. Advantages should be stressed; for example, underground structures, especially those which are built deep on the slope of a mountain

Table II-7. Subterranean House Construction Items to Be Added or Subtracted and the Influence on Construction or Maintenance Cost

TO BE ADDED	SUBTRACTED (COMPARED WITH CONVENTIONAL HOUSE)
• Special study of the site & the soil	• Land cost may be minimal. This may also result from dual use of the site.
• Special grading or blasting	• Some indoor partition walls
• Strong structure of walls and roofs, (mandatory)	• Many design and planning details usually associated with the conventional house are not required
• Special waterproofing (essential)	• Outside painting
• Special ventilation system (forced air) is required	• Most or all windows
• Soil coverage, needed at the end of construction	• Siding (and most material for outside)
• Landscaping & watering of larger lawns due to roof coverage with soil	• Reduction in use of some heating or cooling appliances
• Accessible road (when on slope)	• Most fire resistant materials
• Skylight	• Exterior maintenance, cost of remodelling
	• High insurance rate due to relative immunity to weather and fire destruction

or a hill, have the potential to expand to almost unlimited space within the earth, a great advantage for industry, storage facilities or offices. If the geological formation layers are horizontal, as is the case in Kansas City, such expansion can be relatively easy and routine.

In an interview by Baggs in White Cliffs, Australia, a woman who had lived many years in an underground house in this arid area responded to the question of how she felt in her underground home by saying:

> My husband has been a miner all his life. He loves it, feels at home. The kids love the dugout, in bad weather the kids don't even play outside, they just sit in the kitchen and read or colour—rather than play outside when it is hot or cold or windy.
>
> When we lived aboveground, I had to stay up almost all night in the summer sponging them to cool them down, giving them sips of water continuously—they hardly slept at all as babies, but the twenty month-old girl who came into this dugout as a newborn baby has slept like a top right from the start and has been a totally different type of baby to raise because she has always been comfortable and only needed a minimum of clothes. It's been so much easier for me.[66]

In short, we can summarize the advantages of the subterranean house as follows:

1. Lower planning and construction cost compared with the aboveground house.
2. Land savings because of dual usage.
3. More fire resistance than the aboveground house.
4. Reduction in the length of utility systems and, therefore, reduction of maintenance and repairs. Consequently, initial investment, time required for construction and the tax load for the residents should be reduced.

5. Conservation of heating and cooling energy and minimal maintenance cost due to weathering.
6. Proximity between the various land uses which, therefore:

   a. Saves transportation energy;
   b. Saves energy for utilities supply, such as pumping (sewage) or pressure (water);
   c. Intensifies social interaction between residents and among various age groups;
   d. Increases the use of social facilities and services by the various age groups, especially the elderly and children.

7. Diversified use of the land in more effective ways. Major activities can be carried on underground leaving the surface for special essential land uses that cannot be carried on underground. Examples of such dual usage are:

   a. Industries and manufacturies, nonpolluting, where the underground is used for production, storage, management and the like, while the surface is used primarily for loading and unloading, office space and some parking.
   b. Educational facilities in which study rooms, library and storage areas are underground, with the aboveground area used for recreational open space. The same can be done for public gathering places such as assembly halls and theaters.

8. Lower insurance costs.
9. Protection from surface hazards, noise, vibrations, earthquakes and strong storms.
10. Possible high load-bearing capacity for the floor.[67]
11. Greater security than aboveground buildings (because of limited access), reducing need for guards and cutting vandalism and property losses.

Subterranean living has great advantages, especially in extreme climates, dry-hot or dry-cold. Thus, on the one hand, we have the desert areas with their varieties of climatic conditions or extremes, or, on the other hand, the polar area with its variety of extremes. However, since most human activities are located more in the desert area than in the polar area, our main response here is to view the applicability of the house in the desert, although the same system, with some modification, can apply in the polar areas, also. (Admittedly, the permafrost problem will have to be solved.) However, application in desert regions is particularly appropriate because of the extremes of temperature between night and day; because of solar radiation, aridity and lack of rainfall. Remember, also, that construction of subterranean housing in humid areas brings with it many problems, especially with regard to drainage and to humidity in the soil itself, problems which do not exist in arid areas.

The use of subterranean living is well known throughout the history of mankind. The first shelter used by man in the Paleolithic period was subterranean (caves). Throughout history, the use of subterranean construction was intensified for a variety of reasons. However, the interest in subterranean living increased in the 1970s in the economically developed countries because of the energy crisis and the great concern for environmental and ecological quality.

Notes

1. James Scalise, ed., *Earth Integrated Architecture* (Tempe, Arizona: The Architecture Foundation, 1975), A-12.
2. Baruch Givoni, "Modifying the Ambient Temperature of Underground Buildings," in *Earth Covered Buildings: Technical Notes* (CONF-7805138-P1), Frank Moreland, Forrest Higgs and Jason Shih, eds., Vol. I (Springfield, Virginia: National Technical Information Service, 1979), pp. 123–138.
3. Ibid.
4. Thomas P. Bligh, Paul Shipp and George Meixel, "Where to Insulate Earth Protected Buildings and Existing Basements," in *Earth Covered Buildings: Technical Notes,* p. 256.
5. The Minnesota Energy Agency, *Earth Sheltered Housing Design, Guidelines, Examples, and References.* (New York: Van Nostrand Reinhold Company, 1979, p. 55.
6. Kneeland, William F., "Underground Dwellings—Built in Oklahoma," in *Earth Covered Buildings and Settlements,* Vol. II (CONF-7805138-P2), Frank Moreland, ed. (Springfield, Virginia: National Technical Information Service, 1979), p. 529.
7. Ibid, p. 265.
8. The Minnesota Energy Agency, p. 55.
9. Ibid, p. 63.
10. Bligh, Shipp and Meixel, p. 253.
11. Ibid, p. 254.
12. Barbara Bannon-Harwood, "Earth Shelters are Here to Stay," *Concrete Construction,* **25**/9 (September 1980), p. 648.
13. De Leuw, Catber & Company, *Deep Excavation Techniques for Shelters in Urban Areas, Preliminary Planning and Cost Data* (Arlington, Virginia: Defense Documentation Center, 1963), pp. 6–9.
14. An early version of this concept was introduced by this author in the chapter "Free-energy Cooling Systems for Houses in the Desert," in *Innovations for Future Cities,* Gideon Golany, ed. (New York: Praeger Publishers, 1976), pp. 246–262.
15. Robert L. Lytton, "Soil and Ground Water Considerations," in *Alternatives in Energy Conservation: The Use of Earth Covered Buildings, Proceedings* of a Conference, Fort Worth, Texas, July 9–12, 1975 (Washington, D.C.: National Science Foundation, 1975), p. 261.
16. W. R. van Wijk, *Physics of Plant Environment* (Amsterdam, Holland: North-Holland Publishing Company, 1963), p. 102.
17. Walter F. Spiegel, "Air Quality and Heat Transfer," in *Alternatives in Energy Conservation,* p. 248.
18. "Survival Shelters," *ASHRAE Guide and Data Book Applications 1971* (New York: American Society of Heating, Refrigerating and Air-Conditioning Engineers, 1971), p. 183.
19. Ibid. Recall that the depth to which temperature fluctuation extends varies and is dependent on soil composition and the moisture content.
20. Spiegel, p. 248.
21. Ibid.
22. Thomas P. Bligh, "Energy Conservation by Building Underground," in *Underground Space,* **1**/1 (May/June, 1976), 23.

23. *ASHRAE Guide,* p. 186.
24. Charles Fairhurst, "Introduction," in *Alternatives in Energy Conservation,* p. 1.
25. van Wijk, p. 102.
26. *ASHRAE Guide,* p. 191.
27. Gideon Golany, *Rural Geography of a Traditional Village: A Case Study of Tayiba Village* (Haifa, Israel: Technion-Israel Institute of Technology, The Faculty of Architecture and Town Planning, 1967), pp. 43–57.
28. Henry Orlowski, "Thermal Chimneys and Natural Ventilation," in *Earth Covered Buildings: Technical Notes,* p. 220.
29. *ASHRAE Guide,* p. 184.
30. Lytton, p. 260.
31. *ASHRAE Guide,* p. 195.
32. Sydney A. Baggs, "The Dugout Dwellings of an Outback Opal Mining Town in Australia," in *Underground Utilization: A Reference Manual of Selected Works,* 8 vols., Truman Stauffer, ed., Vol. IV: *Human Response and Social Acceptance of Underground Space* (Kansas City: University of Missouri, Department of Geosciences, 1978), p. 580.
33. Orlowski, p. 221.
34. *ASHRAE Guide,* p. 192.
35. Ibid.
36. The Minnesota Energy Agency, p. 35.
37. Robert B. Bechtel, "Psychological Aspects of Earth Covered Buildings," in *Earth Covered Buildings: Technical Notes,* p. 71.
38. Elizabeth Wunderlich, "Psychology and Underground Development," in *Underground Utilization: A Reference Manual of Selected Works,* Vol. IV, p. 526.
39. Nicholas Chryssafopoulos, "Employee Attitudes," in *Underground Utilization: A Reference Manual of Selected Works,* Vol. IV, p. 530.
40. Birger Jansson, "Terraspace—A Hidden Resource," in *Underground Utilization: A Reference Manual of Selected Works,* Vol. I: *Historical Perspective,* p. 41.
41. Bechtel, pp. 71–72.
42. The Minnesota Energy Agency, pp. 98–99.
43. Daniel G. True, "Economical Structural and Footing Considerations for Buried Structures," in *Earth Covered Buildings: Technical Notes,* pp. 45–46.
44. De Leuw, pp. 10–11.
45. Robert M. Darvas, "Structural Problems in Earth Covered Buildings," in *Earth Covered Buildings: Technical Notes,* p. 72.
46. David J. Bennett, "Notes on the Underground," in *Earth Covered Buildings and Settlements,* Vol. 2, p. 205.
47. C. V. Chester, H. B. Shapira, P. R. Barnes and G. A. Cristy, "An Earth Covered Residential Concept for the Humid Continental Region," in *Earth Covered Buildings: Technical Notes,* pp. 173 and 177.
48. Belinda L. Collins, "Review of the Psychological Reaction to Windows," in *Underground Utilization: A Reference Manual of Selected Works.* Vol. IV, p. 532.
49. Ibid, p. 536.
50. Kenneth B. Labs, "Planning for Underground Housing," in *Underground Utilization: A Reference Manual of Selected Works,* Vol. IV p. 558.
51. Bernatzky, "Performance of Value of Trees," *Anthos,* **I,** (1959).
52. Gary O. Robinelle, *Plants, People and the Environment,* U.S. Department of Interior, N.P.S., 1972.
53. W. J. Flathau and J. P. Balsara, "Soil Structure Instruction—An Overview," in *Earth Covered Buildings: Technical Notes,* p. 16.
54. Donald W. Quigley and James M. Duncan, "Earth Pressures on Shallow Buried Structures," in *Earth Covered Buildings: Technical Notes,* pp. 34–44.
55. Labs, "Planning for Underground Housing," p. 559.
56. The Minnesota Energy Agency, pp. 156–157.
57. Ibid, pp. 169–173.
58. Ibid, pp. 269–276.
59. Melvyn Green, "Building Codes and Underground Buildings," in *Earth Covered Buildings and Settlements,* pp. 25–29.

60. Donald B. McWilliams and Stephen M. Findley, "A Life Cycle Cost Comparison Between a Conventional and an Earth Covered Home," in *Earth Covered Buildings and Settlements,* pp. 90–91.
61. The Minnesota Energy Agency, pp. 191–193.
62. McWilliams and Findley, p. 97.
63. Kneeland, p. 265.
64. McWilliams and Findley, p. 98.
65. Hans R. Isakson, "Institutional Constraints on the Marketing and Financing of Earth Covered Buildings and Settlements," in *Earth Covered Buildings and Settlements,* pp. 7–11.
66. Baggs, pp. 596–597.
67. John D. Rockaway and N. B. Aughenbaugh, "Go Underground for Low Cost Housing," in *Underground Utilization: A Reference Manual of Selected Works,* Vol. IV, p. 616.

Section III:—Integration of Underground Placement within Urban Design

The following section discusses the complexity of the contributing factors to be considered in the design of those innovative urban settlements which are to include underground structures and land uses. Although there have been such settlements throughout history, there is today very little, if any, practical experience in such designs. Historical settlements developed through an evolutionary process rather than through planning; and they instruct us on the small scale rather than on the large: that is, for houses and neighborhoods rather than for cities. Also, most of the historical settlements were agricultural villages rather than urban centers. Certainly, we must continue to discuss theory, but it is practical application which will receive most of our concern.

It seems to us that the primary elements involved in the design of an urban settlement which is at least partially underground are particularly diverse: climatic, physical, social and psychological, economic, educational, management, transportation and landscaping. Although we will focus in this section of our discussion primarily on the physical aspects of designing such a settlement, our analysis will touch upon the other elements as well. Within the area of physical design, we shall consider site selection, new urban design principles and environmental issues. We also shall introduce an urban design case, a neighborhood cell suitable for use in the arid zone of Israel.

Site Selection Considerations

The process of site selection for a settlement in general, and for a settlement with some subterranean parts in particular, is the most vital stage in the success or failure of the settlement. Factors such as soil, orientation, sunshine, wind and many others have much more significance in this case than in any other cases.[1] Moreover, each factor must be looked upon in relation to the others. Further, this author is convinced that subterranean uses must be merged with the supraterranean and special compact forms, making our concept for development unique, complex and promising but requiring special alterations and offering challenges throughout all the detailed design phases. Thus, site selection considerations and methods should anticipate these complexities. It is, however, beyond the ability of a single expert to take such responsibility; a team of diversified experts is required to work jointly on the site selection.

The process of site selection for a combined settlement of supra- and subterranean construction necessitates a thorough investigation and consideration of many criteria. Decisions on the site to be selected involve economic and social aspects as well as matters of health for the population. The development of every city has its own historical mechanism; but in any case, it is crucial to keep in mind that the site once selected involves irreversible decisions. Therefore, in the selection of the site, we must find synthesis among comprehensive treatments; rationale; effective, beneficial output; and, last but not least, intuitiveness.

Although we can generalize about the need for specific land uses in the subterranean settlement, it is essential to consider the needs of the people who will live there and the climatic characteristics of the site. For instance, socializing would require common space specialized for different age groups and perhaps large kitchens in each unit. The open courtyard may need to be dominant in every house design. Thus, these distinctive needs must be kept in mind when choosing a site because they can influence the selection of a proper topographical form. Ancient subterranean settlements evolved as extended processes over many generations, and so did the process of site selection. They were built by people who had a clear understanding of the environment, the climatic conditions, the soil and other factors. Such study was not done on the drafting table, but by learning through experimentation and successful selection. When this author studies the sites and the development of those settlements, he finds that the major difference between them and our development of subterranean space is the availability of advanced technology. However, there is much evidence that their understanding of the potentialities and the constraints of the environment were significantly more developed than ours. If the developers of the ancient subterranean settlements had had our technology to combine with their understanding of their environment, their achievements in subterranean usage might have been even more significant.

Physical criteria to be considered for selecting the subterranean settlement are diversified. We would divide those criteria into primary and secondary levels with primary criteria of:

1. Physiography
2. Climate
3. Hydrology
4. Environmental quality
5. Accessibility and proximity to resources

PHYSIOGRAPHY

We include in physiography a variety of elements: geology, elevation, slopes and soil.

Geological constraints must be considered seriously in selection of a site. Planners of subsurface structures must avoid sites subject to seismic activity such as active faults. A Japanese study concluded that rock caverns are highly safe in earthquake conditions and have many advantages over structures built above the surface.[2] The settlement, however, could be unsafe if it is situated close to a fault. Faults can become active and could create cracks in a building, even though it may be made of solid concrete. Faults, as well as cavities found in rocks, can also be a means for water penetration at their joints. In addition, the house's lower level should be sited above the water table, if any exists in the area. Also, nearby cavities and old mines should be of concern because they may be a source of water leakage or they may weaken the foundation and the structure of the subterranean house. Also dangerous are karst formations, underground hydrological networks of cavities developed mostly in relatively soft limestone.

An ideal geological formation for a subterranean dwelling is composed of alternating hard and soft limestone layers which form horizontal stratification, with each layer 2.5–3.5 meters deep (approximately 8 to 12 feet). Such conditions are ideal for housing and working space because the simple process of destroying the soft layers provides subterranean space without much investment for building conventional walls, ceilings and floors (Fig. III-1). Walls and ceilings can be waterproofed and plastered. These geological conditions are found in the Kansas City valley where, as we have mentioned, underground space left between the geological layers in former limestone quarries has been used for a variety of purposes such as factories, offices, and storage and refrigeration areas.

Elevation is another aspect of the physiography. Absolute elevation is relative to sea level, and it is an indication of temperature and humidity. We can state that, in general, the higher the absolute elevation, the lower the temperature but the higher the relative humidity in the air. As such it has an impact on the macroclimate and the humidity within the soil. Relative elevation, on the other hand, is the height of the site in relation to its sur-

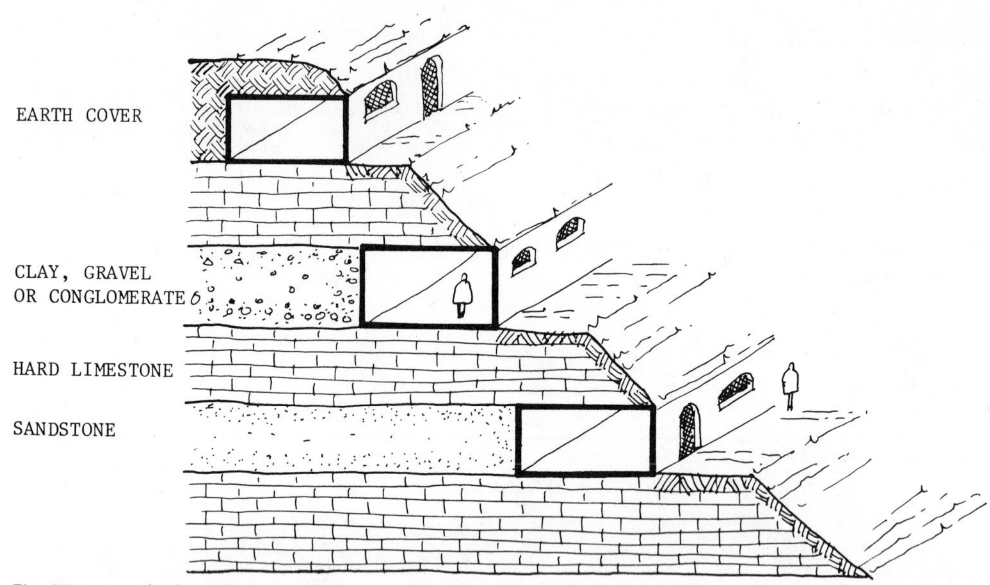

Fig. III-1. An ideal geological structure for subterranean design. Study of the geological sedimentation is necessary. An ideal geological condition is layers of soft and hard stone alternated horizontally, and the soft limestone layers are 2.50–3.5 m. (approx. 8–12 ft). The hard layers can be used as floors and ceilings.

roundings. This interests us because it indicates the possibility of a view to the surrounding environment, an aspect which is crucial for minimizing claustrophobia in subterranean living. In addition, relative elevation is important in relation to the design and maintenance of sewage, water networks and other utilities, and to the accessibility of transportation to the potential site.

Topographical forms are another issue of concern. Forms vary and can be classified, in general, as closed or open. Closed forms include valleys or canyons, craters, playa, or geological depressions. Open forms are plains, hilltops, large slopes, coastal plains, wide valleys, alluvial fans, mesas and plateaus. Each of these forms will affect the diurnal and seasonal microclimate of the site (temperature, relative humidity and wind velocity), making it different from the surrounding, regional area. The form will also affect the range of the view to and from the site, the ventilation of the site, potential pollution and accessibility.

Closed forms tend to limit ventilation; encourage inversion; increase temperature during the second half of the day and decrease it after midnight; support local afternoon wind turbulence; restrict the view to the environs; affect drainage systems; and increase the potentiality for flooding (Fig. III-2). Open forms, on the other hand, encourage ventilation and strong wind movement; support ascending and descending winds (on slopes); create dusty winds when soil cover is absent (especially in arid zones); lower the temperature in general because of strong wind velocity; provide a wide view to the surroundings, and, in some cases, allow for ease of accessibility. The subterranean settlement site should be selected with regard to the optimal form. In practice, it is nearly impossible to find a form which has all the advantages; any site selected is a compromise.

Sewage disposal can become a problem under special topographical conditions where gravity systems cannot be used. Pumping will, of course, mean higher construction and maintenance costs. Serviceability is crucial

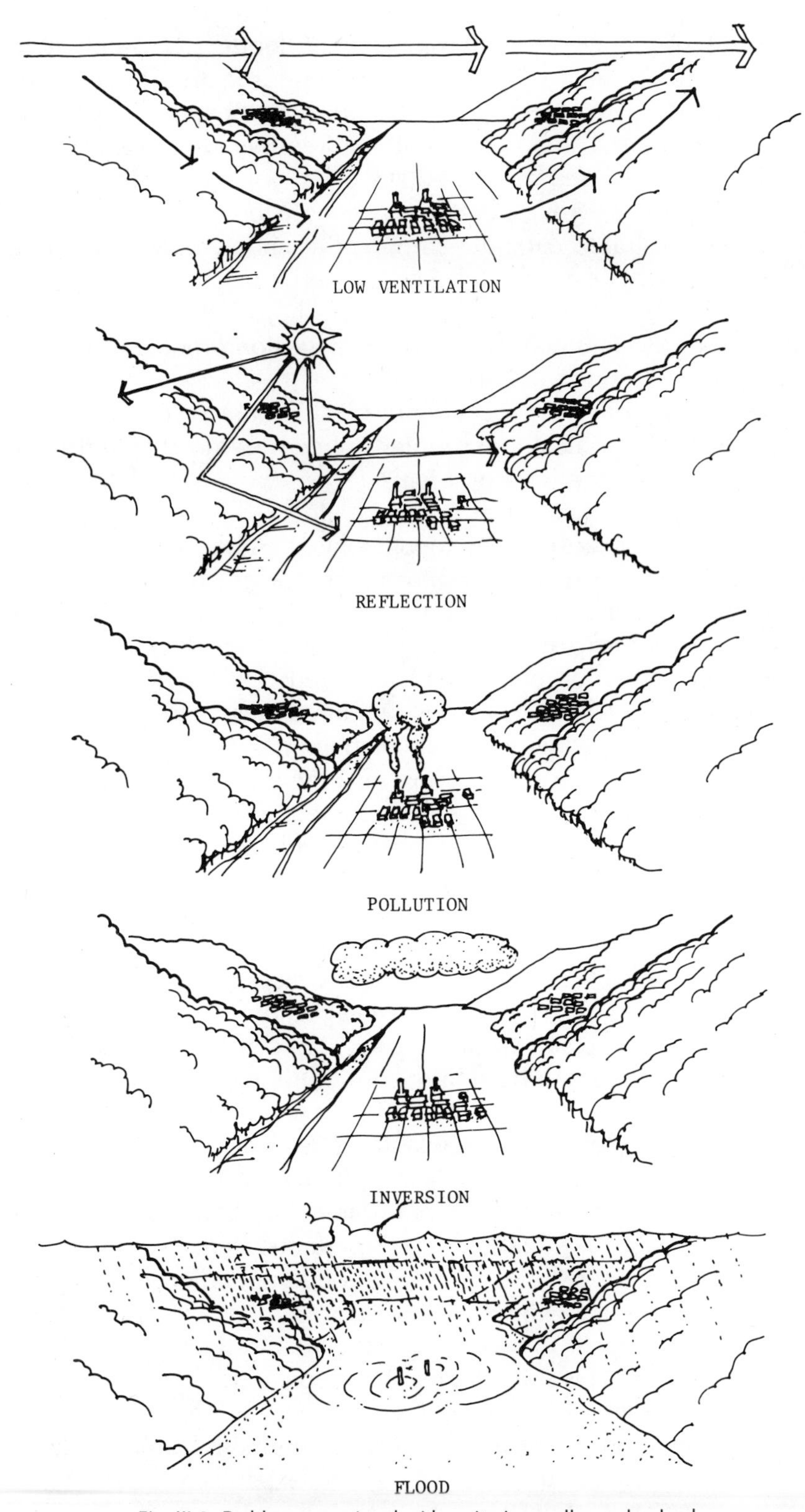

Fig. III-2. Problems associated with a site in a valley or lowland.

here because interrupted operation may cause flooding.[3] One possible solution is the selection of a site on a slope for the structure or neighborhood and the use of a gravity system for the entire development.

Consideration must also be given to the infrequent but torrential and turbulent rain typical of arid zones. A site on a slope can help here, too. Efficient drainage is, however, essential, not only to avoid flooding but also to eliminate undesirable dampness around the underground structures. To this end, it will be necessary to insulate the upper parts of all the structures with impermeable materials. Such insulation, however, will increase surface runoff and consequently accelerate the erosion process, especially at the margins of a clustered development.

As can be seen from the preceding discussion of some of the problems, slopes are the topographical forms which concern us most since they seem to be associated with the least problems.

Most cities in the United States have been located conventionally in valleys and on plains because of the topographical convenience for transportation, for utilities construction and for urban construction. Consequently, those valleys and plains which are usually good agricultural land have been taken over by urban expansion. At the same time, these cities suffer from congestion, high pollution impact and restricted views. Any new subterranean settlements on a flat topography will have these and other problems of existing cities: ventilation, relatively high afternoon temperatures, inversion, sewage pumping, flood risk, loss of privacy, dust and litter—all in addition to the great impact on the environment. On the other hand, sites on the tops of hills have the advantages of wide view, good ventilation, much potential light and sunshine, separation from water tables, ease of sewage treatment, freedom from dust and relatively cool temperatures. Yet, here too there are disadvantages: low water resources require pumping of water; the necessity of blasting for road construction; and, often, high winds.

Although slopes do have some disadvantages (potentials for erosion and landslide, possible necessity for blasting for road construction), with careful design, these problems can be minimized. More importantly, slopes have many advantages. Most offer a direct view over a wide range into what may be a scenic valley, plain or lowland. As we mentioned before, this is significant when possible risk of claustrophobia exists. Also, building on slopes saves the most valuable soil of the lowland. Other positive aspects of the slopes are good ventilation, less interrupted light or sunshine, freedom from dust, relatively good climate, sewage operation by gravity (Fig. III-3), and potentially low water table. Finally, they provide space for possible future expansion of the subterranean structure horizontally or vertically, upward and downward, which a flat topography does not provide to such an extent. It seems to us that this latter aspect can have a far-reaching impact on the environmental quality of the settlement. Residents can expand and improve their house when the family expands without the necessity of moving. Thus, low mobility can be realized and, therefore, neighborhood ties and identity strengthened. At the same time, industry can expand storage or working space with minimal cost under these circumstances.

Another important element of the physiography is soil. The mass and type of the soil will determine construction cost and duration (especially of the

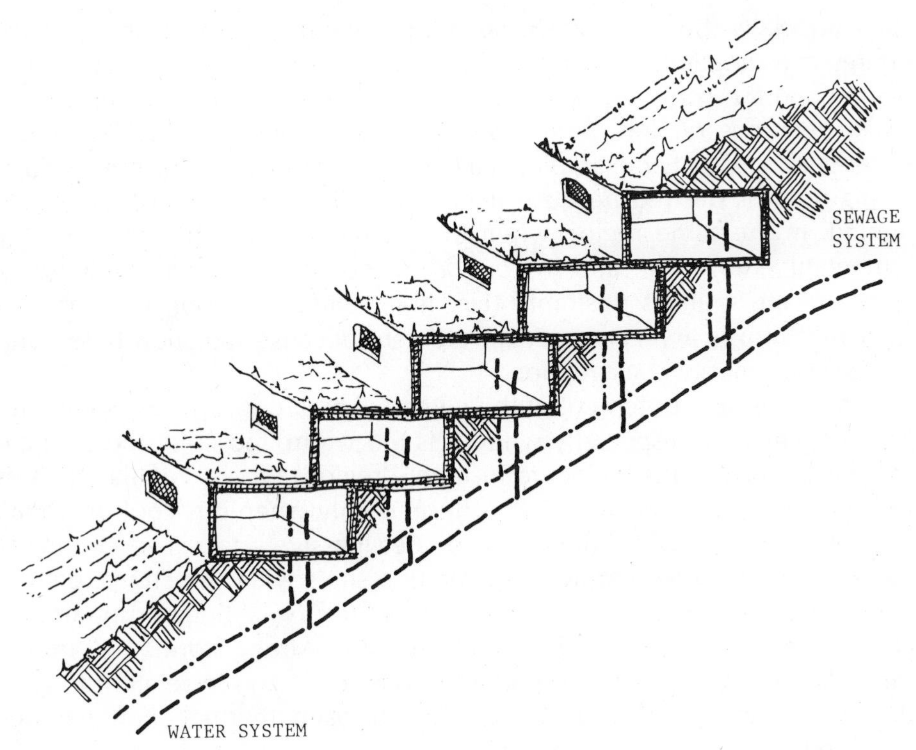

Fig. III-3. Cross section of buildings set into a slope which helps to solve problems of sewage disposal, flooding and drainage in general.

infrastructure); the temperature fluctuation within the soil and the need for insulation of the subterranean structure; the degree of moisture within the soil which will affect the structures directly and determine their needs for waterproofing; resistance of the soil to erosion (very important especially when the settlement is located in the slope); and, finally, soil behavior in expansion and contraction under different conditions (see discussion of soil in Section II). Thus, soil suitability for subterranean space construction is an economic and social matter, as well as one which will reflect the quality of the land uses and of the life and health of the user.

CLIMATE

Climatic conditions are complex, requiring careful study for site selection for a subterranean settlement. We include in the climatic features the criteria of temperature, solar radiation, wind and precipitation. Basically, all the information is required on a diurnal as well as a seasonal basis. Also, in studying a site we should be concerned with the climate on the regional scale (macro) as well as the potential site scale (micro). Generally, data is not available on the microscale, and it may be scarce on the macroscale too. This is especially true in the arid zones where weather stations are sparse, if there are any at all. Therefore, there may be a need to survey the potential site and the region, studying the climate for at least a one-year cycle. Data from a carefully selected similar site could be used with discretion.

A change-in-temperature study combined with a study of the soil will help planners understand the seasonal temperature fluctuation within the soil, especially the lag process. Such a study will determine the probable energy consumption for heating or cooling. In addition, studies of solar radiation and the sun's yearly cycle of movement are necessary in order to make decisions on the orientation of the settlement as a whole and of streets and houses. For example, in a hot-dry climate, site selection on slopes facing north is desired because light is abundant and yet sun radiation is limited. In the same region when the elevation is much higher and the temperatures lowered, and in cold-dry or cold-humid climates, a site on slopes facing south is desirable (Fig. III-4). Site orientation in relation to solar radiation is a very significant element because radiation determines the level of energy consumption.

Wind is the other element of the climate which concerns us; sometimes it is not welcome, especially when it is too warm, too cold, too strong or too dusty. But wind is welcome in humid regions where ventilation is essential. In hot, dry climates wind is desired only when it is cool or when it can be combined with humidity and used passively to cause cooling by evaporation. However, knowledge about a site's wind directions, patterns and velocities is important for the design of the city. Although subterranean houses are protected from tornadoes and other strong storms by virtue of their soil massiveness, it is important to consider wind in relation to street directions, aboveground publicly used open space, thermal chimney directions and all types of house openings.

The origin of the wind will determine the air quality. Thus, any polluting source in the potential site region should be studied in relation to wind. Moreover, wind is not only a source of high or low temperature, but it also may be a carrier of moisture. In addition, the process causes adiabatic wind gains and losses in temperature of the air at different elevations. Descending adiabatic wind gains 1°C for approximately every 100 meters elevation (and loses some of its relative humidity when moisture is present); the process is reversed when the wind is ascending. Further, we can state that sites on the upper part of the slope will be cooler than on the lower part. Wind movement can be a source of more dust and particles in a valley than on a slope, especially when earth cover is minimal or nonexistent.

HYDROLOGY

Precipitation, especially rain, is a major contributor to erosion, landslides, water tables, soil moisture and runoff, all of which affect the construction of the subterranean settlement. In addition, the nature and the pattern of the rain differ from one region to another. For example, although rain is scarce in the arid zone, it is torrential, turbulent, brief and sparse. It often causes flooding since it does not have time to seep into the ground and there is hardly any flora; therefore, it intensifies erosion. On the other hand, in humid regions, such as that of England for example, rain lasts for a long period, but it is comparatively light rather torrential, and much of it seeps into the ground. This creates a relatively high water table but less erosion. If the hydrology pattern of the water catchment area is studied, the network of the watershed system can be influenced and runoff can be properly directed. It may be necessary to decrease or influence the water table in order

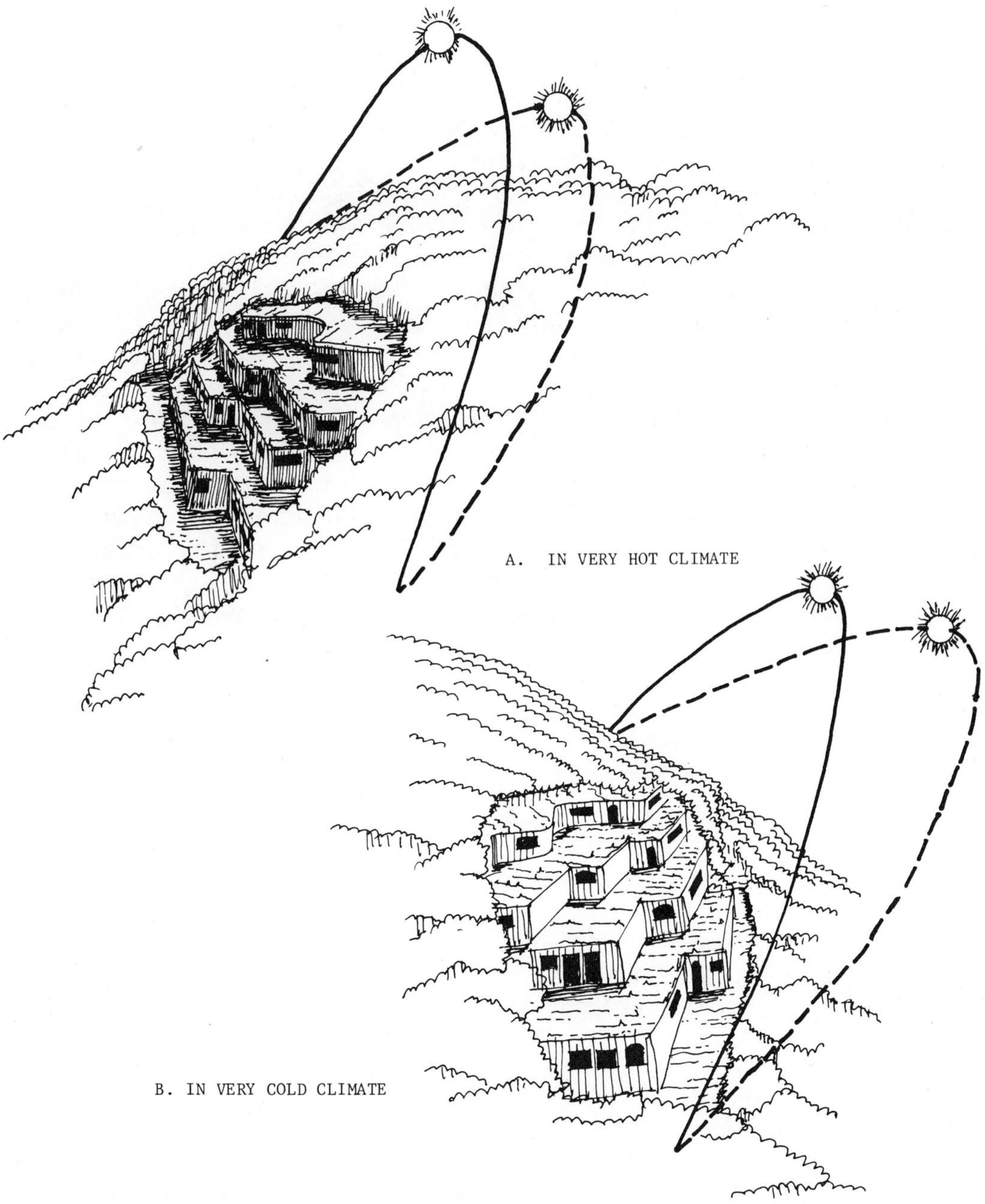

A. IN VERY HOT CLIMATE

B. IN VERY COLD CLIMATE

Fig. III-4. Site selection for a subterranean settlement on the slope in relation to sun orientation. Location in case A minimizes solar radiation and still receives plenty of light. Location in case B maximizes both solar radiation and light.

to minimize moisture in the soil of the site. Here again, slope topography provides better conditions than flatlands.

Rain, snow, fog, dew and relative humidity must be correlated with potential temperature change, visibility and inversion. Availability of water resources is also of concern. If the site is to be selected on the slope or on the top of the hill, it is most likely that pumping will be required; this would increase costs.

ENVIRONMENTAL QUALITY

The study of the environmental quality should be carried out on the regional and on the site scale; it should include pollution sources, pollen, eolian depositions, residuals, noise sources, polluted land, flora and fauna.

Pollution may come from nearby sources such as regional industries, airports, major traffic arteries or soil polluted by chemicals. Also, eolian deposits should be considered: that is to say, dust coming from quarries or unpaved roads or unstable soil resulting primarily from nearby agricultural land intensively cultivated. Another type of pollution is odor, residual or from nearby garbage dumps. Polluted land may result from nearby chemical industry, industrial waste, poorly drained or salty land, open sewage or the like.

The ecological condition of the region and the site should be studied carefully, including flora and fauna of the site and the impact of subterranean site development on the region. It can be expected that, in the long run, a subterranean settlement will have less environmental impact on the site than the equivalent conventional settlement would have. First, such a new settlement will influence the local climate and temperature very little, if any, by virtue of being underground. Wind movement will not be obstructed as in the conventional settlement even though some structures will probably be aboveground. This is not to say that the new settlement will have no impact on the ecological balance of the site and its environs; it will at least disturb the ecology of fauna which may have been living underground. The planner-designer must plan to minimize such disturbances.

ACCESSIBILITY AND PROXIMITY TO RESOURCES

If a new settlement is to be developed, we can expect it to have its own identity. For the sake of our discussion, let us consider a new settlement relatively far from any existing towns or cities. Under such circumstances, we must examine the new site in relation to its degree of isolation, the length of regional infrastructure networks required, its proximity to the hinterland and resources for supplies, year-round accessibility and terrain conditions, elevation and water pumping costs, and, finally, space requirements and site design potentialities. Generally, the rule is: the longer the distance of the new site from existing resources of supply (water, electricity, food, merchandise and transportation), the higher the cost. This higher price includes the building materials which may not be found in the area.

To summarize our discussion of site selection, we have covered a wide range of criteria to be considered comprehensively. We have focused primarily on the physical aspects, without mentioning the social and the economic implications of such development at this time in the belief that physical components are of extreme importance. This does not mean to imply that economic and social considerations are not essential, too.

New Urban Design Principles

Since we have thoroughly discussed the essence of subterranean house placement for different climatic areas, our discussion on the development of new urban settlements will now focus on the combination of aboveground and belowground houses as integrated units within the city. In other words, a house unit or complex of houses can be above- and belowground at the same time. The development of the subterranean house with its energy-free cooling and/or ventilation system should be a part of the design of totally new—and spatial—configurations for urban (or rural) communities in arid zones as well as in humid zones. There must be unity on three levels: (1) house, (2) neighborhood, and (3) city. In our view, the city must become a self-reliant organic unit. To cope with climatic constraints in cases of extreme cold or heat and to cope with high energy consumption, it is desirable to adopt the compact form for the city.

THE COMPACT CITY

Our contemporary city, the largest total form ever produced by mankind, has an impact on its physical environment which goes beyond its real size. For instance, the intrusion of such an immense element as the city into an arid area is a great disturbance to its delicate equilibrium. Furthermore, conventional new town designs have been characterized by horizontal development—low structures, large open spaces, and sprawl over large areas—with resultant high economic, social and energy costs. Our current energy problems require examination of urban design principles with regard to the following problems:

1. Low density, requiring high energy consumption for transportation and maintenance while involving few land uses.
2. Lengthy utility networks, involving expense for their planning, construction, operation and maintenance.
3. Lengthy, often inefficient road networks, including surfaces, lighting, sidewalks and landscaping.

New design approaches should be introduced which respond to these problems directly. Such an approach can be found in a vertical, compact design—not necessarily high-rise buildings, but rather integration of subterranean space with supraterranean forms and elimination of the usually inevitably dense population per living unit which results in a decrease in the desirability of the development.

By compactness, we mean a city that is concentrated and unified with a consolidation of land uses in close proximity. We achieve this by fragmenting the city into multiple residential service units and interrelating them to form the city as a whole. Each residential area is arranged in a neat, orderly manner in a smaller space than that of a standard contemporary city with an equivalent population.

There are three possibilities for physical expansion of the residential area of a neighborhood and of a city:

1. Horizontal development of the residential area, primarily applied in large cities and especially in the United States.
2. Vertical expansion upward, developed throughout the Western world and primarily in the centers of cities where the price of the land has increased immensely.
3. Vertical development downward, a relatively innovative and not yet widely accepted approach.

Underground construction offers desirable conditions for most uses: we suggest an integration of the above three approaches, with first priority given to vertical rather than horizontal expansion. To realize this integrated and compact approach, we have to formulate some basic guidelines:

1. The density of the population in any situation should not be more than the acceptable norms in any given society. These norms should be formulated to allow for future changes in standards and for expansion.
2. Transportation systems and parking spaces should be restricted to the periphery of the neighborhood, thus decreasing noise and air pollution. Of course, emergency transportation networks should be exempt from such restrictions.
3. The maximum distance required for pedestrians to move within the neighborhood should not exceed 500 meters, the distance from the center of the neighborhood to its periphery.
4. Maximum privacy of the residential areas should be provided by individual entrances, variations in the levels of residential units and in the placement of windows, and by the design of dead-end pedestrian paths. Such variety also prevents monotony and unattractiveness.
5. Although transportation systems are restricted, the close proximity of the buildings and other land uses within the residential area can bring noise to the houses. To minimize such noise, we should use technological aids such as insulation.

Some advantages of compact cities are that they:

- Meet the problems posed by a stressed climate condition—intense radiation, diurnal changes in temperatures, extreme dryness, cold or hot winds, dust storms.
- Consume less energy for heating and cooling since they are subject to less heat exchange than the dispersed city.

- Save time and energy for commuting and encourage walking.
- Make social life pleasant, especially for very young and elderly residents.
- Destroy less of the environment, especially the sensitive ecosystem of the arid zone.
- Reduce costs of:

  - planning, construction, maintenance and operation
  - all infrastructure network transportation systems
  - taxes from the citizens.

With reference to the impact of physical patterns of residential areas on the behavior of the people and their degree of interaction, sociologists offer differing theories. Most agree, however, that (1) there is some relationship between the density of population, on one hand, and the standard of living on the other, and that (2) both are associated with some types of social disorders, although such phenomena are also a result of many other influences. In considering density and compactness, the following principles apply:

1. Plan high physical unit density, but not high population density per unit.
2. Physical urban renewal without comprehensive socioeconomic renewal will not solve the problems of deprived neighborhoods.
3. In a compact neighborhood, sociophysical planning can create a service unit which can supply its own daily services and in turn establish community development and strong neighborhood relations.
4. Wide-range residential areas with dense structures can avoid high concentrations of low-income populations by introducing variety, a situation which can foster high standards of education and social services and enrich the community.
5. In combination, building and population density can increase the risks of lack of privacy, noise and high temperatures, while limiting possibilities for movement—all possibly contributing to undesirable behavior in certain groups. However, careful design of the units can improve the physical and psychological conditions of the residents.

Although there may be some social problems associated with the high density of buildings within the compact city, we can make the following assumptions:

1. Homogeneous groups of people, because of better communication, can be more tolerant within a high-density population than heterogeneous groups. In addition, adaptability to space conditions depends on the presence of shared elements—origin, culture and life style.
2. Experience shows that the importance of personal space for the individual and the family is greater within the house than beyond it, as in the street or the market. Furthermore, within the personal space in the house, priorities differ between one family and another and relate to the size of the family and to the age and number of children. We can

assume that similar differences in priorities exist outside in relation to public green open spaces, shopping centers, cultural centers, education services and location of the residences of friends and family.

3. The planner can compensate for the high density of a compact city by giving attention to areas outside the home which are important, such as open and green spaces, safe movement for children and the elderly, proximity to services including education, and distance from noise and pollution.

In the Mediterranean or Middle Eastern cultures, people have adapted to living in close proximity to each other. This adaptability, developed over the centuries, stems from the strong tribal and kinship relations deeply rooted in the societies, from the development of mutual responsibility and personal commitment, and from feelings of security. In Western society, on the other hand, great emphasis is placed on privacy and individuality. Our concern as planners/designers is to examine the feasibility of implementing the concept within different societies. The following issues are important to such an examination:

1. The intensive personal interaction established by living in a physically compact neighborhood may be beyond that which is acceptable in a given society. If basic requirements are provided at the acceptable norms for hygiene, convenience, quiet and privacy, we can expect that the density of compactness will be accepted. Further, we must recognize that norms, whether social, economic or physical, are not static, but dynamic and subject to change within the expectations of the society. As planners/designers, it is our responsibility to introduce some flexibility in our design to meet future changes.
2. Within the compact city, we can expect that the contact of the family with the shopping centers will be greater than it would be in the conventional city, because the compact neighborhood brings together residential areas and shopping areas, creating free movement between these two areas.
3. Given the modern increase in daily consumption, there is also an increase in social contact. In the compact neighborhood, with its concentrated land uses and its convenient green, open spaces, this contact will be accelerated. Different age groups, especially the younger ones, will be brought together because of the freer, safer movement and the proximity of services.

Finally, in planning for optimal land use and minimal energy use, earth-covered or subterranean buildings combined with compact neighborhood development constitute an effective aspect of new design: subterranean use of almost all types of land-use activities, including living. However, we are not suggesting a dominant use of subterranean space, but rather a combination with semisub- and supraterranean units (Fig. III-5).

As we mentioned earlier, the development of any modern city—the most immense project ever created by mankind—is never complete and continues indefinitely; it requires input from everyone, regardless of intellectual

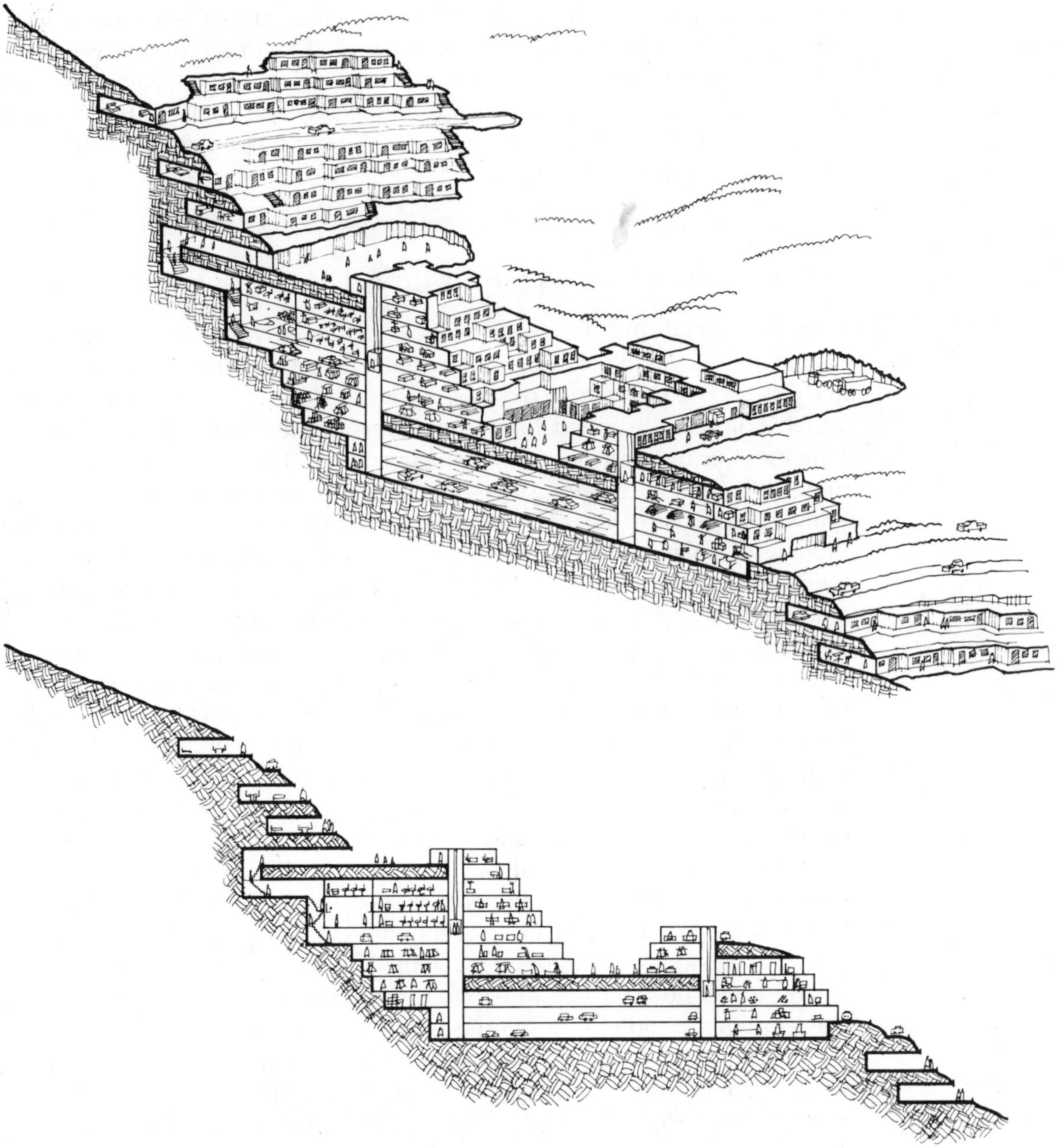

Fig. III-5. Generalized cross section of a combined compact and subterranean city for a stressed climate.

ability, age, race, religion, origin or any other quality. The impact of the conventional city goes beyond its size and the changes it brings in the ecology and the environment of its region. This is the reason we recommend above- and belowground vertical development with physically dense buildings; it is not necessarily associated with high population density per unit. To achieve success, we should establish basic policies for the planning/design of the city and combine compactness with subterranean placement. We must address the subjects of population density, intense social interaction, transportation networks, privacy, future improvements and expansion and open space.

If we compare the three alternatives in urban design—a conventional, aboveground horizontal sprawl, aboveground-compact or aboveground-compact combined with subterranean, we can clearly see that the latter has the most advantages economically, ecologically and socially (Fig. III-6 and Table III-1). We are suggesting here that the third alternative is particularly appropriate in the arid-zone city.

COMPACTNESS IN THE ARID-ZONE CITY

From the social and the psychological points of view, the compact city should be acceptable to persons living in extreme climates such as cold and dry, hot and dry, or cold and humid. In an arid zone, one element which enforces the need for compactness is the feeling of isolation and loneliness in the vast region of a desert area. This is especially true in a new settlement at its early stages. Yet even in the later stages of the community, the sense of vastness of space surrounding the community increases feelings of isolation. There are many aspects, of course, which increase this feeling—infrequent supply of essential goods and the distance between the communities and the centers of supply or between the centers of the country itself, for example. In any case, vastness, isolation, and the feeling of loneliness reinforce the human needs for strong, close relationships in the harsh climate surrounding them. Therefore, we can expect that the need for feeling a sense of community would be strong in communities established in the arid zone. To achieve this sense of community, people must be close to each other; planners can aid in this achievement by designing streets which are short and narrow and many pathways and open spaces or other public gathering centers. All these elements should be in close proximity to each other. This closeness will certainly enforce the sense of community so that it is greater than that in conventional cities with only few land uses and discrete areas of residential living. Thus, the compact form is a response to the problems of arid-zone living. The need to be together, to be close to other people, is answered by the compact city.

Aside from psychological and social advantages, the compact city is ideal for arid zones and provides tangible ways to deal with the problems of intense solar radiation, albedo, dryness, hot winds after noon and great differentiations between day- and nighttime temperatures. We can combine ancient principles and modern technology to build a successful compact arid-zone city.

Likewise in the compact city of an arid zone, transportation has a special role because of the climatic harshness, because of the space it consumes and because of the norms which compactness imposes. An analysis of these causes leads us to recommend that transportation and pedestrian movement should be underground as much as possible, especially within the residential areas. This concept is introduced here in order to suggest an improvement in the overall quality of urban design for an arid-zone settlement. Above all, this concept is consistent with the advantages which the compact city introduces for the arid zone. We shall discuss transportation and traffic patterns more fully later in this section.

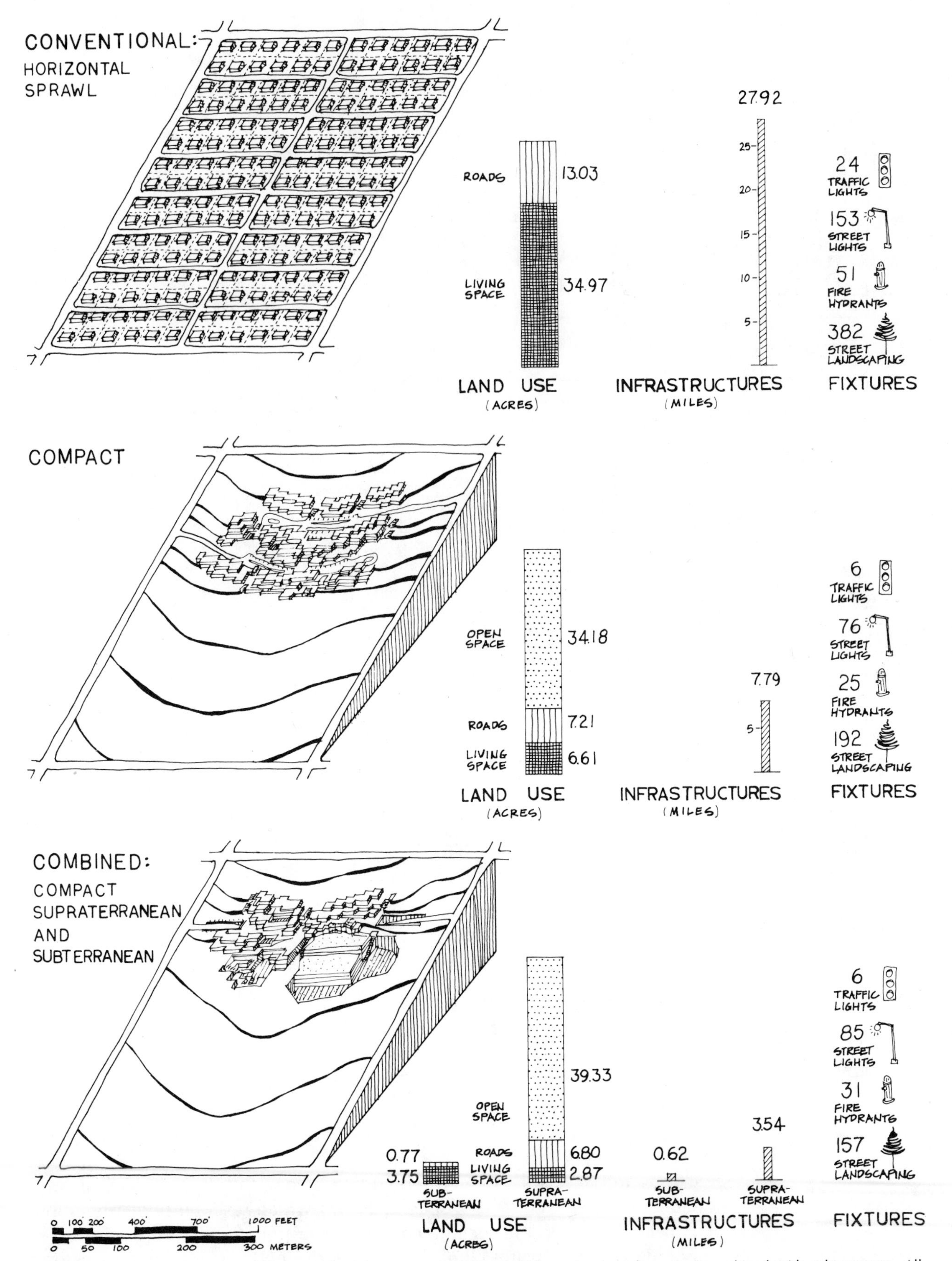

Fig. III-6. Comparative patterns of three residential areas—conventional grid, compact, and compact combined with subterranean. All units are equal in size and number. Total lot size: 194,249 meters (=48 acres) (4,046.72 sq. meters = one acre). Each dwelling is 139 sq. meters (1500 sq. ft.).

Table III-1. Comparative Cost Evaluation of Three Design Alternatives: Conventional Aboveground Sprawl, Compact, and Combined Compact Supraterranean with Subterranean

LAND USE	CONVENTIONAL		COMPACT	
	HORIZONTAL SPRAWL			
	ENGLISH MEASURES	METRIC MEASURES	ENGLISH MEASURES	METRIC MEASURES
1. LAND FORM	ALMOST FLAT		LOW TO MEDIUM GRADE	
2. LAND SIZE	48 ACRES	194,249.04 M^2	48 ACRES	194,249.04 M^2
3. NUMBER OF DWELLING UNITS	192		192	
4. SIZE OF DWELLING UNIT (FLOOR AREA)	1,500 FT^2	139.35 M^2	1,500 FT^2	139.35 M^2
5. SIZE OF ONE LOT UNIT	7,933.92 FT^2 = 18.21%A.	737.08 M^2	NONE	
6. LAND USE:				
6A. NET FOR DU*	6.61 A = 288,000.00 FT^2	26,755.20 M^2	6.61 A (288,000 FT^2)	26,755.20 M^2
6B. NET LIVING SPACE**	34.97 A = 1,523,312.64 FT^2	141,519.36 M^2	6.61 A (288,000 FT^2)	26,755.20 M^2
6C. ROAD AND STREET SURFACES:				
6CA. INTERIOR ROADS (25' WIDE)	5.52 A = 240,512.50 FT^2	22,344.33 M^2	1.15 A (50,000.00 FT^2)	4,645.15 M^2
6CB. PERIPHERAL STREETS (40' WIDE)	5.22 A = 227,355.20 FT^2	21,121.98 M^2	5.22 A (227,355.20 FT^2)	21,121.98 M^2
6CC. PEDESTRIAN WAYS (4' WIDE)	2.29 A = 99,699.52 FT^2	9,262.38 M^2	0.38 A (16,600.00 FT^2)	1,542.19 M^2
6CD. PARKING SPACE	NONE		0.46 A (20,000.00 FT^2)	1,858.06 M^2
6CE. OPEN SPACE	NONE		34.18 A(1,488,924.80 FT^2)	138,326.46 M^2
TOTAL LAND USE	48.00 A = 2,090,880.00 FT^2	194,249.04 M^2	48.00 A(2,090,880.00 FT^2)	194,249.04 M^2
7. LENGTH OF INFRASTRUCTURES:				
7A. INTERIOR ROADS	9,620.50 FT.(1.82MI)	2,932.33 M(2.9KM.)	2,000 FT.(0.38 MI)	609.60 M.(0.6 KM).
7B. PERIPHERAL STREETS	5,683.88 FT.(1.08MI)	1,732.45 M(1.7KM.)	5,683.88 FT.(1.08 MI)	1,732.45 M.(1.7 KM).
7C. PEDESTRIAN***	24,924.88 FT.(4.72MI)	7,597.10 M(7.5KM.)	4,150.00 FT.(0.79 MI)	1,264.92 M.(1.2 KM).
7D. SEWAGE	15,304.38 FT.(2.90MI)	4,664.78 M(4.6KM.)	3,589.00 FT.(0.68 MI)	1,093.93 M.(1.0 KM).
7E. STORM SEWER	15,304.38 FT.(2.90MI)	4,664.78 M(4.6KM.)	7,683.00 FT.(1.46 MI)	2,342.05 M.(2.3 KM).
7F. WATER	15,304.38 FT.(2.90MI)	4,664.78 M(4.6KM.)	3,589.00 FT.(0.68 MI)	1,093.93 M.(1.0 KM).
7G. TELEPHONE	15,304.38 FT.(2.90MI)	4,664.78 M(4.6KM.)	3,589.00 FT.(0.68 MI)	1,093.93 M.(1.0 KM).
7H. GAS	15,304.38 FT.(2.90MI)	4,664.78 M(4.6KM.)	3,589.00 FT.(0.68 MI)	1,093.93 M.(1.0 KM).
7I. TV CABLE	15,304.38 FT.(2.90MI)	4,664.78 M(4.6KM.)	3,589.00 FT.(0.68 MI)	1,093.93 M.(1.0 KM).
7J. CENTRAL HEATING	15,304.38 FT.(2.90MI)	4,664.78 M(4.6KM.)	3,589.00 FT.(0.68 MI)	1,093.93 M.(1.0 KM).
TOTAL	147,359.92 FT(27.91MI)	44,915.34M(44.91KM)	41,050.88 FT.(7.77 MI)	12,812.60 M.(9.26KM).
8. TRAFFIC LIGHTS	18-27 FIXTURES		6 FIXTURES	
9. STREET LIGHTS (ONE FOR EVERY 100 FT.)	153 FIXTURES		76 FIXTURES	
10. FIRE HYDRANTS (ONE FOR EVERY 300 FT.)	51 FIXTURES		25 FIXTURES	
11. STREET LANDSCAPING (ONE TREE FOR EVERY 40 FT.)	382 TREES		192 TREES	

* ONE FLOOR STRUCTURE WITHOUT THE LOT

** LIVING SPACE IS THE LOT SIZE WHICH INCLUDES WITHIN IT THE FLOOR STRUCTURE

*** TWICE THE LENGTH OF THE INTERIOR ROADS + ONE LENGTH OF THE PERIPHERAL ROADS IN THE CONVENTIONAL CASE. IN THE CASE OF THE COMPACT CITY: THE LENGTH OF THE INTERIOR ROADS + ONE LENGTH OF THE PERIPHERAL ROADS.

Thus, the new arid-zone city should be planned in special physical configurations and according to specific norms which respond primarily to the climatic stress and the social, economic, transportation and environmental problems caused by this climatic stress. To this end, a few important lessons should be kept in mind by the planners-designers of arid-zone cities. First, most, if not all, traditionally acceptable historical or contemporary urban

Table III-1 continued

C O M B I N E D					
SUPRATERRANEAN		SUBTERRANEAN		TOTAL	
ENGLISH MEASURES	METRIC MEASURES	ENGLISH MEASURES	METRIC MEASURES	ENGLISH MEASURES	METRIC MEASURES
MEDIUM TO HIGH GRADE				MEDIUM TO HIGH GRADE	
48.00 ACRES	194,249.04			48.00 ACRES	194,249.04 M^2
192				192	
1,500 FT^2	139.35 M^2	1500 FT^2	139.35 M^2	1,500 FT^2	139.35 M^2
NONE				NONE	
2.87 A (125,000 FT^2)	11,612.88 M^2	3.75 A (163,000 FT^2)	15,143.19 M^2	6.61 A (188,000 FT^2)	26,756.03 M^2
2.87 A (125,000 FT^2)	11,612.88 M^2	3.75 A (163,000 FT^2)	15,143.19 M^2	6.61 A (188,000 FT^2)	26,756.03 M^2
0.36 A (15,500 FT^2)	1,440 M^2	0.17 A (7,500 FT^2)	696.77 M^2	0.53 A (23,000 FT^2)	2,136.77 M^2
5.22 A (227,355.20 FT^2)	21,121.98 M^2	NONE		5.22 A (227,355.20 FT^2)	21,121.98 M^2
0.22 A (9,600 FT^2)	891.87 M^2	NONE		0.22 A (9,600 FT^2)	891.87 M^2
NONE		0.60 A (25,536 FT^2)	2,372.37 M^2	0.60 A (25,536 FT^2)	2,372.37 M^2
39.33A(1,713,424.80 FT^2)	159,182.30 M^2	NONE		39.33A(1,713,424.80 FT^2)	159,182.30 M^2
48.00A(2,090,880 FT^2)	194,249.04 M^2	4.52 A (196,036 FT^2	18,212.33 M^2	52.52A(2,090,880 FT^2)	212,461.37 M^2
620 FT (0.12 MI)	188.00 M(0.18 KM)	300 FT (0.06 MI)	91.44 M	920 FT (0.18 MI)	279.44 M(0.27 KM)
5,683.88 FT (1.08 MI)	1,732.45 M(1.7 KM)	NONE		5,683.88 FT (1.08 MI)	1,732.45 M(1.7 KM)
2,400 FT (0.45 MI)	731.52 M(0.73 KM)	NONE		2,400 FT (0.45 MI)	731.52 M(0.73 KM)
1,400 FT (0.27 MI)	426.72 M(0.42 KM)	400 FT (0.08 MI)	121.92 M	1,800 FT (0.34 MI)	548.64 M(0.54 KM)
1,400 FT (0.27 MI)	426.72 M(0.42 KM)	400 FT (0.08 MI)	121.92 M	1,800 FT (0.34 MI)	548.64 M(0.54 KM)
1,400 FT (0.27 MI)	426.72 M(0.42 KM)	400 FT (0.08 MI)	121.92 M	1,800 FT (0.34 MI)	548.64 M(0.54 KM)
1,400 FT (0.27 MI)	426.72 M(0.42 KM)	400 FT (0.08 MI)	121.92 M	1,800 FT (0.34 MI)	548.64 M(0.54 KM)
1,400 FT (0.27 MI)	426.72 M(0.42 KM)	400 FT (0.08 MI)	121.92 M	1,800 FT (0.34 MI)	548.64 M(0.54 KM)
1,400 FT (0.27 MI)	426.72 M(0.42 KM)	400 FT (0.08 MI)	121.92 M	1,800 FT (0.34 MI)	548.64 M(0.54 KM)
1,400 FT (0.27 MI)	426.72 M(0.42 KM)	400 FT (0.08 MI)	121.92 M	1,800 FT (0.34 MI)	548.64 M(0.54 KM)
18,503.88 FT (3.50 MI)	5,639.01 M(5.6 KM)	3,100 FT (0.59 MI)	944.88 M	21,603.88 FT (4.09 MI)	6,583.89 M(6.58 KM)
6 FIXTURES		NONE		6 FIXTURES	
63 FIXTURES		22 FIXTURES		85 FIXTURES	
21 FIXTURES		10 FIXTURES		31 FIXTURES	
157 TREES		NONE		157 TREES	

configurations in humid areas are not valid or operable in arid zones and, therefore, should not be transferred to them. Second, the study of contemporary and ancient arid-zone cities can provide models and indicate possible solutions to modern problems, models which can be used—with or without special adjustment—in modern city designs in arid zones. Third, modern urban design configurations cannot (and in some cases must not) be transferred to the arid-zone city overnight for economic, social and energy reasons. Planners should keep in mind that advanced technology is not always the answer to the problems facing the new arid-zone city.

URBAN CELL FORMS

To achieve satisfactory social, economic and physical results in the city, the new urban form should be made of chains of independent but interconnected units, here called urban cells. These cells can be arranged in linear, radial, concentric, polycentric or grid forms, as Figure III-7 depicts. However arranged, each urban cell forms an independent territorial unit and a relatively independent service unit (perhaps sharing some services with a neighboring cell), connected to its neighbors by a pedestrian and/or transportation network.

In response to the arid-zone's stressful climate, the new urban cells and cities should exhibit compactness and subterranean construction. The design of each urban cell should minimize the distance between land uses, maximize shadow during the day, and minimize the influence of cold winds at night and hot winds during the day (Fig. III-8). Services should be provided on a small scale, with frequent mass transportation reserved for connection with the more distant urban cells. Compactness, however, necessitates careful design in order to plan for growth and to achieve good health conditions, reasonable social and individual privacy, minimal noise and a socially pleasing environment. To achieve compactness and still allow for urban growth with a minimal use of motor vehicle transportation, the city can be built in the territorially and socially independent cells described above. City growth can then be accommodated by adding new urban cells to the existing configuration.

The proximity of places of work to places of residence is essential for cell compactness (Fig. III-9). To achieve the necessary proximity of land uses and also to minimize the impact of climatic stress, it is essential to use underground spaces extensively for nearly all types of land use. As a corollary of this principle, the new urban form should expand vertically (above- and belowground) rather than horizontally.

Sloped locations can be most suitable for subterranean urban cell development. Figure III-10 depicts a possible design. Such sites will offer ease of excavation, good drainage, an attractive view, alternative design opportunities (combining under- and aboveground forms) in accordance with local topographical features, increased climatic comfort and minimal ecological disturbance.

Whether the site is on a slope or not, the climatic advantages of the suggested urban concept are important and should be briefly discussed. Meteorologists have noticed that conventional urban centers create a microclimate known as a heat dome as a result of the immense tracts of land laid bare of vegetation and other natural heat-absorbent features. The urban cell concept will lessen the heat dome phenomenon because each compact cell will exist in relative isolation from its neighbors and because most of the heat is absorbed within the subterranean space.

The conventional aboveground city, especially in arid zones, retards strong prevailing winds which sweep across vast open spaces and increase heat

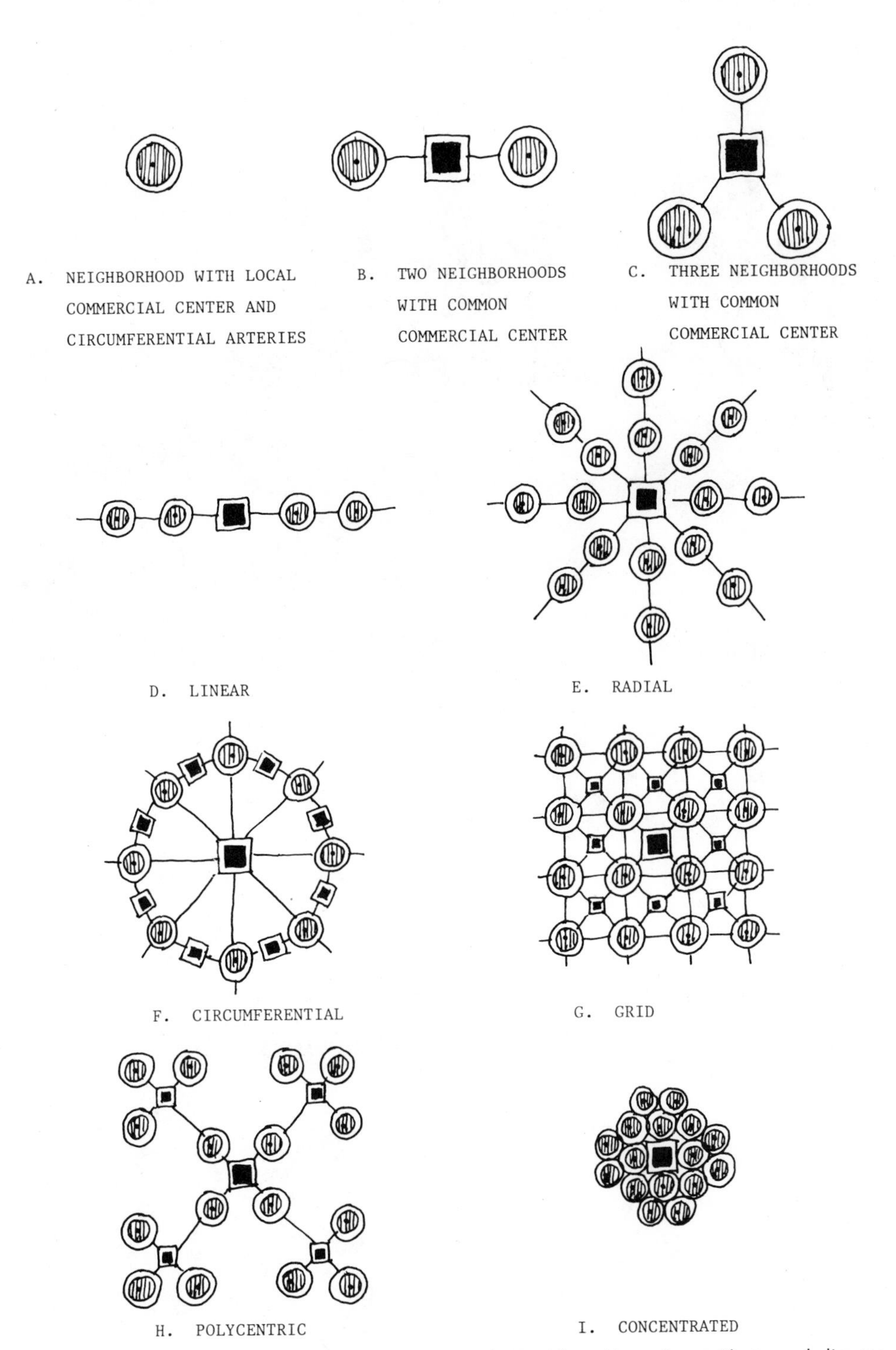

Fig. III-7. The concept of urban cell configurations can be introduced in regions with stressed climates which require compact forms. Transportation routes would surround or penetrate into the cell but not cross it.

gain and heat loss. However, temperature increases generated by the solar reflection of the city itself can result in physical turbulence at ground level even when the wind speed is very low. Underground structures will neither inhibit nor cause wind movement. Thus, to maximize protection from turbulence, the integration of both above- and belowground construction is

Fig. III-8. Cross section of subterranean house units on terraced topography for a warm, dry climate. Outdoor spaces are paved to minimize dust movement, while green areas are primarily kept for indoor space. Windows are designed to establish direct eye contact with an outdoor lower level (or valley) and to permit plenty of light penetration. The pedestrian alley is narrow and partially shadowed for protection from solar radiation and strong winds. Within the compact residential area, privacy is preserved. Car movement and parking are on the periphery.

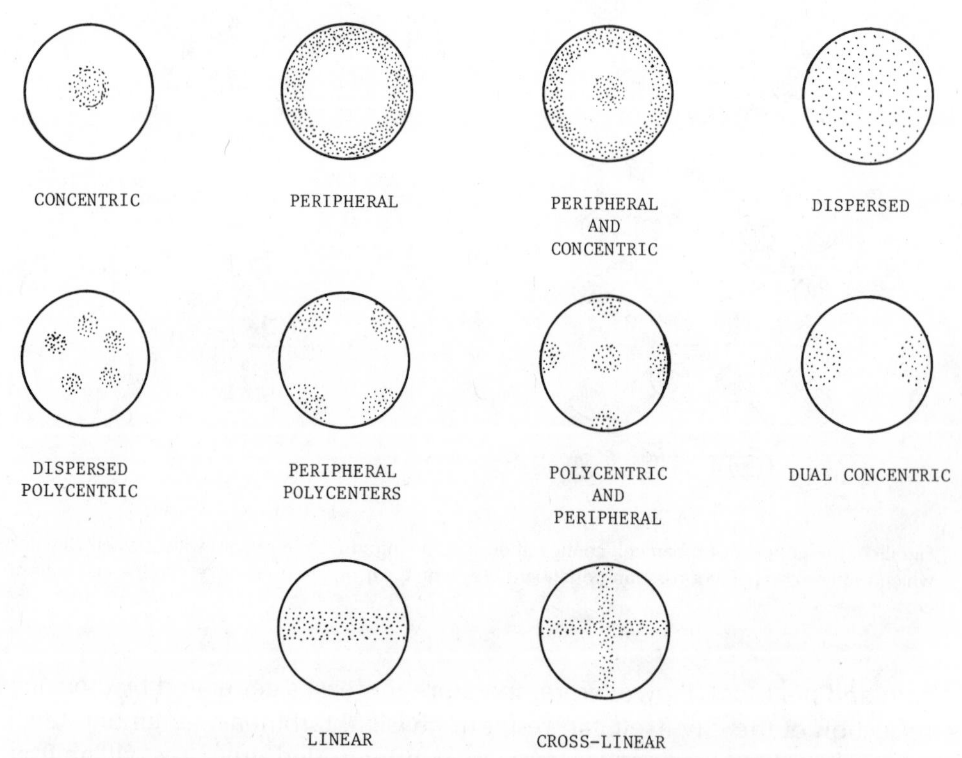

Fig. III-9. General concepts of alternative employment dispersion (dotted) configurations within the city. Each one results in a land-use pattern which influences the transportation pattern, energy consumption, degree of congestion and environment.

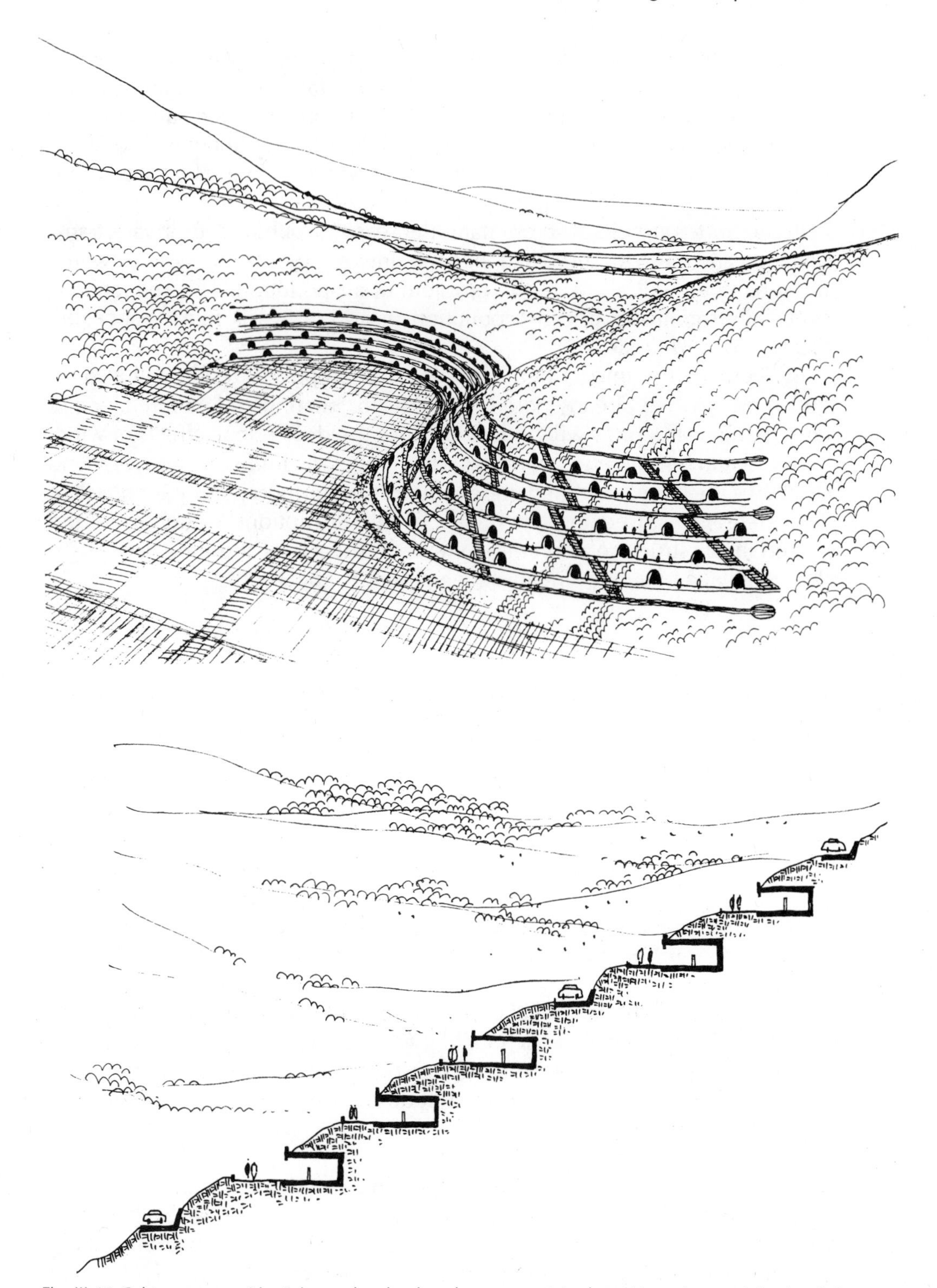

Fig. III-10. Subterranean residential complex developed on a mountain slope. Note alternated dead ends for cars.

desirable. The compactness of the proposed urban cell design will help reduce another related arid-zone climatic problem: covering the selected site with urban cell configurations will minimize the dust pollution so common in arid-zone cities and, consequently, improve health conditions and ease the housekeeper's cleaning burden.

Built-up areas in general and the proposed urban cell in particular will increase water runoff and therefore require special storm drainage systems. Sloped sites, as we already noted, should support the necessary drainage systems conveniently. The more acute problem facing underground structures will be posed not by runoff but by the water table in flat areas, especially because drainage expenses are considerable.

In the arid-zone city, the dramatic decrease of public and private transportation and the widespread adoption of energy-free cooling systems within the urban cell will alleviate the problem of air pollution. This reduction is especially significant because pollution in arid zones is more noticeable than in nonarid areas.

By reducing the impact of the climate, the proposed design will also reduce costs. As we already noted, the costs for heating and maintenance of underground structures will be low because the deleterious effects of wind, moisture and temperature extremes which regularly influence aboveground structures are largely avoided underground. Furthermore, utility supplies are protected against freezing and other climatic disruptions. If the cost of the subterranean structure and cell is viewed in terms of the life-cycle costs, the advantage of subterranean development is apparent.[4]

LAND USE

We have discussed the way our new urban settlement should follow three land-use principles:

1. Compactness, a unified consolidation of land uses in close proximity.
2. Urban cell configurations as physically integrated service forms.
3. Accommodation of man's residences, activities and space needs in both supraterranean and subterranean development.

Therefore, we are now ready to present two additional land-use principles:

1. Synthesis of the historical approach to land use calling for integration, and the modern approach calling for total segregation of land uses.
2. Total separation of the vehicle transportation network from other land uses and greater emphasis on a pedestrian system which is integrated within all land uses.

Synthesis of Integration and Segregation in Land Uses. In general, there have been two land-use models, with the first developed throughout history: integrated land uses and most daily uses and services within walking distance of the residential area. The city in this model was relatively small and formed primarily from conglomerates of neighborhoods based on social identity within a physically defined entity. Thus, residents had access to daily services and places of work on a human scale of integrated land-use patterns. Accordingly, mixed land use was essential to operate such a city effectively and to create social integration. This model is usually favored by social scientists. The second model is the conventional one widely used in modern design in which the city is viewed as a unit of economic cells with

Table III-2. Suggested Land-Use Division for Combined Subterranean and Supraterranean Compact City (in Percentages)

TYPE	COMBINED: COMPACT SUPRATERRANEAN	SUBTERRANEAN	TOTAL	BRITISH NEW towns*
Residential	30	30	60	43
Industry	2	10	12	12
Open space	14	0	14	22
Education & health	2	2	4	4
Commercial	0	2	2	2
Streets	4	4	8	17
Total	52%	48%	100%	100%

*G. Golany, *New-Town Planning: Principles and Practice.* New York: John Wiley & Sons, 1976, pp. 330–331.

its major function the trading of goods within the city and between cities. According to this concept, the city operates on the basis of efficient transportation networks which necessitate segregated land uses. People who strongly favor economic activity as the core of the city existence have been the decisionmakers—politicians and financiers.

We believe that a synthesis between the two models may be found in our new urban settlement which combines compactness and a configuration of urban cells, an integration of compact supra- with subterranean placement (Table III-2). In the conclusion of this section we shall review and expand our discussion of underground land uses, for we recommend increased usage of this form in our new urban settlement.

Traffic Network Design. The unique city design we are calling for requires a nonconventional traffic pattern for vehicles and pedestrians. To coincide with the urban cell structure of the city, the policy for the vehicular network (public and private) should require the following:

1. Transportation shall penetrate deeply within the compact neighborhood as a *terminal point,* not for the purpose of crossing the neighborhood. All transportation networks shall be circumferential to the neighborhood with none crossing it.
2. The penetration applies to public as well as private transportation. Public transportation may have its circulation within underground terminals. In these terminals, stairways or elevators should be planned and built for access to housing or shopping areas.
3. Public transportation shall take the form of subways, electric cars or buses. The private car underground penetration shall be terminated in cul-de-sacs from which stairways and elevators will lead to the upper ground.
4. The underground penetration of transportation shall be extensive when approaching the built-up area. However, the underground section will not necessarily be covered from all sides; one side can be exposed vertically (to the sky) or horizontally (to the view) to admit light or sunshine.

5. If some sections of the transportation network must be constructed aboveground, they should be developed as a suspended network above all pedestrian paths and not intersecting them. Also, the design of the landscapes for this aboveground transportation becomes crucial to the overall appearance of the neighborhood and shall be considered carefully.
6. The transportation network shall connect one cell with another linearly with a common shopping area and offices to serve the two cells.

The policy for pedestrian design, above- and belowground, should be guided by needs for variety, social space, hierarchy, protection and safety, pleasure and convenience. Variety in design has the potential of introducing aesthetics as well as enriching the environment and pleasing the users. Variety can also introduce possibilities of combined functions (Fig. III-11). This variety can be accomplished, for example, by offering occasional wider spaces along the line, in the intersections or at its terminal dead end. It can also be introduced by making the connecting path a curvilinear or zig-zag one. The intervening space along the path or at its end should take different forms (Figs. III-12 and III-13).

Since social space is a positive element to be considered in the design of a compact city, expecially when it is combined with subterranean living, pedestrian paths should be viewed not only as connecting lines between two points, but also as meeting spaces, as places for social interaction, as playgrounds for different age groups and as recreational spaces, especially for the elderly. Also, since parents need to watch their very young children when they are playing, resting space should be provided. Terminals of the pedestrian network receive special attention as social spaces for the complex of houses related to them. The pedestrian terminals can be formed of multiple combined dead ends which can stimulate more social interaction and still retain privacy (Fig. III-14). The cell is a physical unit as well as a social unit. Children, women and the elderly will develop ties with their peer groups during the day. Thus, subunit and subsubunit systems are to be developed and should be supported by preplanned physical forms. The design of the pedestrian network must be viewed to meet this development. In addition, the network should have direct connections with the community open spaces provided within the cell (Fig. III-15).

Hierarchy in the pedestrian network is related to the concept of cell subdivisions. Traffic is also hierarchical when a full network of pedestrian ways is considered. The hierarchy should be designed to avoid traffic flow in an area when it is not necessary. It would be proper to have three levels in the hierarchy which coincide with the dead-end patterns (Fig. III-16).

The design of the alley should protect the pedestrian against the effects of climate by considering orientation in relation to the sun; strong, cold or hot winds; dusty wind; or noise. Such considerations become especially important in an extreme climate (Fig. III-17). In a hot-dry climate, the alley should receive minimum sunshine and maximum shadow because pedestrians would use it frequently (Fig. III-18). A pleasing environment is essential there. Variation of the view along the alley and between one alley and another, landscaping with greenery in intersections and along the alley, the location of benches occasionally for rest, recreation and relaxation, pav-

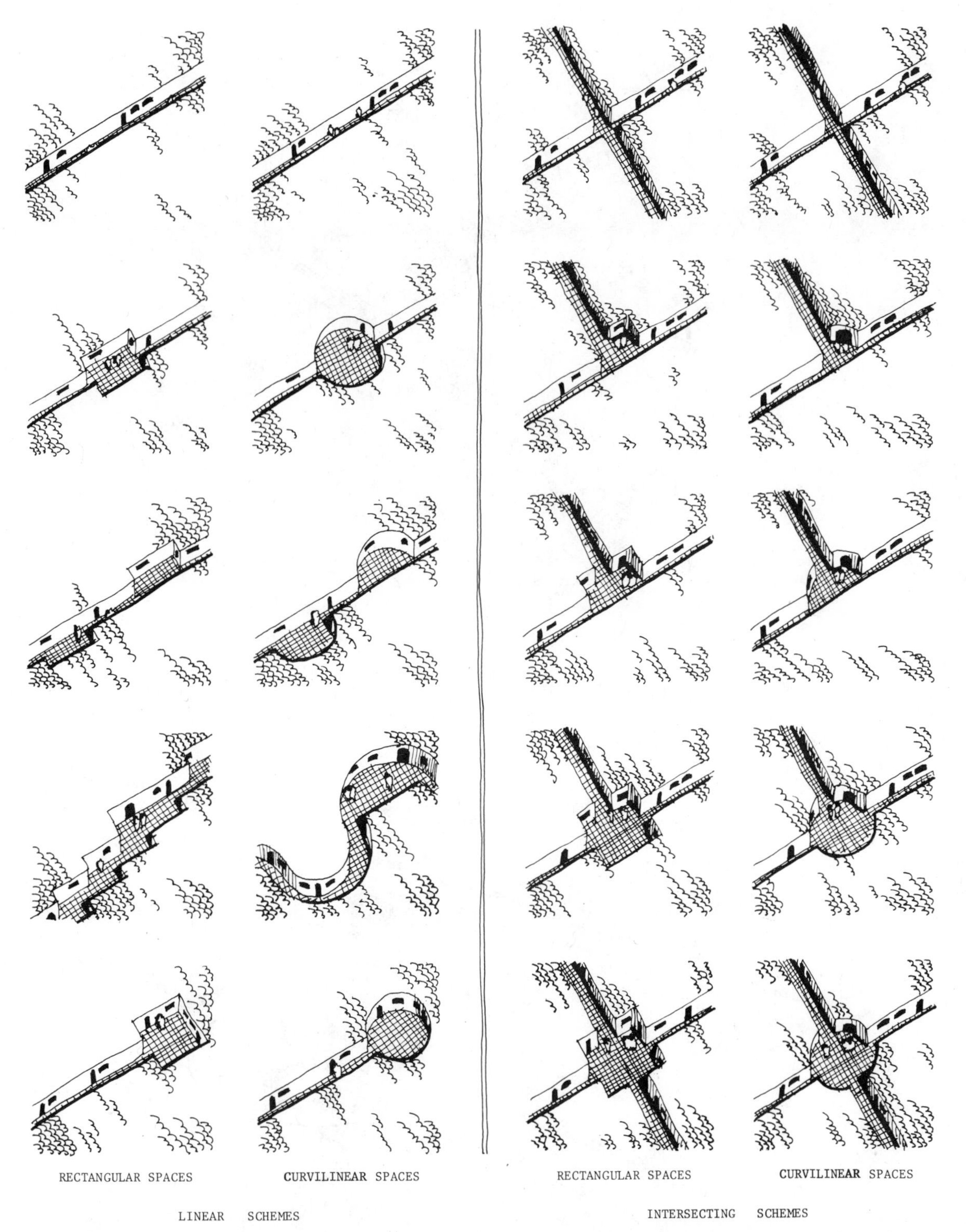

Fig. III-11. Variety of general subterranean pedestrian system schemes combined with small open spaces. The pedestrian network should be designed for movement, for social gatherings, for interaction, for visual pleasure and for expression of personal preferences and local privacy.

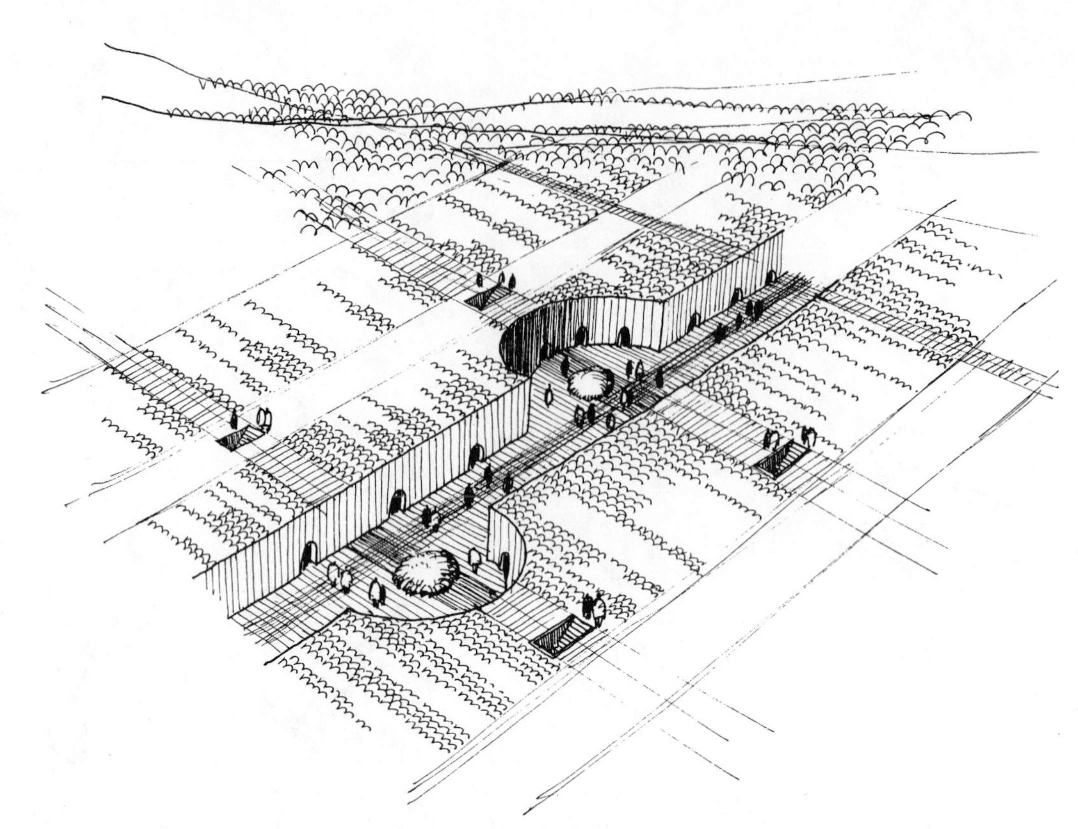

Fig. III-12. Design of belowground alleys.

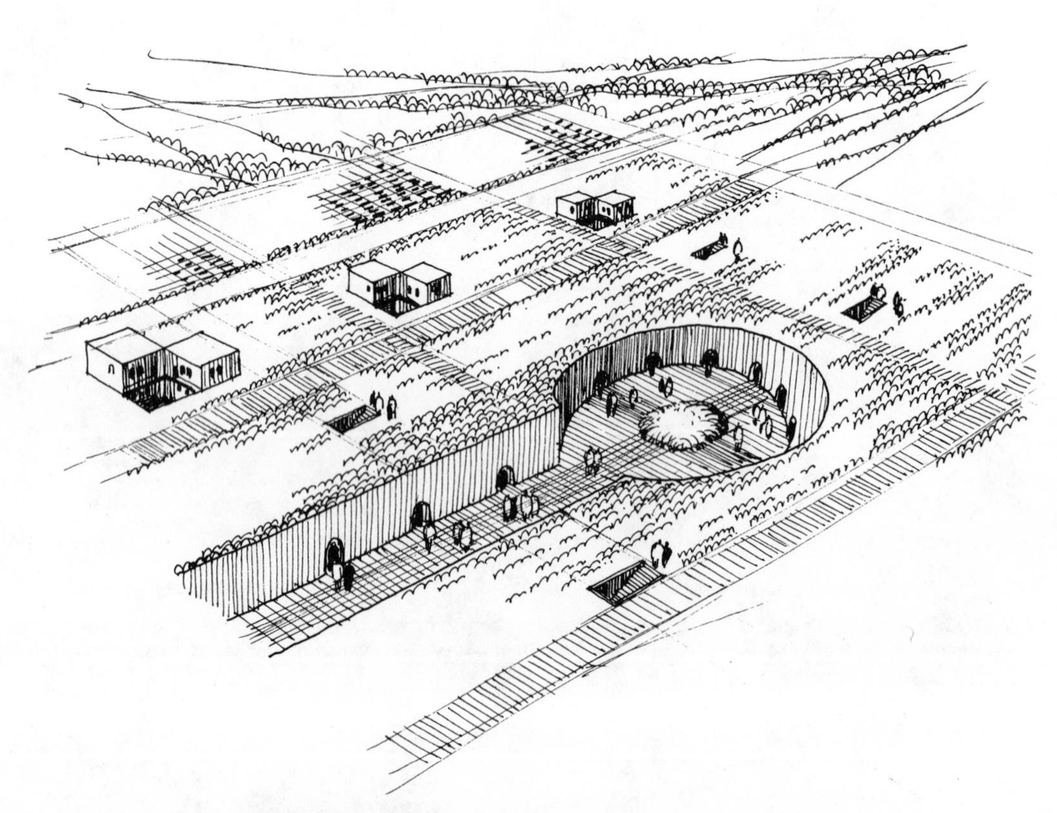

Fig. III-13. Belowground dead-end alley for the subterranean neighborhood.

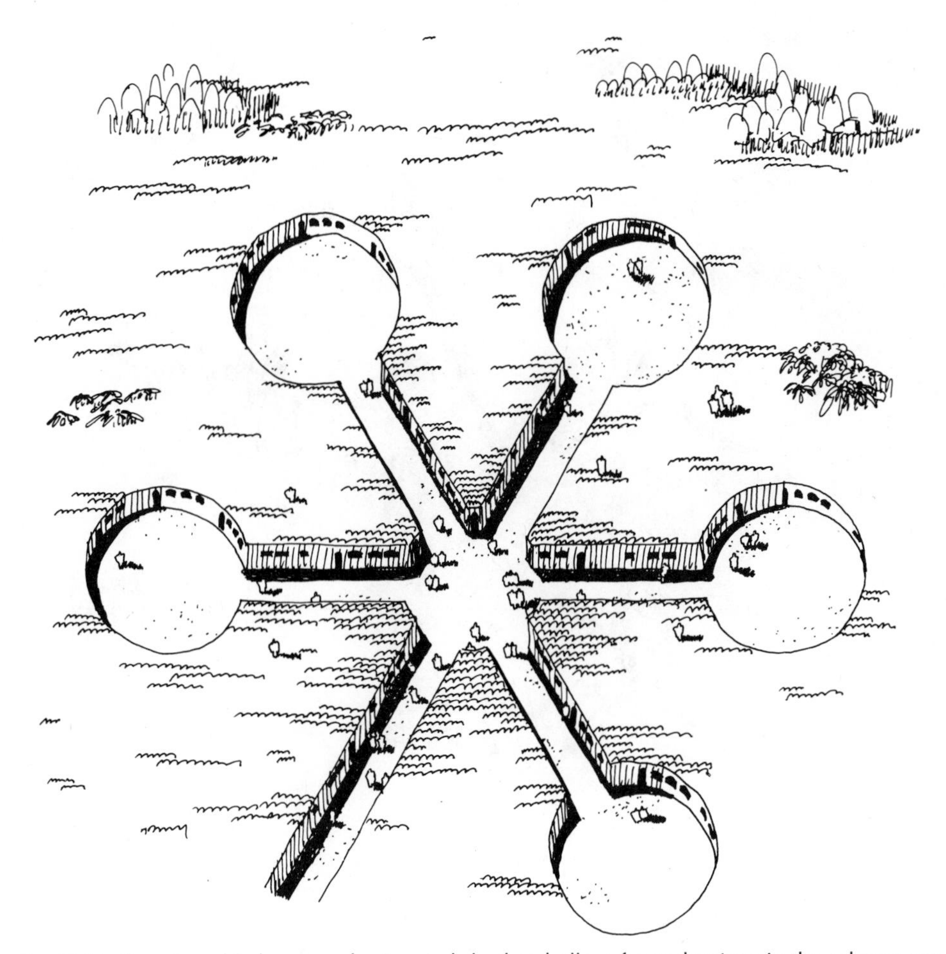

Fig. III-14. Complex of belowground integrated dead-end alleys for pedestrians in the subterranean neighborhood.

Fig. III-15. Multiple and concentrated dead-end configurations of belowground alleys to provide privacy within the community.

Fig. III-16. Hierarchy of belowground pedestrian networks can be used especially effectively in the arid-zone city. Such a pattern allows for elimination of unnecessary traffic within the neighborhood and coincides with the neighborhood structure.

ing—all can contribute to offering a pleasing environment. Also, the view from the alley to its environment is to be considered; it can be most enriching when the alley is along slopes (Fig. III-19). The path can be made pleasant by the creation of points of unobstructed view to the open landscape. In cold and snowy weather, it may be necessary to create covered spaces along the pedestrian path for protection.

Convenience is another element to be considered in the design of the pedestrian network. The paths must be as short as possible, they must provide access to the park or the open space of the community and to the service centers, they should be lighted at night and they must be safe. All parks should be able to accommodate the peak hours of traffic.

Pathways, if built below the ground level, raise two problems which have to be considered and solved:

1. Pathways are apt to become systems for rain drainage and are flooded under heavy rain, especially in arid zones where rain is torrential. Therefore, a special drainage system should be designed for those paths.
2. Dust and garbage deposits carried by wind movement tend to be accumulated in the lower levels and brought to the pedestrian paths. To minimize such a development, the edges of the paths can be elevated

Fig. III-17. Alley shadowing. Belowground alley shadowed by semidome (A); by semicovering arches (B); by tree plantings (C) for hot, dry climates.

and curved to trap the garbage. Also, when the path is belowground, a fence should be constructed at the ground-level edges of the path for safety and protection.

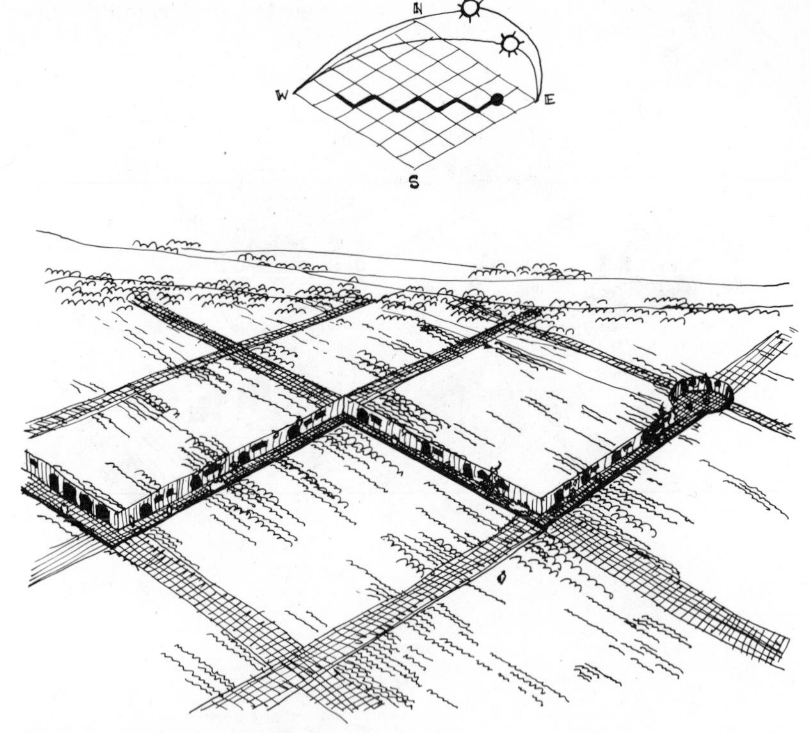

Fig. III-18. Belowground pedestrian alleys of proper orientation, an intensive communication system within an underground arid-zone community. The variation in the solar orientation of the sections of the alley creates shadow through most of its course and increases pedestrian comfort.

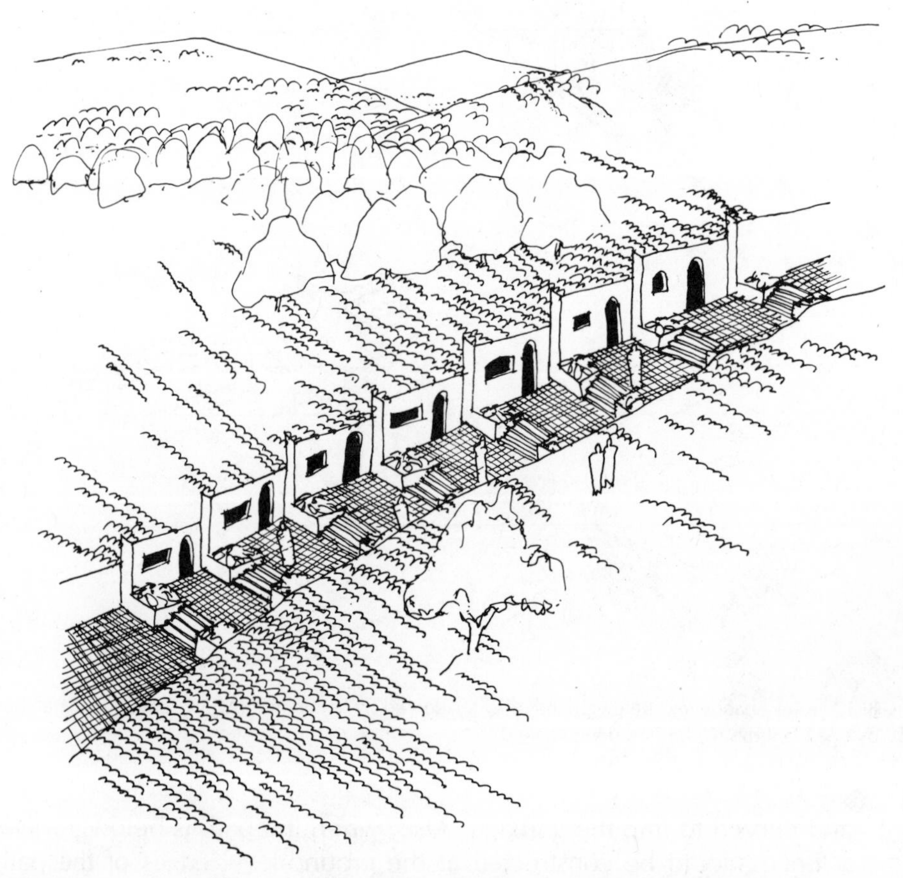

Fig. III-19. An alley and house facade along a moderate topographical rise and open to the view.

DESIGN OF UNDERGROUND HOUSING COMPLEXES

Housing complexes may constitute the majority of urban land use and thus become the dominant feature of the community landscape. Here again, design treatment should be innovative and imaginative. As we explained, the design of settlements on slopes will offer many advantages to the community, such as easy sewerage, an attractive view, plenty of light, deep penetration of sunshine when it is desired, possible future house expansion horizontally and vertically downward, good ventilation, safety from the traffic burden and much privacy. Likewise, housing locations on slopes require water pumping, topographical adjustment and possible investment for access and utilities; but when the total advantages are weighed against the possible slightly higher investment, the slope still proves to be the best location. Housing complex design, if planned for slopes, should be guided by policies addressing the following: adjustment to the topographical forms, orientation, ventilation, accessibility and social cohesiveness (Fig. III-20 and III-21).

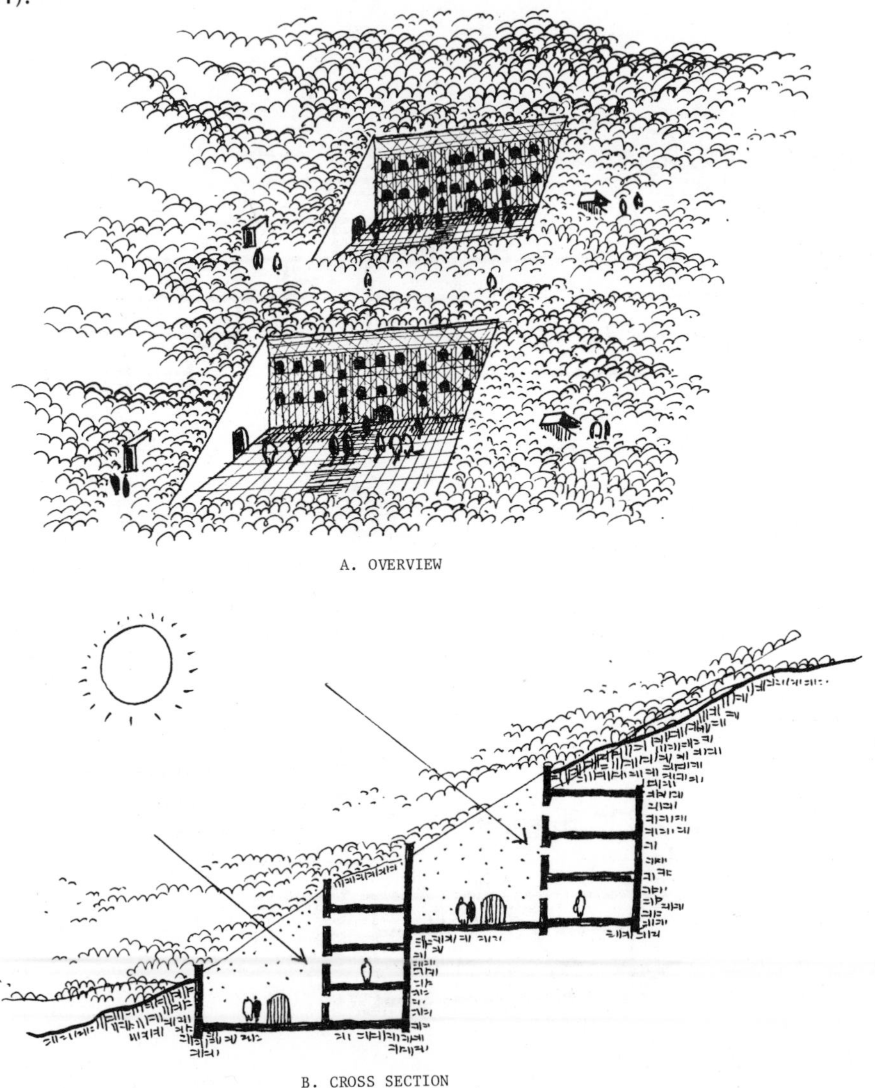

Fig. III-20. A subterranean three-story apartment built on a slope so that light can penetrate to most sections of the structure.

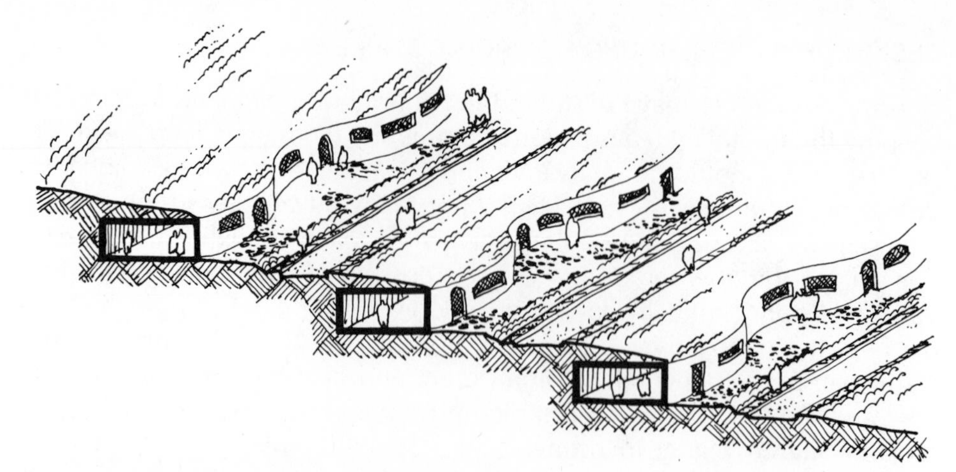

Fig. III-21. Cross section and view of terraced subterranean units on slopes with private patios on the slope area to provide a view to the environs.

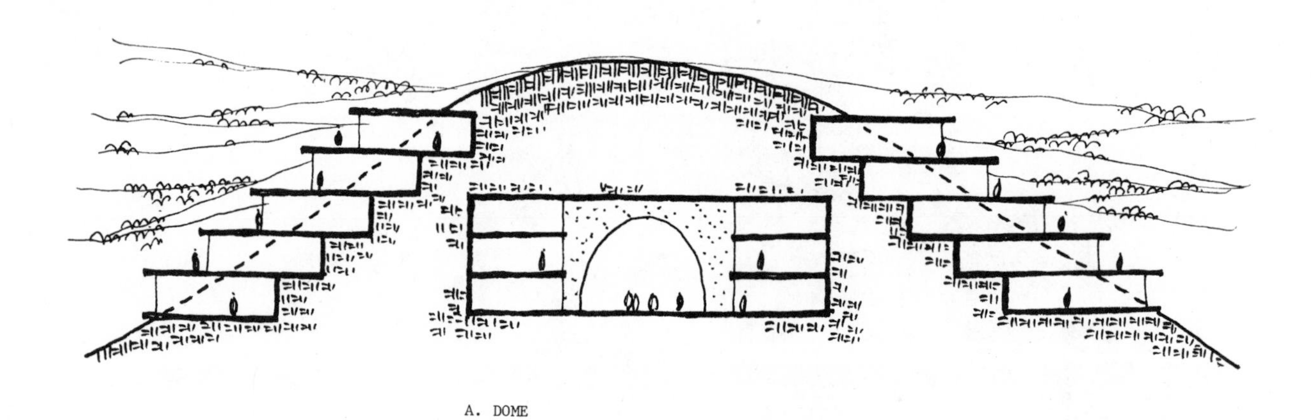

A. DOME

B. VALLEY

Fig. III-22. A cross section of two schemes for integrating a semisubterranean and subterranean residential area with an underground community center composed of social and educational facilities, shopping areas and places of work.

Housing complexes can be harmoniously integrated within the topographical dome form or into the valley, and with the location of the community center (Fig. III-22). The dome housing form has the advantages of an attractive and wide view, privacy, light and good ventilation (Fig. III-23). Also, unique slopes with curved forms can be selected since they tend to integrate housing and develop physical semienclaves to support social interaction among various age groups (Fig. III-24). Some communities, such as the Mediterranean cultures, are more open than others to socializing, with little or no formality; these cultures would be receptive to such enclosed forms. Housing complexes can also be integrated with steep slopes, especially when an ideal geological formation is found; such slopes have even more advantages than the dome form (Fig. III-25). In any case, the use of the slope requires terraced housing designed along the contour line of the topography (Fig. III-26). Low slopes would require little investment for grading and topographical adjustments (Fig. III-27). In all three slope types, the advantages make development feasible. The design, however, should keep the direct view to the environs unobstructed (Fig. III-28).

Although the least desirable alternative is the flat topography, housing complexes can also be adjusted to such land (Fig. III-29). A patio open to the sky can provide privacy and safety for the residents (Fig. III-30). As we have already mentioned, the view to the environs is obstructed, however. Also, sewage pumping will be required. Special designs should be considered to avoid the hazard of flooding and to minimize the dust and litter falling into the patio.

Ventilation is important for the subterranean house, especially in humid regions. On the slope, ventilation can be passive if careful design is considered after study of the site's microclimate. If air movement exists at the site, ventilation can also be achieved by diverting the wind to the interior spaces.

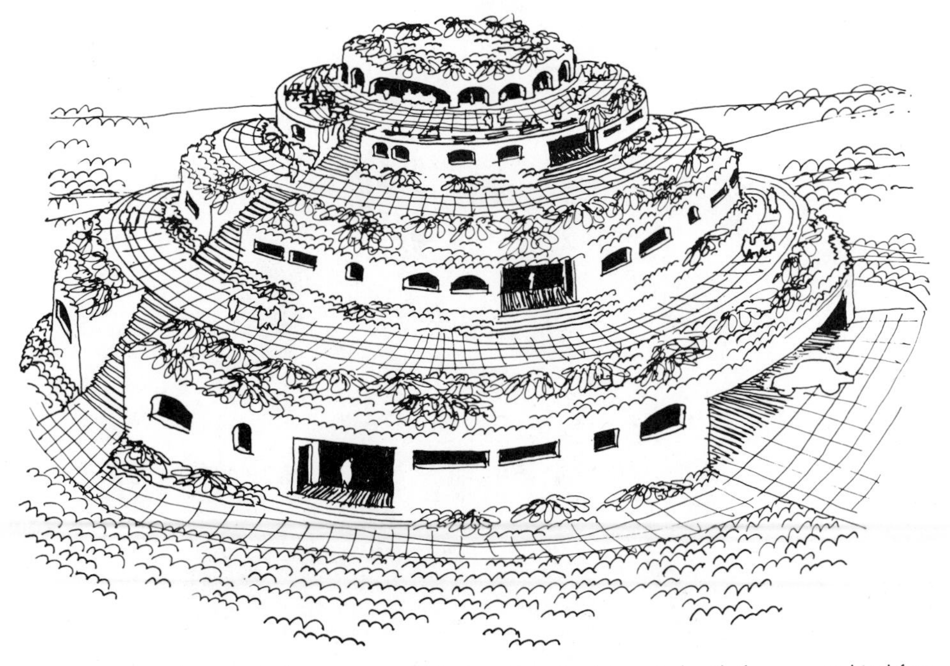

Fig. III-23. General view of subterranean neighborhood housing integrated with the topographical form of a dome. Public center is at the top.

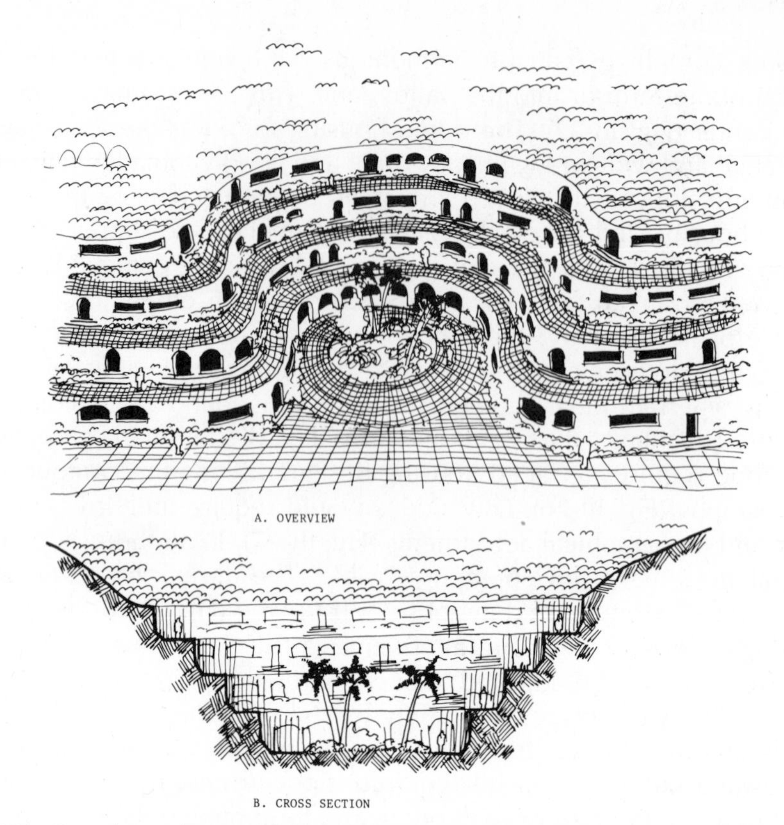

Fig. III-24. Circumferential terraced housing complex on a slope, adjusted to special topographical configurations and with a common patio.

Fig. III-25. Terraced houses on high, steep slope.

Fig. III-26. Terraced subterranean houses on topography with a steep slope usually ignored by conventional development. Note curved, narrow horizontal windows positioned to obtain maximum daily light and sunshine.

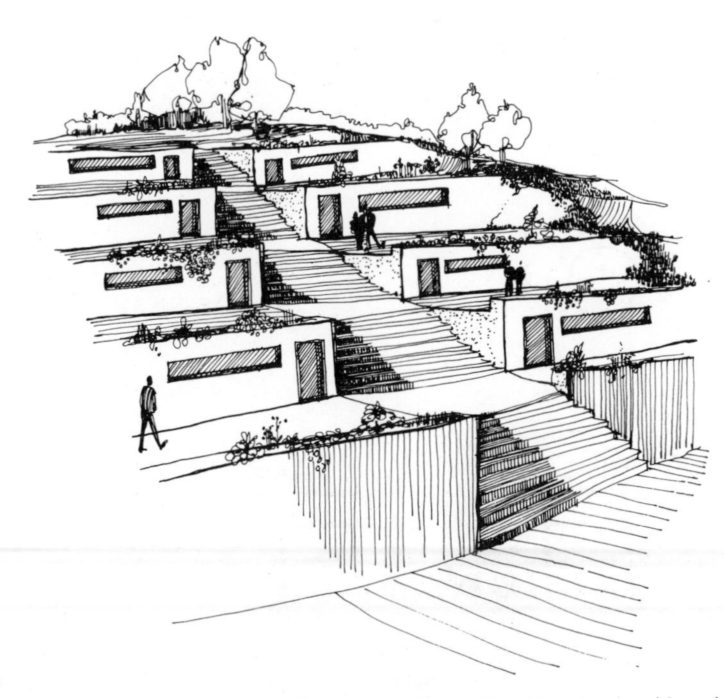

Fig. III-27. Attached subterranean housing complex units with unique and low slope configurations.

Fig. III-28. Linear subterranean complex using relatively low-grade slopes.

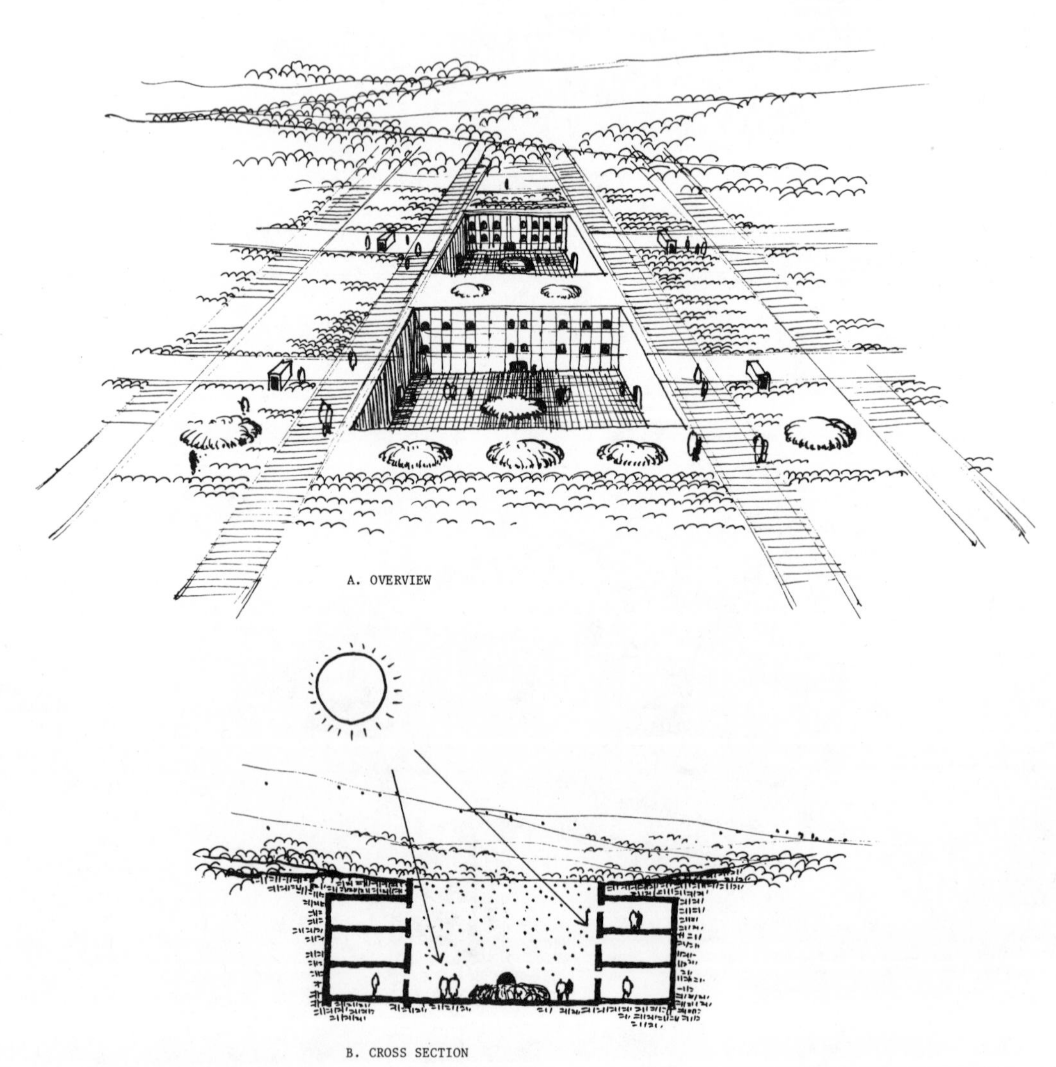

Fig. III-29. An underground apartment complex featuring an open-air patio which makes light penetration into the apartments possible. Garages can be built aboveground with a direct underground connection between the garages and the apartment units.

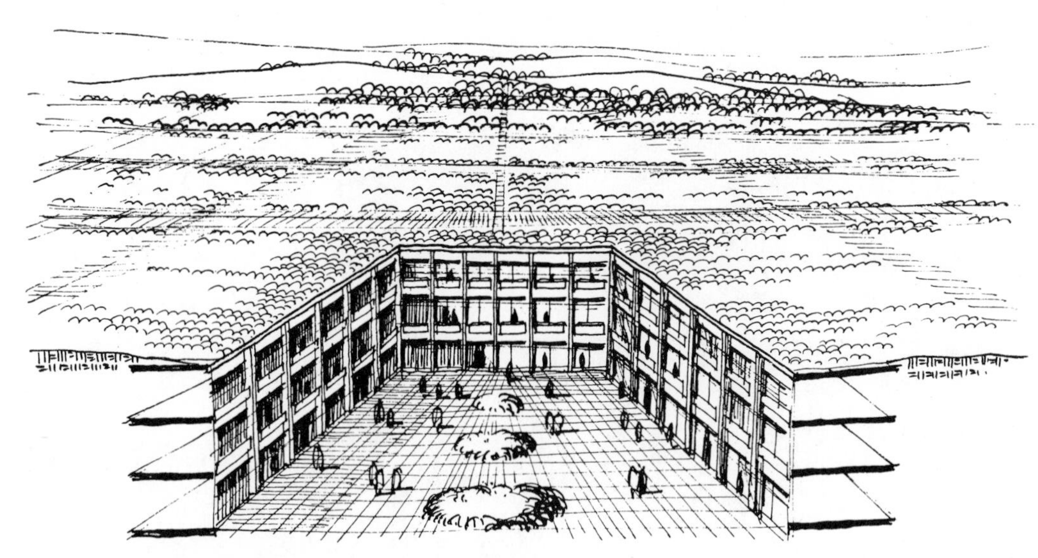

Fig. III-30. Three-sided apartment building underground with common open-to-the-sky patio. One side of the patio can be open partly or fully to the view.

Orientation of the house openings is to be considered in relation to sunshine, light and view. As we have said, sunshine cycle studies should be made of both the seasonal and diurnal movement. In cold regions, sunshine is most desirable, especially when dampness is associated with the cold. In hot climates, the penetration of solar radiation should be avoided, unless, again, dampness is a problem. Both light and direct view to the environment are essential in all climates, and the housing complex design should take them into consideration (Fig. III-31).

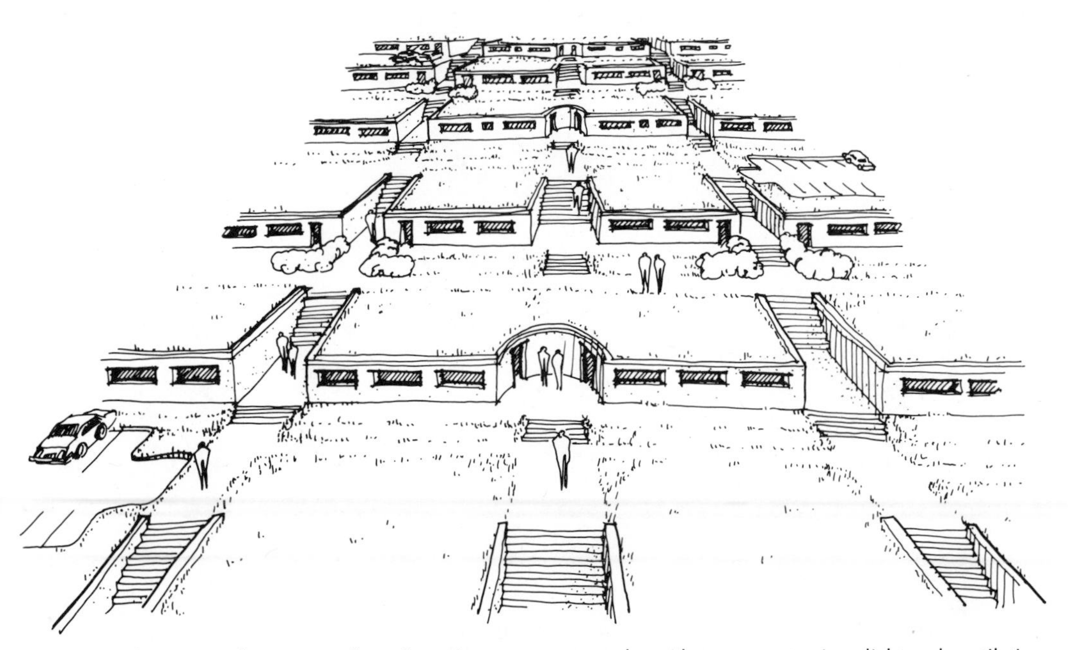

Fig. III-31. Subterranean house complex adjusted to steep topography with exposure to view, light and ventilation. Closed car parking spaces are in alternate levels.

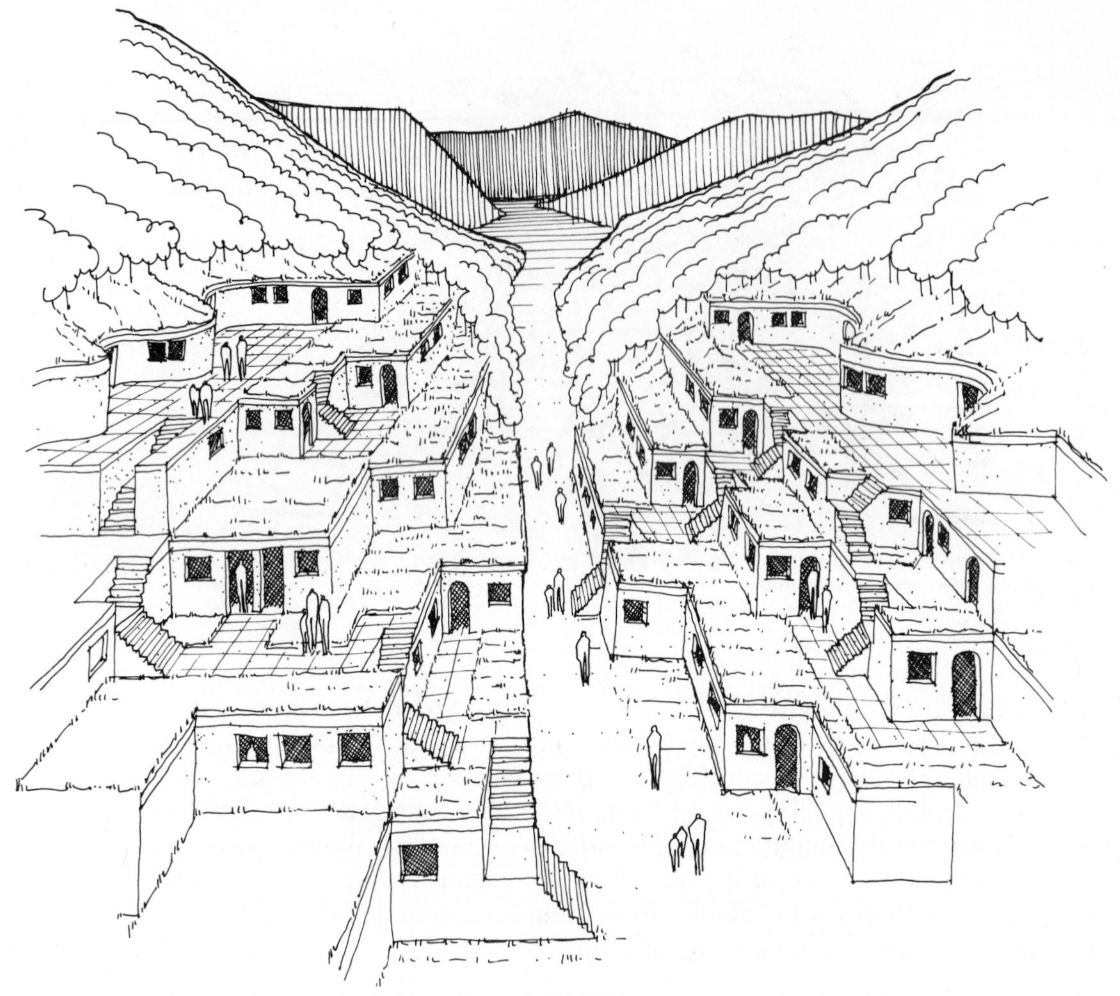

Fig. III-32. Subterranean and compact housing complex adjusted to special steep topographical forms but still facing the view, light and ventilation. The complex can also be related to the sun cycle. The design introduces privacy and plenty of space for pedestrians.

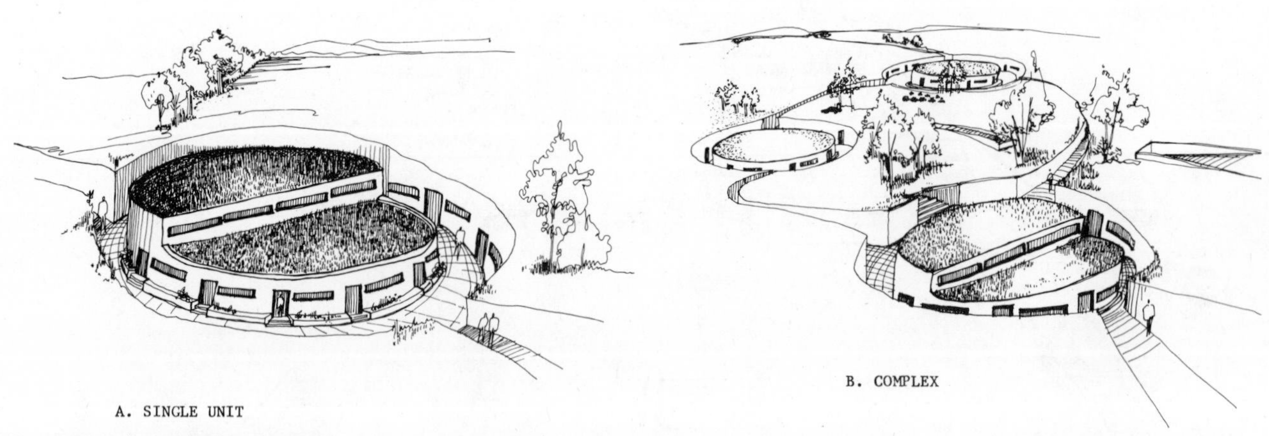

Fig. III-33. Subneighborhood units:

A. Unit of circumferential, concentrated housing combined with daily service units.

B. Complex of circumferential, concentrated subterranean housing units located on low-grade slopes with centers, common parking and separated pedestrian paths. Houses can face light or sunshine or both, while nonresidential sections do not need them.

Accessibility to and within the housing is essential, and plans should provide for easy and convenient movement without loss of privacy. Cars should have close access to the complex without interrupting pedestrian movement. However in slope sites, cars can approach the complex from the upper parts so that residents can descend through stairways, especially after shopping, to their houses.

Finally, social cohesiveness is another aspect of the housing complex design. Subterranean living provides a great deal of privacy within each house. Therefore, the physical form of the complex as a whole should introduce and foster social cohesiveness for the residents (Fig. III-32). The housing complex can be circumferential on a low grade and provide social cohesiveness within the area. Figure III-33 shows single and complex units for housing, with car movement entirely separated and enough space for groups of residents to interact.

Environmental Issues Involved in Going Underground

Undoubtedly, subterranean or earth-covered buildings make more radical changes in their sites than equivalent conventional houses do. These changes affect the geomorphological form, soil deposit, slope, topographical configuration and soil structure; and they occur on the site and in its immediate environment. The end result of these changes does, however, go beyond the immediate environs. For instance, slope grade, slope length, soil cover, percentage of impervious surface area and general rainfall characteristics are factors that influence runoff. The first four factors will be significantly changed with the development of a subterranean house or earth-covered building. We already know how rain in the arid zone can have a special impact on runoff and change the equilibrium of the soil.

The first consequence of subterranean usage is a hydrological one. The new development will change the historically developed balance of the site. At the beginning stage, following the development of the site, the permeability of the soil there is higher than in its surroundings because the ground is not yet packed hard. The result is that much rain will seep into the ground. Because of this condition, the runoff will be more erosive, carrying with it higher quantities of soil. Therefore, large areas recently developed with earth-covered buildings may be subject to intensive soil erosion. The development of methods to hold the soil, such as terracing or planting grass or other coverage, becomes necessary. This problem is further compounded by the fact that earth removal for grading creates steeper slopes than before, a condition which intensifies soil erosion. This process will continue until such time as the soil and the geomorphological form achieve a new equilibrium which is balanced with the older surrounding environment. However, the new site equilibrium will never be the same as it was before the development. Moreover, the new equilibrium will influence the hydrological pattern of its surrounding area to some extent.

The second development resulting from the site balance disturbance is eolian erosion. When soil is removed, especially in a dry climate, the ground becomes subject to wind erosion during the construction and for some period afterward until the soil reaches stability again. In addition, the new changes in the soil order will result in the loss of the upper layer of soil, which may have been of high quality, and its replacement by a lower quality soil. Thus, soil fertility may be reduced. Therefore, new soil may have to be imported for top soil to support the landscaping.

The third result of the site development is the change in the balance of the ecology. The flora and fauna of the site will be entirely destroyed, and new flora will have to be brought to the site. In newly developed areas, the fauna will never be the same as before because of the human activities. This is especially true of the jerboa in arid zones or the mole and the groundhog in nonarid regions—fauna which lives in cavities belowground.

Another result concerns the general runoff pattern throughout the city as a whole. In a conventional city, the permeability of the city ground is low due to the asphalt and built-up coverage of the city. Consequently, the runoff generated is higher than in an equivalent nonbuilt-up area under the same condition. This large runoff may run through stormwater systems, if they are well designed, or make the streets a stream-bed and eventually destroy them. Such a process is more acute in an arid zone where rain is torrential than in a more moderate rainfall area. However, in a subterranean neighborhood or city, the soil is more permeable and the ratio of runoff is expected to be less, a condition which can enrich the aquifer, but still necessitates efficient and lasting structures, waterproofing and an effective treatment against erosion.

Another environmental change is related to the heat island (dome) phenomenon associated with cities. It is clear now that earth-covered buildings save energy for heating and cooling, whether they are built in cold or in warm climates. For instance, the earth-covered Terraset School in the new town of Reston, Virginia, an area of moderate climate, has an energy savings rate of 70 to 75 percent.[5] In a neighborhood made up of a large number of subterranean houses, we can anticipate that most of the home heat is kept within these houses and there is minimal, if any, heat loss. Thus, the heat island usually found above the conventional city is minimized in the subterranean city, resulting in less impact on the environment of this city.

Application of Urban Design Principles to a Specific Arid Israeli Site

To help the reader envision the applicability and the feasibility of the concepts introduced in this chapter, we briefly present here the planning of a specific site in the northern part of the Negev of Israel. It seems to us that the stressed climate here makes it a most appropriate choice because the effectiveness of the subterranean house relates to the climate in general and to the extreme climate in particular. On the other hand, some of the solutions to be introduced here can be applicable in another type of climate with some modification and adjustment by the planner-designer.

THE SITE

Solely as an academic exercise, the author selected a site located near the ancient city of Avdat, Israel. The author believes that this typically arid site would be an appropriate place to build a subterranean cell (not necessarily introducing compact principles). The site's elevation is 400 meters above sea level, and it is close to a major national highway running north-south. Although the climate is hot and dry in summer and moderate to cool and dry in winter, there is some water available in the area. The topography is characterized by a slope of 3 to 15 percent, and the geological conditions are stable. The area is made of limestone layers.

The soil of the selected site is loess, which is quite appropriate for subterranean settlement. As we have already noted, when the loess surface becomes dry, it forms a hard crust which retains and protects the moisture found in the deeper layers of the soil.[6] The energy-free cooling system for subterranean structures suggested in this book requires just such soil characteristics.

Microscopic examination reveals the presence of minerals (quartz, feldspar, calcite or mica) together with binding materials which harden the texture of the soil when dry.[7] "This binder can give loessial soils considerable dry strength, sufficient to sustain the overburden of 100 m. or more of dry soil."[8] Loess, however, loses this strength when water destroys the binding and causes serious erosion on the surface. Nevertheless, loess is very porous and well-drained vertically.[9]

THE PLAN

A primary goal of this modern plan of an urban cell is to minimize the impact of the climatic stress of the arid zone on urban living and thereby

Table III-3. Israeli Population by Sex and Age

AGE GROUP	FEMALES		MALES	
	THOUSANDS	PERCENTAGE	THOUSANDS	PERCENTAGES
Total	1481.5	100.0	1477.9	100.0
0–9	305.5	20.7	321.4	21.7
10–19	255.7	17.2	269.9	18.3
20–29	271.2	18.4	277.0	18.7
30–39	157.9	10.6	154.4	10.4
40–49	153.2	10.3	140.5	9.6
50–59	138.2	9.4	128.4	8.7
60–69	122.7	8.2	115.9	7.8
70+	77.1	5.2	70.4	4.8

Data from *Statistical Abstract of Israel*, no. 27 (1976), table ii/15, p. 40.

to increase the comfort of the designed residence area in social as well as in climatic terms. To achieve this defined goal, the following secondary goals were established: (1) a compact physical environment; (2) an emphasis on vertical development rather than on the more conventional horizontal; and (3) an inclination toward semisub- and subterranean rather than high-rise construction. Adhering to these principles will bring noticeable savings in the costs of utilities and the infrastructure in terms of both initial investment and in maintenance and operation.

Population size and land size were computed on the basis of the average national norms of Israel. Table III-3 divides Israel's 1976 population by sex and age (in 10-year groups); Figure III-34 is a population pyramid presenting graphically the current demographic age distribution. In Table III-4 the same 1976 population is divided by using different age increments. This data will be used throughout this model to calculate labor force, school-age population, etc., for the proposed site we have selected. Our assumption is that the demographic structure of the community we are planning will mirror that of the nation. Thus, most of the residents will be young, and there will be fewer elderly residents. Most of the young adult residents will be married, some with children and some without. As the community develops, the average family size will gradually rise to 4.0 members.

Each neighborhood unit will consist of 5000 to 6000 residents. This range will result in the necessity for one elementary school per cell within easy

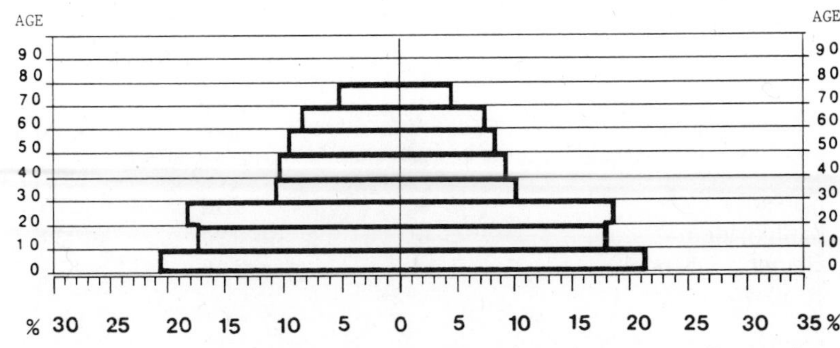

Fig. III-34. Population pyramid for the Israeli state and the proposed urban cell. Data from *Statistical Abstract of Israel*, No. 27 (1976), table ii/15, p. 40.

Table III-4. Israeli Population by Sex and Age for Use in Calculating Labor Force, School Attendance, etc.

AGE GROUP	FEMALES		MALES	
	THOUSANDS	PERCENTAGE	THOUSANDS	PERCENTAGE
Total	1481.5	100.0	1477.9	100.0
0–4	165.8	11.1	174.3	11.7
5–9	139.7	9.4	147.1	9.9
10–14	125.3	8.4	132.3	8.9
15–19	130.4	8.8	137.6	9.3
20–24	141.7	9.5	146.1	9.8
25–34	211.3	14.2	212.1	14.3
35–44	152.2	10.2	145.3	9.8
45–54	154.7	10.4	139.4	9.4
55–64	127.4	8.5	118.4	8.0
65+	133.0	8.9	125.3	8.4

Data from *Statistical Abstract of Israel*, no. 27 (1976), table ii/15, p. 40.

walking distance of all residents. (See Table III-5 for enrollment data, using both the 5000 Alternative (1) and the 6000 (2) upper limits.)

Table III-6 distributes the proposed cell population (5000 and 6000) according to the national demographic profile. Following the same statistical procedure, we drafted Table III-7 showing the distribution of family sizes across the proposed cell populations. The circular cell's size will be 196.25 acres, based on a radius of 500 meters. Given this, we computed the population densities for the two proposed cell populations. They were 25.5 persons/acre for a 5000-person cell and 30.6 persons/acre for a 6000-person cell.

As for the economic structure, for the most part, each cell will be relatively self-contained as far as the necessary daily and weekly services and facilities are concerned. A reasonable proportion of the population will work within the cell itself. There will be some population mobility, making necessary the construction of some rental housing units in the cell. As is the case in the rest of Israel, the middle-income group will dominate the community (75 percent); therefore, there will not need to be much low- or upper-income housing built.

Table III-5. Ultimate Attendance Figures for the Educational Facilities of the Proposed Urban Cell*

TYPE OF FACILITY	AGE OF STUDENTS	NUMBER OF ATTENDANTS	
		ALT. 1 5000 RESIDENTS	ALT. 2 6000 RESIDENTS
Nursery	3–4	230	276
Kindergarten	5–6	194	234
Elementary School	7–14	728	872
		1152	1382

*High school will not be provided within the urban cell itself.

Table III-6. Population in the Proposed Urban Cell by Sex and Age

	FEMALES			MALES		
AGE GROUP	%	ALT. 1*	ALT. 2**	%	ALT. 1*	ALT. 2**
Total	100.0	5000	6000	100.0	5000	6000
0–4	11.1	279	335	11.7	294	353
5–9	9.4	237	284	9.9	249	299
10–14	8.4	212	254	8.9	224	269
15–19	8.8	221	266	9.3	234	281
20–24	9.5	239	287	9.8	246	296
25–34	14.2	357	428	14.3	359	430
35–44	10.2	256	308	9.8	246	295
45–54	10.4	261	314	9.4	236	283
55–64	8.5	214	256	8.0	201	241
65+	8.9	224	268	8.4	211	253

*Alt. 1 is based on a population of 5000 for the urban cell.
**Alt. 2 is based on a population of 6000 for the urban cell.
Data from *Statistical Abstract of Israel,* no. 27 (1976), table ii/15, p. 40.

The labor force can be fairly easily and accurately calculated. To do this, we must first distinguish between the potential labor force and the actual labor force. We considered every male and female person between 25 and 65 years old as part of the potential labor force. In addition to that, an estimated 20 percent of the age group between 15 and 20 was considered as having left the school system to become part of the potential labor force (Table III-8).

The potential labor force was expected to be the same as the actual labor force among the men. Among the women, the actual labor force was computed as the potential labor force minus those females living in households with at least one child (Table III-9). We are aware of the inevitable inaccuracy of this calculation since mothers of grown-up children, who are nevertheless still members of the household, might return to the labor force. Making the adjustments described above to the potential labor force and allowing for this slight inaccuracy, the actual labor force would be 1781 persons in Alternative (1) or 2138 persons in Alternative (2). The industrial

Table III-7. Number of Households by Family Size for Proposed Urban Cell

		ALT. 1 (5000 RESIDENTS)		ALT. 2 (6000 RESIDENTS)	
TOTAL PERSONS/ HOUSEHOLDS	% 100.0	HOUSEHOLDS 1450	PERSONS 5000	HOUSEHOLDS 1740	PERSONS 6000
1	12.2	178	178	215	215
2	24.4	354	708	425	850
3	17.7	263	789	314	924
4	19.7	287	1148	345	1380
5	12.6	183	915	219	1095
6	5.7	82	492	99	594
7+	7.6	110	770	132	924

Data from *Statistical Abstract of Israel,* no. 27 (1976), table xi/17, p. 272.

Table III-8. Potential Labor Force of Proposed Urban Cell

	FEMALES		MALES	
AGE/GROUP	ALT. 1*	ALT. 2**	ALT. 1*	ALT. 2**
15–19	44	53	47	56
20–24	239	289	246	296
25–64	1327	1593	1288	1545
Total	1610	1935	1581	1897

*5000 Residents
**6000 Residents

labor force is a special economic category within the urban cell. We computed it by taking 50 percent of the actual labor force left after subtracting those employed in transportation, infrastructural services, security, health, education, social services, communications, recreation and commercial services.

Some housing units will include aboveground, semisubterranean and subterranean structures, and some of them will feature open-to-the-sky patios. There will also be a few apartment buildings of between 80 and 120 m^2. All roofs will be flat and suitable for summer sleeping, and energy-free cooling systems will be used in all units. Housing for the handicapped will be provided close to and on the same level with shopping centers and other public areas.

The urban cell will be compact to a limited extent; however, it will be designed so as to ensure the maximum possible privacy (Fig. III-35). Within it, land uses will be mixed with each use spatially integrated and all uses within no more than 500 meters' walking distance. Around the cell's circumference, land would be used for residential purposes in order to concentrate nonresidential uses at the cell's center.

The housing, as noted above, will be multilevel. Most offices, public meeting spaces, "clean" manufacturing plants and similar facilities will be subterranean; however, many of these will have access to the sun and sky via patios and belowground, open-air pedestrian alleys. Green areas will be located aboveground in a central park as well as in scattered parks throughout the cell. In both subterranean and supraterranean space, shadows will be maximized by careful design and strictly enforced land-use and land ownership regulations covering this and other important points.

Table III-9. Calculation of Actual Female Labor Force in Proposed Urban Cell

	ALTERNATIVE 1	ALTERNATIVE 2
Potential Labor Force	1327	1593
Household with One Child or More	−925	−1109
	402	484 = 30.5% of Potential Labor Force

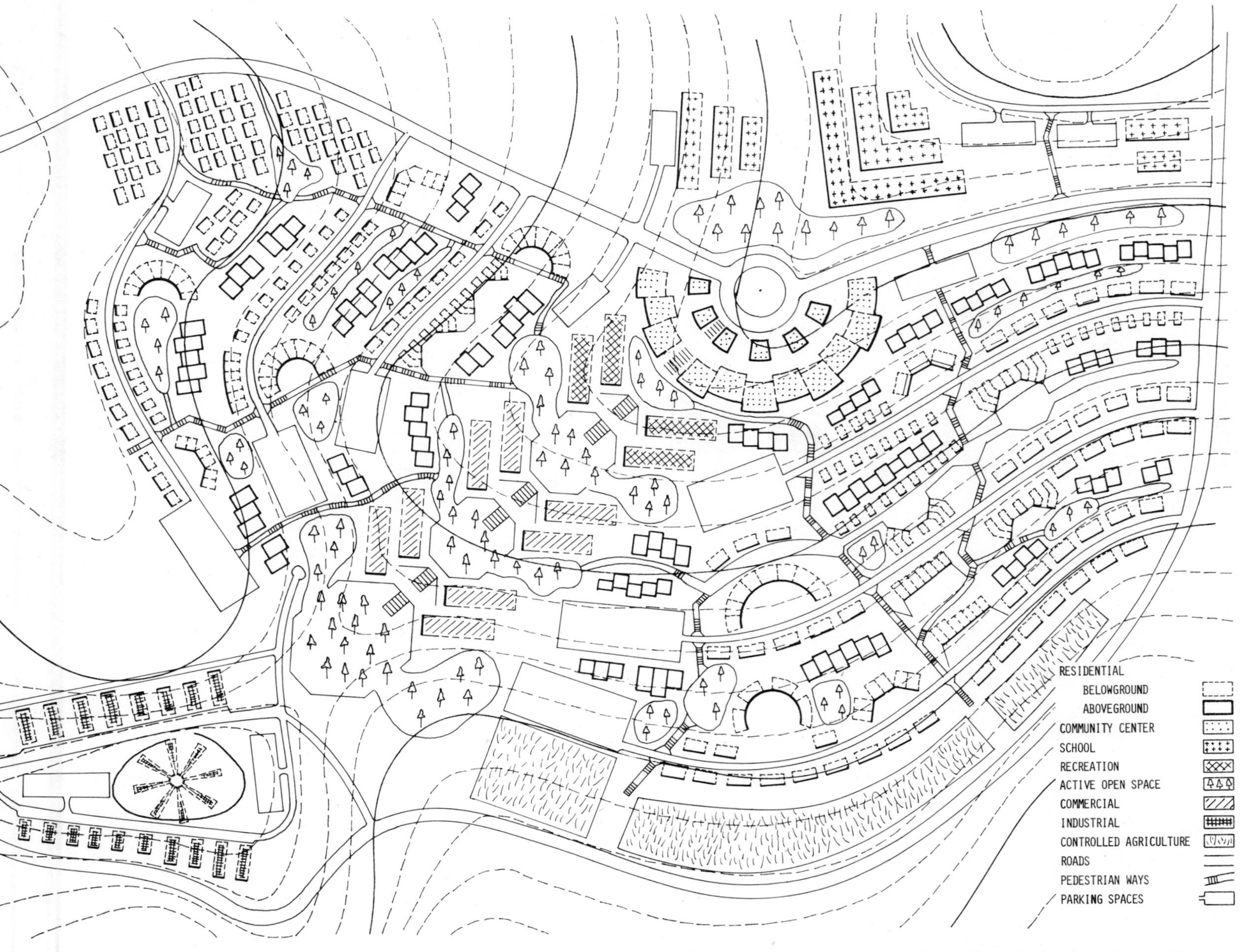

Fig. III-35. Plan used at the proposed Negev arid-zone site in Israel with exposure to the south for penetration of light and sunshine. The altitude of the site is 400 meters above sea level.

Table III-10. Land-Use Distribution in the Proposed Urban Cell

CATEGORY OF LAND USE	ABSOLUTE AREA IN M^2 ALT. 1 5000 RESIDENTS	ABSOLUTE AREA IN M^2 ALT. 2 6000 RESIDENTS	PERCENTAGE OF TOTAL AREA IN M^2 ALT. 1 5000 RESIDENTS	PERCENTAGE OF TOTAL AREA IN M^2 ALT. 2 6000 RESIDENTS
1. Transportation	132,844	135,344	16.5	17.5
2. Services	1,000	1,000	0.2	0.2
3. Security	400	400	0.1	0.1
4. Health	420	470	0.1	0.1
5. Education	58,890	66,640	7.8	8.7
6. Social Services	4,750	5,450	0.8	0.9
7. Communication	600	600	0.1	0.1
8. Recreation and Leisure	23,000	23,800	3.2	3.4
9. Residential	121,390	145,710	15.5	18.3
10. Commercial	2,825	2,895	0.6	0.6
11. Industrial	41,400	49,850	5.6	6.5
Total	396,519	442,159	50.5	56.4
12. Open Land	388,481	342,841	49.5	43.6
Total	785,000	785,000	100.0%	100.0%

Table III-10 divides the land available for our alternative cell designs into eleven uses. Table III-11 further elaborates on the ninth land-use category, residential. A planner would have to compile tables for all eleven uses. Quite possibly, the tables offered here can serve as models.

Also as a model, still more detailed information is presented here for the first listed land use, transportation. This land use can be divided into five categories: external access road(s), internal access roads, parking areas, pedestrian routes, miscellaneous (including bus stops, taxi stands and routes used in serving the residents and in transporting goods), and bicycle routes.

No motor vehicle traffic (except for emergency, sanitation, maintenance and home-delivery) crosses the urban cell. The primary transportation modes are bicycling and walking; for these, we allow in our model cell (either

Table III-11. Calculation of Residential Area within Proposed Urban Cell

PERSONS/HOUSEHOLD	AVERAGE SIZE OF DWELLING	ALT. 1 (5000 RESIDENTS) NUMBER OF HOUSEHOLDS	ALT. 1 (5000 RESIDENTS) AREA	ALT. 2 (6000 RESIDENTS) NUMBER OF HOUSEHOLDS	ALT. 2 (6000 RESIDENTS) AREA
1	65 M^2	178	11,570 M^2	215	13,975 M^2
2	65 M^2	354	23,010 M^2	425	27,625 M^2
3	80 M^2	263	21,040 M^2	314	25,120 M^2
4	95 M^2	287	27,265 M^2	345	32,775 M^2
5	95 M^2	183	17,385 M^2	219	20,805 M^2
6	110 M^2	82	9,020 M^2	99	10,890 M^2
7+	110 M^2	110	12,100 M^2	132	14,520 M^2
Total			121,390 M^2		145,710 M^2

alternative) 30,000 m^2. Within the cell, the Radburn concept is used: the cell is one cohesive physical block, and the roads in it are dead-end, beginning at the cell's circumference and ending toward the center. For six of these roads, each one 240 meters long and 20 meters wide, we allot 28,800 m^2. Around the cell, a major circumferential highway serving distribution and collection functions marks the boundary between the cell and the open space which separates it from neighboring cells. For this 20-meter wide road, we allot 61,544 m^2. Parking space will be supplied, assuming that there are 100 cars per 1,000 people and that 25m^2 of space is needed per car. If the cell's planned population is 5,000, we alot 12,500 m^2 for the 500 vehicles; if the cell's population is 6,000, we allot 15,000 m^2 for the 600 vehicles. Finally, for the miscellaneous transportation needs, we allot 20,000 m^2. The totals are 132,844 m^2 (Alternative 1) and 135,344 m^2 (Alternative 2).

This is, of course, only a model, although we believe it is a viable one. In moving from the calculation of the land needed for transportation to actual construction, we must consider two other points: the need to shadow pedestrian space and the need to provide areas where pedestrians are protected from rain both at the ground level and below (when walkways are still largely open to the sky). The recognition of these issues will help make the design responsive to human needs as well as to calculated spatial demands. Similarly, the design meets the need for private and mass transportation and fixed rail intercell transportation. Along these narrow linear routes, facilities common to two or more neighborhood cells can be located both above and below the ground surface.

Open space is an important element here in urban cells planned for the Negev, as well as in all such neighborhoods. This importance is reflected in the following figures: 388,481 m^2 of open space to 396,519 m^2 of used land for Alternative 1 and 342,481 to 442,159 m^2 for Alternative 2. This generous allotment of open space is necessary to help provide a sense of community through the shared use of public playgrounds and parks (as well as indoor community facilities). Granting each cell at least partial self-government would further engender such a sense of community.

Conclusion

It is certainly feasible for many of our urban space needs to be filled by subterranean forms. Again, this can be more easily achieved on slope forms, lands which are usually abandoned, than on flat topography. We surely have made it clear that nearly all facets of life have been and can be accommodated within the subterranean form. In fact, diversification has been accelerated, especially recently (Fig. III-36). The energy crisis has caused a resurgence of interest in subterranean housing. In the past, the motivation for use of the underground was diversified: saving land for agriculture, need for multiland usage, cost of land, cooling or escape from outside harsh climate (igloo), security, avoidance of alien surveillance, reduction of evaporation (water storage), and the need for proximity and compactness of land uses. Judging by the existing tendency throughout the world and especially in the United States, we can expect that the usage for living space will increase in the future through public as well as private enterprise.

Subterranean settlements have been widely used throughout history; the first section of this book discussed areas such as Cappadocia in central Turkey, provinces in northern and eastern China, and North Africa. Regions with extreme climates, either very cold or very hot and dry, prove to be most suitable for the development of subterranean living quarters as protection from the harsh climate. In highly populated regions as well, subterranean settlements can offer an alternative solution, the utilization of both above- and belowground space. We have mentioned many other specific cases, of course.

Shopping centers, developed in some large cities in Japan and in Montreal, Canada, and Stockholm, Sweden, accommodate hundreds of thousands of shoppers a day. Some shops, because of equipment which requires stable temperature and humidity, can be more properly built using subterranean spaces. Others go underground to acquire additional space for storage. Security and safety problems tend to be reduced in these cases. Shopping centers are a major focal land use in the community. Here again we suggest the use of the slope for the development of the community center or the shopping center (Fig. III-37). The structure can also take a terraced form when it is on flat topography (Fig. III-38). In cold climates, it may be necessary to cover all the shopping center with glass, thus providing a completely controlled climate (Fig. III-39). In warm, dry climates, it may be necessary to design central large patios (Fig. III-40).

Power plants are conveniently located underground and include electric power generation and transmission, nuclear power plants and power plants

Fig. III-36. Terraced subterranean hotel.

generating hydroelectric power near waterfalls. An example of this type is the Churchill Falls hydroelectric power installation in Canada, which is the largest man-made excavation and supplies 20 percent of the total electricity consumed in Canada. Frequently, subterranean space is used for electric power cables. Italy has 60 underground power plants, and they can also be found in Sweden, Norway, Australia and several other countries. There are over 100 subterranean power plants operating in the world.[10] The construction of such subterranean power plants is possible year-round, especially in areas with severe winter weather which makes surface construction very difficult if not impossible. The advantages are numerous: no weather change impact, no damage by storms, no effect on landscape, the capacity of the ground to carry heavy loads, lower maintenance cost, freedom from limitations imposed by topographical conditions, and lower costs overall.[11] However, planning and construction require thorough investigation and testing of the rocks and their geological characteristics which can be time consuming.

Tunneling has been used throughout history for pedestrian passage or for water supply. In Jerusalem, the water tunnel of Hezekiah, the king of Judea, carried water to the city from a rural stream. Similarly, tunnels were used in the ancient city of Megiddo (Israel) many centuries B.C. The Romans used tunneling to aid in quarrying copper and gold. In the United States, tunnels were dug to provide passage, for example, under the Hudson River in New York and under the Chesapeake Bay between Virginia and Maryland. Underground utilities were also used widely in Europe during the Middle Ages or before, especially for sewage.

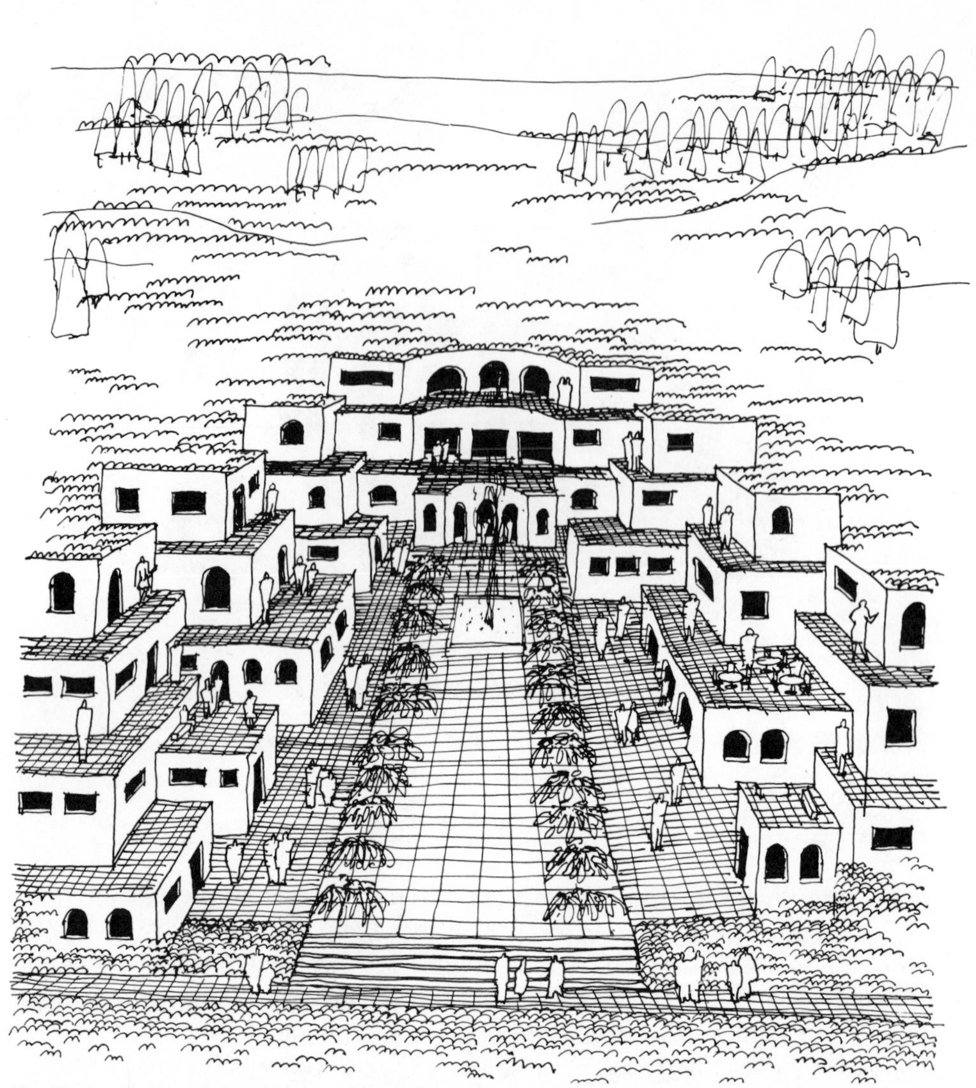

Fig. III-37. Subterranean, terraced, compact, comprehensive shopping center with central open patio located on a slope overlooking the lowland or the valley. Outer walls are thick, double or earth-covered.

Contemporary usage in many large cities and university campuses includes telephone cables, electricity, centralized heating pipes, sewage and water pipes and others. In new communities in the United States, the concept of Utilidor was introduced to encompass all the systems of utilities in one planned and organized tunnel.[12] There are many advantages in the use of the underground for utilities: low maintenance cost, few power failures, no storm hazards and lower fire hazards which reduce insurance rates, high capacity of the ground for heavy load, and easily maintained stable humidity and temperature conditions.[13] Also, space for utilities can be multilevel.

Manufacturing and industry have profitably used the underground, especially industries which require stable temperature and humidity or minimum vibration (e.g., Brunson Instrument Co., a manufacturer of precision surveying and optical instruments in Kansas City). Other advantages in working underground include greater efficiency in production since the work is not influenced by weather conditions, more efficient operation of the

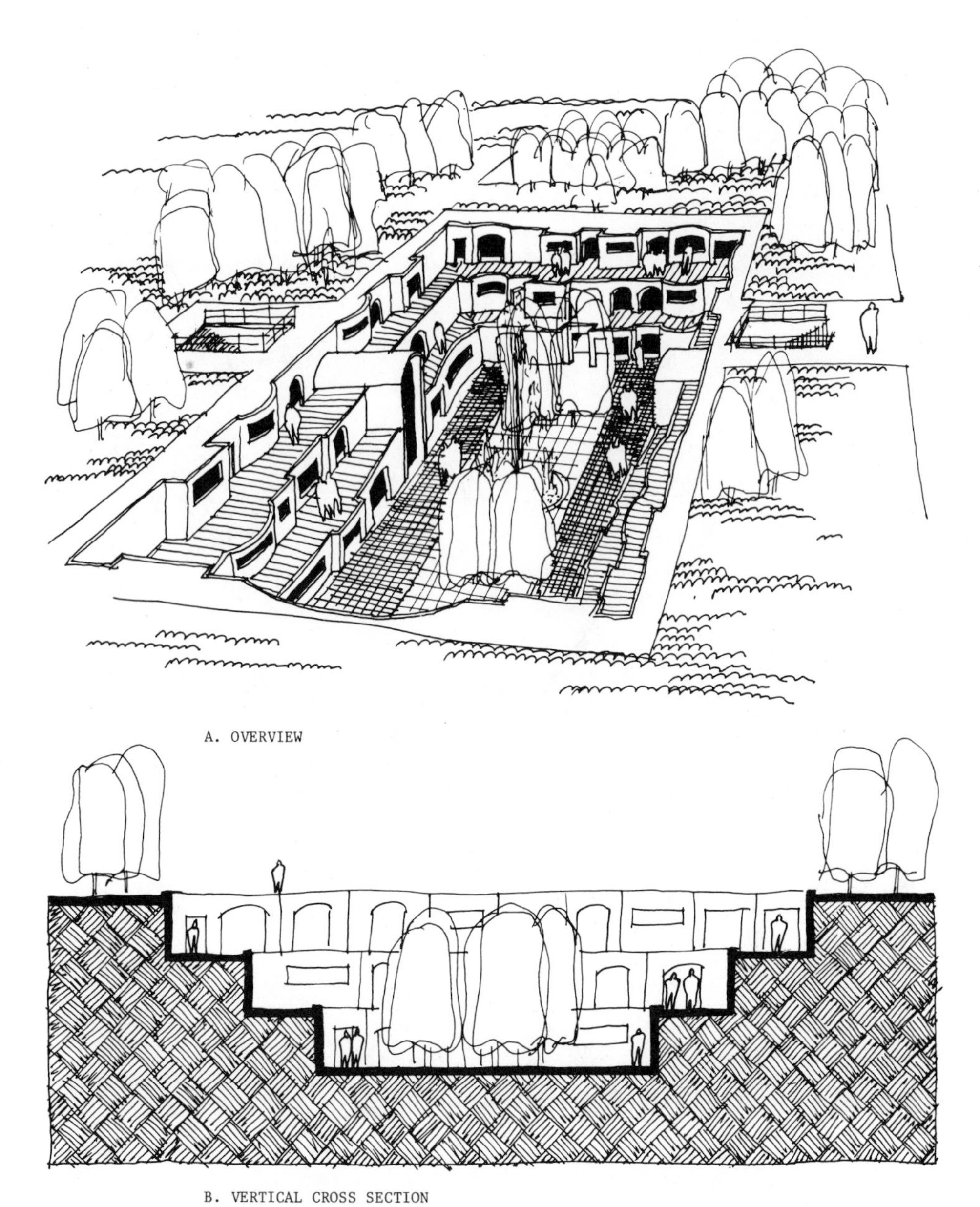

Fig. III-38. Terraced subterranean shopping center with open patio located on flat area.

equipment, little effect on working hours which could be influenced by outside light.

It would seem as if other manufacturing processes which require stable temperature and humidity (such as the processing of foods and spirits) and manufacturing processes and storage businesses needing refrigeration facilities would find distinct advantages in naturally insulated underground structures. Yet, we must consider some of the basic requirements for underground industrial storage and manufacturing structures which differ from underground residential structures:

1. Accessibility and circulation for heavy trucks loading and unloading, and in some cases the need for railroad extension to the site itself;
2. Need for a large parking space close by, above or below the ground;

Fig. III-39. Subterranean shopping center promenade covered with skylight for cold climate site.

3. Space for temporarily storing large amounts of garbage;
4. Utilities systems, such as high voltage power, water supply and storage and sewage. Also fire protection equipment is necessary.

Undoubtedly, research would prove that nonpolluting manufacturers and storages could be accommodated within the complex neighborhood since the noise will not affect the residential section.

Refrigeration and cooling is another usage. The stable and relatively low temperature saves energy for refrigeration. In Iran, people have been cooling or icing water by channeling outside nightime air to the subterranean section. Subterranean refrigeration is also widely used in Kansas City today.

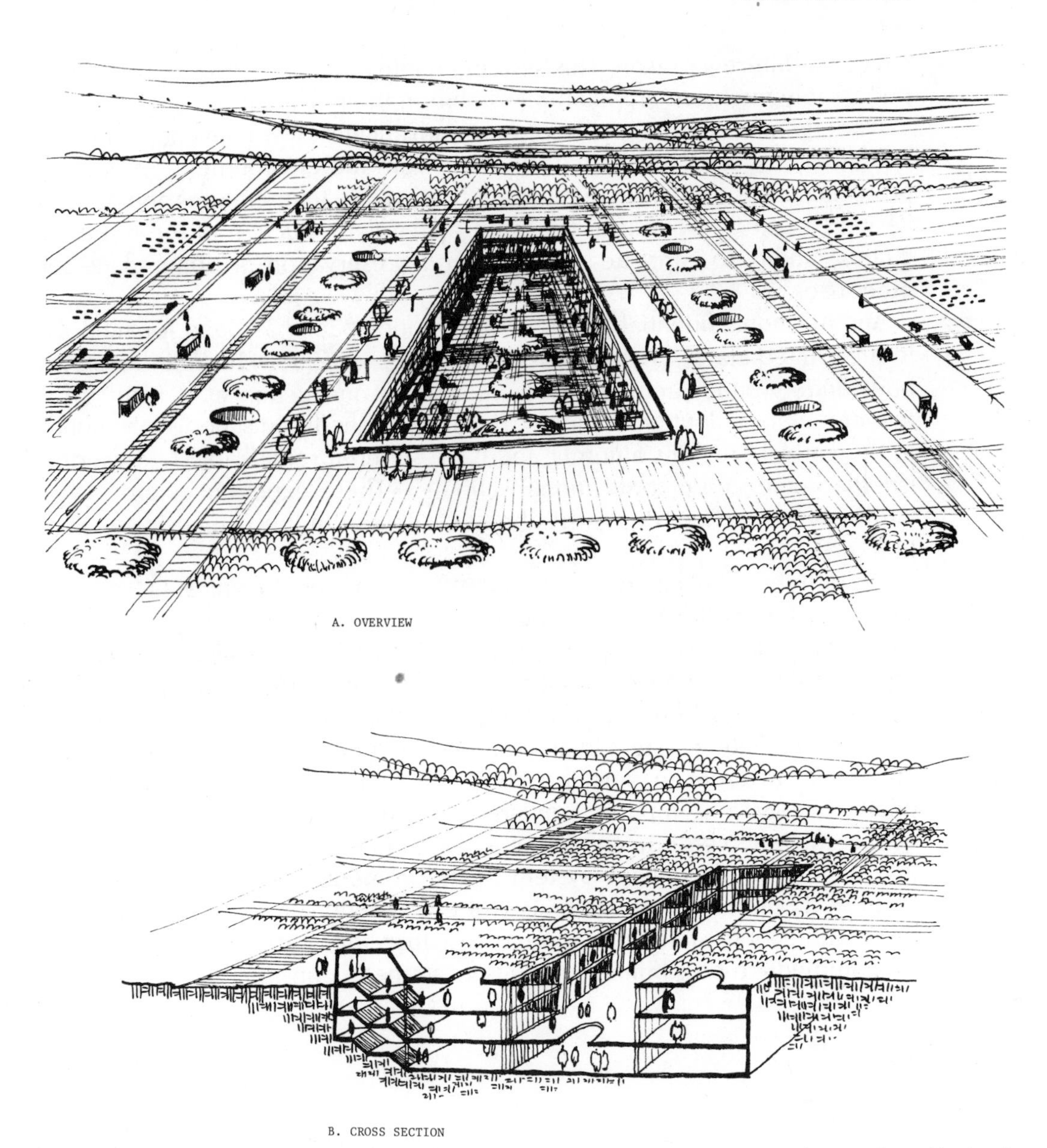

Fig. III-40. Land saving. The aboveground space saved by the construction of an underground neighborhood center (commercial and manufacturing) in an arid zone can be landscaped and turned into parks, playgrounds and social gathering areas.

Storage of oil has been primarily subterranean in Norway since World War II and now this method is used by many countries and in military installations. It has been found that certain rock formations which have minimal or no leakage make ideal sites for storage because they are secure against fire and storm damages, they require minimum, low-cost construction, and they are hidden and protected. Underground water storage has been used to avoid water loss by evaporation, especially in warm, dry climates where water is scarce. The most remarkable case is that of the Nabatean settlement in the Negev (2nd century B.C. to 6th century A.D.). The runoff water was diverted to a large excavated limestone cave to be used by forcing it through another controlled opening at the lower side of the cave in a different season. This system enabled the Nabateans to develop a remarkably intensive argricultural system in their arid zone.

Underground warehouses have many advantages over conventional ones, including low-cost land; less destruction from the environment; safety and protection from vandalism or adjacent fires; stable temperatures and humidity which can be beneficial for some materials; reduction of the need for guards; ability to carry heavy loads; and preservation of open space for other uses.

Another nonresidential use of underground space has been for schools. In the United States, by the mid-1960s, there were 96 schools in 23 states built underground; another 25 were under construction as a result of a movement for fallout protection.[14] Three earth-sheltered schools were built in the early 1970s in California and have suffered remarkably less vandalism than other more conventional schools.[15] These experiences prove that earth-covered school design has been successful in the United States (Fig. III-41).

Despite these successes, we feel that the concept of using subterranean structures for shelter and working space requires laboratory and practical experimentation before large-scale implementation can begin. The physical and psychological impact of subterranean living on different age and sex groups must be studied more closely. Furthermore, we must develop an efficient ventilation system in an underground house, office, or manufacturing facility in order to meet health and fire protection standards without undue energy consumption. Similarly, we must meet the need to develop quick, efficient, versatile and low-cost excavation techniques and an effective underground drainage system in the immediate vicinity of development. Although at this time some of these problems may seem outwardly irresolvable for the fully subterranean structure, there seem to be very promising immediate possibilities in semisubterranean construction.

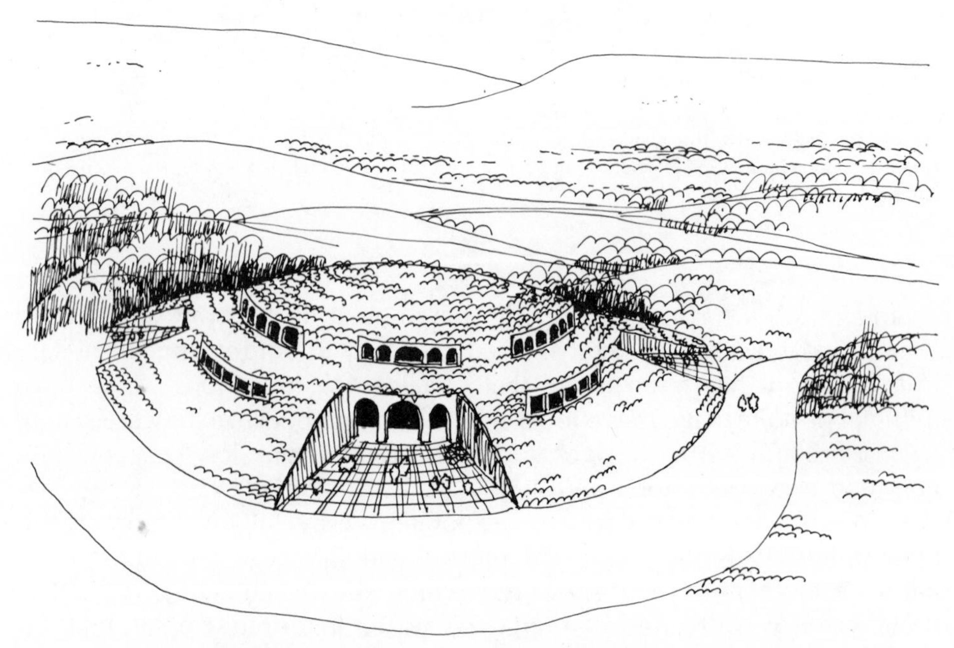

Fig. III-41. The earth-covered structure as an alternative design for school or public library.

A partially underground structure (shelter, working place or city) must respond to climatic stress, save energy, meet low-income family housing demands, meet physical and mental health standards (light, ventilation and good acoustics), provide direct contact with the natural environment, feature an efficient infrastructure and adequate utilities, introduce a good quality living environment, and be safe.

In this author's opinion, an integrated semiunderground settlement can indeed respond favorably to stressed climates, in particular that of some particular arid zones where a combination of above- and underground environments can, in fact, better solve climatic problems such as strong and dusty winds, solar reflection and high amplitude of diurnal temperature than conventional aboveground construction. Such a combination allows for the use of the house in a seasonal cycle when temperatures move from one extreme to another and introduces a quite livable landscape. The proposed urban center would have less impact on the landscape and the fragile ecology than a conventional city. In addition, we feel that joint development of semisubterranean or totally subterranean space, the proposed compact urban cell and open spaces aboveground will result in a semirural environment. This urban configuration can realize that old dream of many planners: combining urban living with a rural environment while reducing or eliminating the tyranny of the automobile within the urban cell.

Greenhouses can be incorporated within proposed urban cell design to bring about a totally atypical urban environment with several advantages (Fig. III-42): (1) It will establish agricultural production for the community, make the urban cell a more self-supporting and self-contained community, and result in less commuting to obtain food supplies; (2) it will provide job opportunities for the local residents; (3) it will make possible the recycling of the cell's sewage; and (4) it will introduce additional green space into the urban cell. It is also necessary to develop a visually pleasing design for the greenhouses in order to integrate them harmoniously into the urban environment.

The combined aboveground compact and underground form for urban settlements in stressed climate zones is, in our opinion, an unconventional planning concept. It promises to alleviate climatic stress in these regions and diminish the tyranny of conventional transportation in the residents' daily lives. As a result, this new urban pattern can help ease both the financial and ecological burdens of energy consumption in today's world.

In our discussion of this endeavor, we have tried to introduce the importance of the lessons which could be learned from the historical experience. We have suggested that combining the ancient experience with our modern technology will bring an innovation that can meet our modern standard of living. It is our belief that a comprehensive treatment of the issue, as introduced here, can ultimately offer housing possibilities for low-income families.

On the other hand, because of the public bias against subterranean habitation, it is necessary to conclude with an analysis of the pros and cons of subterranean structures. Such analysis can offer more awareness of the potentialities, from the comprehensive point of view, and the possible hazards which subterranean development may bring. In any case, the examination

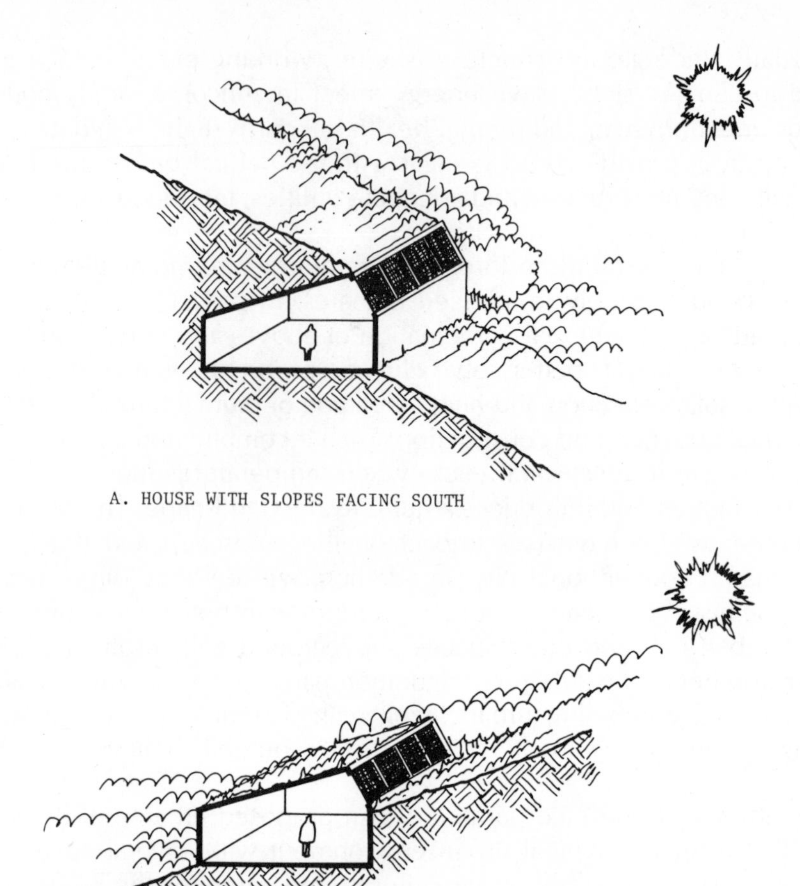

Fig. III-42. A simplified sketch of a solar collector integrated with the subterranean house design.

of such an endeavor must be comprehensive. Thus, we must deal with the physical, social, economic, environmental and other variables of such a project. It is also believed that in order to treat the subject objectively we should approach it without prejudice. The following is a summary attempting to offer such treatment.

PROTECTION AGAINST CLIMATE

1. Weatherproofing: The subterranean structure is weatherproofed against extreme climate, such as solar radiation, extreme cold or heat, dusty winds and storms and tornadoes, and benefits both the structure and its inhabitants.
2. Temperature Fluctuation: The use of subterranean structures can minimize the impact of the daily temperature fluctuation and offer a constant temperature within the structure. This will introduce more comfortable conditions for human beings as well as for certain industries.
3. Moderated Climate: Underground dwellings offer a comfortable, moderated microclimate compared to the outdoor harsh macroclimate. This may apply to arid zones as well as to cold polar zones.

4. Survival: In a very cold climate the subterranean house protects residents in case of a breakdown in electricity for heating. In such a case, temperatures will remain tolerable for a human being.

On the other hand, some reservations should be noted here. It is possible that during the dust storms, which occur in arid climates, the openings of the subterranean structure may be buried, since the dryness, the frequent change in the temperature and the coverage make dust particles more abundant on the ground surface. Thus, even small turbulences can have some negative impact by polluting the house air, making life difficult for the housekeeper. However, preventive measures can be taken in the design to minimize or to eliminate such hazards. Extensive and intensive planting around the house and the settlement environs, minimization of open space and proper design of the house openings can provide effective solutions. The risk, of course, is increased in sandy soil, especially in settlements located close to shorelines. Of course, consideration of wind direction and site selection for the house and the settlement will make such a preventive approach more efficient.

LAND USE

1. Land Saving: It is evident that subterranean housing utilizes new space which conventional settlements have not used. This new plan, in our opinion, does not eliminate the traditional system; but rather it adds a new, innovative dimension to it. However, the saving of land is an economic as well as a social factor.
2. Social Interaction: The subterranean land use, when it is combined with the supraterranean, offers proximity which supports, in general, the intensification of social interaction. In the modern, traditional urban setting where social alienation is common, the implications for new land use which brings social proximity are most welcome.
3. Traffic: The new, combined land use supports to a large extent the old dream of planners, social scientists and others to minimize or eliminate the use of the motor vehicle, especially the private one, within the settlement. Consequently, social and physical health will improve tremendously.
4. Utilities: The new design will reduce the length of all types of utility networks and minimize land uses. It will save energy, planning, construction and maintenance.
5. Landscape: The introduction of the subterranean house will not greatly disturb the natural environment, and restoration of the landscape is possible.

It could be argued that this combined land-use settlement with strong emphasis on subterranean structures leads to compactness, and such compactness is usually associated with nuisances, like lack of privacy (and some cultures, such as the American, are sensitive to it), noise of collective and individual activities, and the feeling of lack of "visual" space. Although such phenomena are possible, it is the belief of this author that such disadvantages can be overcome or eased through a proper and careful design.

Provision of private dwelling-unit entrances, segregated common playgrounds for age groups, and segregated pedestrian networks can contribute to such improvement.

ENERGY COSTS

In general, the subterranean settlement will have a far-reaching, positive impact on energy savings in dry, arid regions as well as in nonarid ones.

1. Fuel: The use of subterranean structures for habitation and for working can reduce the need for air conditioning and heating by 80 percent or more and, in some cases, even eliminate such needs.
2. Heat Loss: Subterranean structures have minimal or no heat loss because of the massive soil enveloping the house. The conventional supraterranean house loses or gains heat through the building materials of the walls and ceilings, through the windows and doors, and any other openings.
3. Refrigeration Costs: Subterranean structures, because of the minimal temperature fluctuation, require less electricity for refrigeration. This applies to large storage refrigeration for commercial use and to that for housing needs.

It can be argued that subterranean housing will consume more electricity for lighting than the conventional aboveground house does. In general, it is possible to reduce such consumption through innovative devices or designs (such as those we discussed in Section II). An additional solution is the exposure of one side of the structure (in a sloping topography) to the optimal light direction with thick walls and large windows which can bring direct light to the inner part of the house. Also, a centered patio can bring light through partially or fully glassed walls facing the patio. Thus, natural lighting can be brought to the interior parts of the house when sun and light cycles are well studied.

CONSTRUCTION COSTS

This factor can be disputed and depends to a great extent on the features of the specific site. Our following statements are general in nature and rely primarily on previous research concerning costs for:

1. Land: Due to the multiple uses of the land, its cost will become minimal in relation to its many uses.
2. Design: The simplicity of the planning and design reduce the cost of the project investment.
3. Materials: There is less material required for the construction of the subterranean structure since such things as windows, exterior finishing, roofs and the like are used minimally.

It can be argued that the excavation cost is higher than in the conventional aboveground structure, and it is possible that this argument is valid when excavation is made in rocks or in large quantity. Here again, improvement of methods can be introduced to minimize costs under such special conditions. One possibility is the use of abandoned mines or quarries, or preplanning done for the dual use of the excavated land, with the ultimate goal being subterranean settlement—a concept which can improve the negative landscape of abandoned quarries. Also, the achievements in mine techniques and machinery can be improved and applied in excavation for subterranean settlements. One further consideration is that large-scale development reduces the per-unit investment for excavation.

An intensive geological survey will be required to study the potential sensitivity of the area to vibration and earthquakes and to discover geological faults, landslide and erosion potentialities and the like. There is also a need for comprehensive information on soil temperature at different depths and on soil dynamics. We can anticipate, again, that large-scale development reduces the price-impact of such surveys.

MAINTENANCE COSTS

1. Exterior Maintenance: For a subterranean house, there will be almost none of the exterior maintenance costs usually associated with the conventional aboveground structure such as windows, exterior walls, roofs, exterior painting, repairing and remodeling. The only exterior maintenance will be the outdoor landscaping which, obviously, becomes easier than that for the conventional house.
2. Durability: The minimal impact of the outdoor weather on the structure minimizes the deterioration of the structure and, therefore, prolongs its life and lowers amortization costs.
3. Housekeeping: There may be less housekeeping expenditures due to minimal dust and outdoor weather influence. It is this author's experience that conventional houses in arid zones require above-average housekeeping treatment because of the large quantities of dust coming into the house through the windows. Subterranean settlements will minimize this housekeeping burden and cost.
4. Insurance: Fire insurance of the structure is expected to be lower than for the conventional house.

Critics argue that subterranean structures will require high expenditures to maintain the utilities system. Here again, this cost may be reduced in settlements built on a sloping topography where pumping is minimized and the utility network can be tunneled as part of the excavation. The Utilidor system of conduits can be introduced here: all the networks are in tunnels allowing for ease of repair. It is imperative, especially in nonarid zones, that special and reliable insulation of the exterior walls be introduced to avoid dampness.

SAFETY

Subterranean settlements offer a comprehensively and uniquely safe living and working environment.

1. Fire: Fire expansion from one unit to another is minimized or even eliminated due to the unique features of the subterranean settlement. Consequently, the subterranean neighborhood is safer than the conventional one.
2. Earthquakes: Due to the secure enveloping of the structure by its massive surroundings, the structure will be less subject to collapse in an earthquake than the supraterranean one. This condition may be different, however, when a fault exists nearby. In general, areas subject to earthquakes can accommodate subterranean settlements and still be safe.
3. Tornadoes: Subterranean structures are safer in tornadoes than supraterranean ones. Subterranean settlements can, therefore, be a good solution for regions subject to such occasional hazards.
4. Nuclear Plants: Such plants are safer underground than aboveground.
5. Defense: In countries subject to potential hostilities, subterranean settlements will be safe against enemy attacks and the shelling of civilian areas. Such construction will eliminate the need for special underground shelters and will save special expenditures.
6. Radiation Protection: A subterranean home can provide a shield to protect inhabitants from radiation fallout, if designed to meet such needs.
7. Noise: The subterranean structure protects against outside noise or vibrations.
8. Freezing: In very cold climates, the subterranean house provides protection against pipes freezing in the low winter temperatures.

Subterranean housing can be subject to hazards such as flooding, especially when utilities are malfunctioning, or when settlements are located on or by geological faults. However, flooding hazards can be eliminated with careful design, particularly when structures are located on a sloping topography. Settlements located on flat ground are more likely to be subject to such hazards. Codes and regulations of construction should require more than one entrance and exit in every dwelling and working unit for easy evacuation of the inhabitants.

HEALTH

1. Comfort: In addition to the comfortable climate, subterranean housing provides an isolated and quiet environment due to the massive soil around the development. It, therefore, can be healthful. For instance, sleeping in subterranean structures is not interrupted by outdoor noise.
2. Creativity: The isolation of the dwelling facilitates (for certain age groups and for specific types) creativity. This applies to scholarly work, painting, writing and other artistic work.
3. Dampness: Subterranean structures are subject to dampness and associated odor and mildew, especially in nonarid zones. Avoidance of these problems will require a careful design of the drainage system and proper selection of insulating materials.
4. Pollution: The structure will be less subject to pollution, pollen, dust or inversion problems than the conventional house.

Since ventilation is essential in the subterranean structures, as we have discussed before, it should be carefully designed and properly controlled and more than one alternative should be available. In any case, an exchange of fresh air must be introduced. It is our belief that a careful design, such as one with patios or one combining under- and aboveground sections, will alleviate any problem with lack of sunshine. Also, watering landscaped areas must be carefully controlled, or preventive action against dampness will be needed.

Claustrophobia is the major argument often raised against subterranean dwellings. Such claustrophobia may exist; but it could be minimized or eliminated by proper education about the advantages of subterranean living and by good design: for example, dwellings with one side open to the outdoor view, especially if the dwellings are on a slope.

ENVIRONMENTAL IMPACT

We can conclude that, generally, subterranean settlements have a positive impact on the environment:

1. Ecosystem: Subterranean construction has a minimal impact on the ecosystem of the arid zone after construction is completed and the site has regained its equilibrium.
2. Nuisance: Such construction reduces outside nuisance, especially noise, because of soundproofing and minimizes or eliminates vibration caused by roads, vehicles or nearby airports. Residential areas built underground can be even closer to airports than now. However, strong vibrations from railroad systems and other heavy transportation forms may be transferred across large distances through the ground.
3. Wind: Subterranean construction minimizes the impact of the wind, especially important when such an impact is negative.

All in all, it seems to this author that subterranean settlements offer a variety of advantages; in addition, they solve some diversified problems which the supraterranean structures have not solved. Some of these problems are not exclusively arid-zone problems, but rather universal ones. Moreover, most of the limitations are resolvable. It is this author's belief that the major resistance is psychosocial in nature rather than technological or economical.

If we analyze the trends of the contemporary underground movement we can outline some final remarks about its future development:

1. Underground housing will be integrated in combination with aboveground housing. The major motivations will be its energy benefits and the improvement in the quality of life.
2. Environmental and economic pressure will accelerate the movement of a nonresidential underground usage, especially in industry, storage, offices, educational centers and schools, and libraries.
3. Countries with a shortage of space, with agricultural land vital to survival and limited in area—as in Israel, Japan, Belgium and Britain—will sooner or later discover the potentialities of underground devel-

opment, especially for nonresidential use. The highly sophisticated technology for excavating certainly makes this endeavor possible.

4. Underground space can be planned in conjunction with quarry development, especially near cities where quarries are planned, and cut for the future alternative use of underground space. This preplanned cooperation makes both projects economically feasible and worthwhile.

In the final analysis, the subterranean housing movement will not be evaluated by its quantity or its aesthetic value only, but rather by two important issues: its ability to solve or ease problems, especially the energy problems which are most pressing; and the quality of its design and construction so that existing uncertainty is eliminated and marketability is expanded.

The construction of subterranean structures in the city center introduces new variation and choice to the center which should be thoughtfully and comprehensively designed. This opportunity will give flexibility to the city center for expansion, meeting the pressing demands for space in addition to bringing new social and economic dimensions to the center.

Our concept of urban subterranean housing is meant to introduce the development of a system to cope with climatic stress. The overall urban environment, therefore, should be a mixture of conventional and subterranean structures. The combined system will benefit the subterranean system by reducing the loss or gain of temperature within the subterranean environment.

The city will be designed so that a large portion of it is subterranean, a development which can improve the urban environment, especially if compactness is introduced. This improvement applies to:

- Energy savings
- A decrease in transportation and, therefore, in pollution sources
- An increase in social interaction
- Time savings
- Safety for children and adults
- Saving of lengthy utility networks
- Land savings

As we have tried to show, the subterranean house introduces all of the advantages of its conventional aboveground counterpart and can be designed to reflect the culture and the personality of its owners.

In conclusion, our analysis shows that subterranean placement is most effective in regions with climatic extremes:

1. Very cold, such as those located in the northern United States or northern Europe; or cold, dry climates such as those found in central Asia and central Turkey;
2. Very warm, especially the dry and warm found in the arid zones of the world.

Subterranean placement can suit moderate climates but would not be as effective nor as necessary. It must also be acknowledged that going subterranean does not necessarily guarantee all the benefits introduced here. Only careful study, design and implementation will determine the effectiveness of such development.

Notes

1. For more details on methods and criteria of site selection of a settlement see: Gideon Golany, "Arid Zone Settlements Site Selection: The Case of Egypt." Paper presented at the 4th Annual Conference of the International New Towns Association, Cairo, Egypt, October 6–14, 1980. 36 pages.
———, "Selecting Sites for New Settlements in Arid Lands: Negev Case Study." Paper presented at the International Symposium on the Impact of Climate on Planning and Building, Hertzelia, Israel, November 4–7, 1980. 34 pages.
———, "A Quantitative Method for Site Selection of a New Town." Paper presented at the International Congress on New Towns: *How to Build a New Town*, Tehran, Iran, December 9–15, 1977. 69 pages.
———, "A Quantitative Method for Site Selection of a Regional Shopping Center," in *Proceedings* of the International Conference on Mathematical Modeling, St. Louis, Missouri, August 29–September 1, 1977. 25 pages.
———, "A Quantitative Method for Land-Use Planning," *International Technical Cooperation Center Review*, **6**/2 (22)(April 1977), pp. 11–37.
———, "Site Selection: Process, Criteria and Method," in *New-Town Planning: Principles and Practice*, G. Golany, ed. (New York: John Wiley and Sons, 1976), pp. 60–97.
———, *New Community for Virginia in the Roanoke Valley: Site Selection and Feasibility Study* (State College, Pennsylvania: Gideon Golany Associates, January 1972); idem, "Site Selection and Feasibility Study: A New Town for Roanoke Valley," in *Strategy for New Community Development in the United States*, G. Golany, ed. (Stroudsburg, Pennsylvania: Dowden, Hutchinson and Ross, 1975), pp. 129–154.
2. H. Yamahara, Y. Hisatami and T. Morie, "A Study on the Earthquake Safety of Rock Cavern," in *Storage in Excavated Rock Caverns, Rockstore* **77,** Vol. 2 (Oxford, England: Pergamon Press, 1978).
3. Kirby T. Mayer, "Utilities for Underground Structures," in *Alternatives in Energy Conservation: The Use of Earth Covered Buildings*, *Proceedings* of a Conference, Fort Worth, Texas, July 9–12, 1975 (Washington, D.C.: National Science Foundation, 1975), p. 165.
4. John E. Williams, "Comparative Life Cycle Costs," in *Alternatives in Energy Conservation: The Use of Earth Covered Buildings*, pp. 50–59.
5 Steven J. Foute and Douglas B. Cargo, "Earth Covered Housing: Hydrological and Pollution Considerations," in *Earth Covered Buildings and Settlements*, Vol. II (CONF-7805138-P2), Frank L. Moreland, ed. (Springfield, Virginia: National Technical Information Service, 1978), pp. 112–113.
6. Cultivation, tillage and irrigation modify the porosity of the soil and its profile, and this in turn affects its thermal properties. In loess, as in many other soils, moisture and thermal behavior will vary with depth. See Marcel Fuchs and Amos Hadas, "The Heat Flux Density in a Non-Homogeneous Bare Loessial Soil," *Boundary Layer Meterology*, **3**/2 (1972), 191–200, for specific information on loess behavior.
7. A soil profile of the northern Negev examined at the Gilat Experimental Farm reveals a texture homogeneous to a depth of about 2 m of 18% clay, 27% silt, and 55% fine sand. See Fuchs and Hadas, p. 192.
8. Robert F. Legget, *Cities and Geology* (New York: McGraw-Hill, 1973), p. 469.
9. Douglas S. Way, *Terrain Analysis: A Guide to Site Selection Using Aerial Photographic*

Interpretation (Stroudsburg, Pa.: Dowden Hutchinson and Ross, Inc., 1973); and Daniel Hillel, *Studies on Loessial Crusts* (Beit-Dagon Israel: State of Israel, Agricultural Research Station, Division of Publications, 1959).

10. Ellis L. Armstrong, "Underground Space as a Resource," in *Underground Utilization: A Reference Manual of Selected Works*, 8 vols., Truman Stauffer, ed., Vol. I: *Historical Perspective* (Kansas City: University of Missouri, Department of Geosciences, 1978), p. 29.
11. Ibid.
12. Walter A. Lyon, "Innovations in Environmental Control," in *Innovations for Future Cities*. G. Golany, ed. (New York: Praeger Publishers, 1976), pp. 170–174.
13. Armstrong, p. 27.
14. "Going Underground," in *Underground Utilization: A Reference Manual of Selected Works*, Vol. I, pp. 50–65; Douglas N. Carter, "Community and Building Official Reaction to Earth-Covered Buildings: A Case Study, Terraset Elementary School, Reston, Virginia," in *Earth Covered Buildings and Settlements*, p. 81.
15. Lloyd S. Jones, "Thinking Down Through the Earth," in *Underground Space Utilization*, Vol. I, p. 46.

Selected Bibliography*

*Articles from *Alternatives in Energy Conservation* (Frank Moreland, ed.), *Earth Covered Buildings and Settlements* (Frank Moreland, ed.), *Earth Covered Buildings: Technical Notes* (Moreland, Higgs and Shih, eds.), and *Underground Utilization: A Reference Manual of Selected Works* (Truman Stauffer, Sr., ed.), 8 vols. are cited in brief. Complete data on those volumes are included in their listings under "General and Basic Information." Journal sources are abbreviated per *Chemical Abstracts List of Periodicals* or *Art Index* with some journals cited in full.

General and Basic Information

1. Armstrong, Ellis L. Underground space as a resource. In *Underground Utilization,* Truman Stauffer, Sr. (ed.). Vol. 1, pp. 27–30.
2. Aughenbaugh, N. B. Development of underground space. In *Earth Covered Buildings and Settlements,* Frank Moreland (ed.), pp. 152–168.
3. Bacon, Vinton W. Overview of underground construction in congested areas. In *Underground Utilization,* Truman Stauffer, Sr. (ed.). Vol. 2, pp. 176–179.
4. Birkerts, Gunnar. Subterranean systems. *Archit. For.* **135**/4: 58–59 (1971).
5. Bligh, Thomas. *Building underground. Build. Syst. Des.*: 1–22 (October/November 1976).
6. Brown, G. Z. and Novitski, B.-J. Climate responsive earth-sheltered buildings. *Underground Space* **5**/5: 299–305 (1981).
7. Campbell, Stu. *The Underground House Book.* Charlotte, Vt.: Garden Way Publishing, 1980.
8. Coogan, Alan H. Classification and valuation of subsurface space. *Underground Space* **3**/4: 175–186 (1979).
9. Costello, Michael J. Reclaiming minelands for earth-sheltered housing. *Underground Space* **5**/5: 279–286 (1981).
10. Dempewolff, Richard F. Underground housing. *Science Digest* **78**/5: 40–53 (1975).
11. DeSaventhem, E. M. Insuring risks underground—Some general considerations. *Underground Space* **2**/1: 19–25 (1977).
12. DiMatteo, L. P. and Stewart, M. W. Prepare to go 100% underground. *Am. City* **82:** 62, 64, 66 (1967).
13. Fairhurst, Charles. Going under to stay on top. *Underground Space* **1**/2: 71–86 (1976). Also in *Underground Utilization,* Truman Stauffer, Sr. (ed.). Vol. 7, pp. 942–954.
14. ——— et al. *Potential Use of Underground Space.* Minneapolis: Department of Civil and Mineral Engineering, University of Minnesota, 1975.
15. Garbutt, P. E. *How the Underground Works.* London: London Transport, 1963.
16. Green, Bruce. Basic principles of underground house construction. In *Underground Utilization,* Truman Stauffer, Sr. (ed.), Vol. 4, pp. 608–613.
17. Hagman, Donald G. Planning the underground uses. In *Legal, Economic, and Energy Considerations in the Use of Underground Space.* Washington, D.C.: National Academy of Sciences, 1974, pp. 52–67.
18. ———. Planning the underground uses. In *Underground Utilization,* Truman Stauffer, Sr. (ed.). Vol. 6, pp. 845–851.
19. Harrison, Lloyd. Should the city be interested in an underground subdivision? In *Underground Utilization,* Truman Stauffer, Sr. (ed.). Vol. 6, pp. 860–862.
20. Hollister, Hugh. The development and operation of underground space. In *Underground Utilization,* Truman Stauffer, Sr. (ed.). Vol. 1, pp. 67–69.
21. Horsbrugh, Patrick. Geospace: The concept and definition of subterranean accommodation. In *Underground Utilization,* Truman Stauffer, Sr. (ed.). Vol. 2, pp. 155–160.
22. ———. *Geotecture, Concept, Design, Construction and Economy of Geospace—The Creation of Subterranean Accommodation.* Minneapolis: Department of Civil and Mineral Engineering, University of Minnesota, 1973.
23. ———. Urban geotecture: The invisible features of the civic profile. In *Alternatives in Energy Conservation,* Frank Moreland (ed.), pp. 151–164.

24. Jansson, Birger. Terraspace—A world to explore. *Underground Space* **1**/11: 9–18 (1976).
25. ———. The urban development and the demand on the underground during the year 2000. Washington, D.C.: Paper presented at the Annual Meeting of the Geological Society of America, 1971, 10 pp.
26. ——— and Wingvist, Torbjörn. *Planning of Subsurface Use*. Oxford, England: Pergamon Press, 1977.
27. Kommendant, August E. Earth covered structures. In *Earth Covered Buildings: Technical Notes,* Moreland, Higgs & Shih (eds.), pp. 1–12.
28. Labs, Kenneth B. Performance of earth covered development: Planning issues. In *Earth Covered Buildings and Settlements,* Frank Moreland (ed.), pp. 125–140.
29. Mason, Roy. Projections on the future of underground development. In *Earth Covered Buildings and Settlements,* Frank Moreland (ed.), pp. 169–173.
30. Minnesota. University. Department of Civil and Mineral Engineering. *Potential Use of Underground Space*. Minneapolis, February 1975.
31. Minnesota. University. Underground Space Center. *Earth Sheltered Housing Design: Guidelines, Examples and References*. Minneapolis, 1978. Also published in New York: Van Nostrand Reinhold, 1979.
32. Moreland, Frank L. Earth covered habitat—An alternative future. *Underground Space* **1**/4: 295–307 (1977).
33. ——— (ed.). *Alternatives in Energy Conservation: The Use of Earth-Covered Buildings* (NSF/RA-760006). Washington, D.C.: National Science Foundation, 1975.
34. ———. *Earth Covered Buildings and Settlements*. (CONF-7805138-P2). Vol. 2. Springfield, Va: National Technical Information Service, 1979.
35. ———. Higgs, F. and Shih, J. (eds.). *Earth Covered Buildings: Technical Notes*. (CONF-78005138-P1). Vol. 1. Springfield, Va: National Technical Information Service, 1979.
36. Randall, (Honorable) Wm. J. Space age cave dweller. *Congressional Record—Appendix,* Vol. 109, part 10, July 18, 1963: A4569–A4571.
37. Reyner, J. F. Analysis of Several Surveys Relative to Problems of Shelter Habitability. Washington, D.C.: National Academy of Sciences, January 1960.
38. Riechers, Maggie. Federal activities in earth-sheltered and underground construction. *Underground Space* **5**/5: 275–278 (1981).
39. Roberts, A. *Applied Geotechnology: A Text for Students & Engineers on Rock Excavation and Related Subjects*. Elmsford, N.Y. Pergamon Press, 1981.
40. Roy, Robert L. How to build an underground home. *Farmstead,* March 1979.
41. ———. A log-end cave. *The Mother Earth News* **67:** 110–113 (1981).
42. Scalise, James W. (ed.). *Earth Integrated Architecture: An Alternative Method for Creating Livable Environments with an Emphasis on Arid Regions*. Tempe, Ariz.: Architecture Foundation, College of Architecture, Arizona State University, 1975.
43. ———. *Terratecture: The Environmental Benefits of Earth Integrating Architectural Design Techniques*. Tempe, Ariz.: College of Architecture, Arizona State University, May 1977.
44. Seeley, Barrett (ed.). *Earth-sheltered Housing: The Comprehensive Bibliography*. Washington: Eco-Terra, March 1981.
45. Stauffer, Truman, Sr. (ed.). *Underground Utilization: A Reference Manual of Selected Works*. 8 Vols. Vol. 1: *Historical Perspective;* Vol. 2: *Uses for Underground Space;* Vol. 3: *Space Construction Underground;* Vol. 4: *Human Response and Social Acceptance of Underground Space;* Vol. 5: *Advantages in Underground Space Use;* Vol. 6: *Regulations and Policy in the Use of Underground Space;* Vol. 7: *The Future of Underground Development;* and Vol. 8: *Index*.
46. van der Meer, Wybe J. Underground and earth covered housing deserves consideration. In *Proceedings IAHS Symposium,* Clemson, S. C., May 1976. Vol. 2, Parviz Rad et *al.* (eds.). Coral Gables, Fla.: International Association for Housing Science, 1976, pp. 1137–1150.

Section I: Historical Lessons

Tunisia:

47. Haan, Herman. Matmata. In *Architects Yearbook,* Vol. 11, David Lewis (ed.). London: Elek Books, 1965, pp. 126–128.
48. Hallet, Stanley. Mountain villages of southern Tunisia. *Journal of Architectural Education* **29**/2: 22–25 (1975).
49. Williams, Maynard O. Time's footprint in Tunisian sands. *National Geographic* **71**/3: 344–386 (1937).

China:

50. Boyd, Andrew C. H. *Chinese Architecture and Town Planning 1500 B.C.—1911 A.D.* Chicago, Ill.: University of Chicago Press, 1962.
51. Cressey, George B. *Land of the 500 Million.* New York: McGraw-Hill, 1955.
52. Fuller, Myron L. and Clapp, Fredrick G. Loess and rock dwellings at Shensi. *Geog. Re.* **14:** 215–226 (1924).
53. Wu, Nelson I. *Chinese and Indian Architecture.* New York: George Braziller, 1963.

Cappadocia:

54. Andolfato, U. and Zucchi, F. The physical setting. In *Arts of Cappadocia,* L. Giovannini (ed.). Geneva: Nagel Publishers, 1971, pp. 51–65.
55. Erguvanli, A. K. and Yüzer, A. E. Past and present use of under-ground openings excavated in volcanic tuffs at Cappadocia area. In *Storage in Excavated Rock Caverns, Rockstore 77,* Magnus Bergman (ed.). Vol. 1. Oxford, England: Pergamon Press, 1978, pp. 31–36.
56. Fahrner, Rudolf. *The Face of Anatolia, Caves and Khans in Cappadocia.* Vienna: Austrian National Printing Press, 1955.
57. Giovannini, Luciano. The rock settlements. In *Arts of Cappadocia,* Luciano Giovannini (ed.). Geneva: Nagel Publishers, 1971, pp. 67–83.
58. ——— (ed.). *Arts of Cappadocia.* Translated from French; produced in collaboration with the Instituto Internazionale di Arte Liturgica, Rome, Chicago. London: Barrie & Jenkins, 1971; Geneva: Nagel Publishers, 1971.
59. Hazer, Fahriye. Cultural-ecological interpretation of the historic underground cities of Göreme, Turkey. In *Alternatives in Energy Conservation,* Frank Moreland (ed.), pp. 21–36.
60. Kostof, Spiro K. *Caves of God: The Monastic Environment of Byzantine Cappadocia.* Cambridge: MIT Press, 1972.

United States:

61. Bennett, David. University of Minnesota Bookstore. In *Alternatives in Energy Conservation,* Frank Moreland (ed.). pp. 117–130.
62. Boyer, L. L. (ed.). *An Earth Sheltered Guest House.* Stillwater, Okla.: Oklahoma State University, 1979.
63. ———, Grondzik, W. T., and Bice, T. N. Energy usage in earth covered dwellings in Oklahoma. *Proceedings: Earth Sheltered Building Design Innovations,* Lester L. Boyer (ed.). Stillwater, Okla.: Oklahoma State University, April 1980, pp. IV. 17-IV. 31.
64. ———, Grondzik, W. T., and Weber, M. J. Passive energy design and habitability aspects of earth-sheltered housing in Oklahoma. *Underground Space* **4**/6: 333–339 (1980). Also published in *PreConference Proceedings of the Solar Heating & Cooling Systems Operational Results.* Department of Energy, Colorado Springs, Colorado, November 27–30, 1979, pp. 183–189.
65. ———, Weber, M. J., and Grondzik, W. T. Energy and habitability aspects of earth-sheltered housing in Oklahoma. Project Report, Presidential Challenge Grant. Stillwater, Okla.: Oklahoma State University, March 1980.

66. Citation: Campus store is put underground to preserve trees and sloping contours of the site (Cornell University, Ithaca, NY). *Progres. Arch.* **50**/1: 108–109.
67. Dean, Lester. How underground space use started in the Kansas City area. In *Underground Utilization,* Truman Stauffer, Sr. (ed.). Vol. 1, pp. 65–66.
68. Exhibition Hall Goes Underground. *Eng. News-Rec.* **174:** 87–88 (1965).
69. Goleman, Harry. University of Houston Student Center. In *Alternatives in Energy Conservation,* Frank Moreland (ed.), pp. 131–134.
70. Hall, Charles L. Architecture of the Anasazi Pueblo culture. *New Mexico Architecture,* May–June 1967.
71. Hiding out in Harvard Yard, the New Nathan Marsh Pusey Library. *Int. Des.* **47:** 144–147 (1976).
72. Largest earth shelter in San Francisco. *Earth Shelter Digest and Energy Report,* No. 9: 37 (May/June 1980).
73. MacCurdy, George G. American caves and cave dwellers. *Am. J. Archaeol.* **41**/3: 383–387 (1937).
74. Machowski, Barb. Earth shelters enrich American history. *Earth Shelter Digest and Energy Report,* No. 11: 48–51 (Sept/Oct 1980).
75. Martin, Paul and Plog, Fred. *The Archaeology of Arizona.* New York: Doubleday/Natural History Press, 1973.
76. Morgan, William. Buildings as landscape: Five current projects by William Morgan. *Arch. Rec.* **152**/3: 129–136 (1972).
77. ———. Molding our man-made world. *The Florida Architect,* May/June 1975.
78. Sanoff, Henry. Seven acres of underground shelter. *AIAJ* **47**/2: 67–68 (1967).
79. Saving by going underground. Ecology House—Architect, John Barnard, Jr. *AIAJ* **61**/2: 48–50 (1974).
80. Scaltro, Louis P. Subsurface design and construction techniques in Manhattan. In *Underground Utilization,* Truman Stauffer, Sr. (ed.). Vol. 3, pp. 383–387.
81. Schmertz, Mildred F. In deference to its environment the Pusey Library was built beneath Harvard Yard. *Arch. Rec.* **160:** 97–102 (1976).
82. Smith, Watson. *Prehistoric Kivas of Antelope Mesa, Northeastern Arizona.* Cambridge, Mass.: Peabody Museum, Harvard University, 1972.
83. Welsch, R. L. The Nebraska Soddy. *Nebraska History* **48**/4 335–342 (1967).
84. ———. *Sod Walls; The Story of the Nebraska Sod House.* Broken Bow, Neb: Purcells, 1968.
85. Your society to seek new light on the cliff dwellers. *National Geographic,* January 1959, pp. 154–156.

Australia:

86. Baggs, Sydney A. The dugout dwellings of an outback opal mining town in Australia. In *Underground Utilization,* Truman Stauffer, Sr. (ed.). Vol. 4, pp. 573–599. Also published in Mimeo, Sydney: University of New South Wales, Department of Landscape Architecture, 1977.
87. ———. The lithotecture of Australia: With specific reference to user health factors. In *Proceedings Earth Sheltered Building Design Innovations,* Lester L. Boyer (ed.). Stillwater, Okla.: Oklahoma State University, April 1980, pp. 11.19–11.26.
88. ———. Underground architecture. *Architecture Australia:* 62–69 (Dec. 1977/Jan. 1978).
89. Earth-sheltered design wins award in Australia. *Underground Space* **5**/2: 70 (1980).
90. Lang, Thomas A. Underground experience in the Snowy Mountains—Australia. In *Protective Construction In A Nuclear Age.* J. J. O'Sullivan (ed.). Proceedings of the Second Protective Construction Symposium, Santa Monica, Calif., 1959. Vol. 2. New York: The Macmillan Company, 1961, pp. 688–765.

Japan:

91. Brown, R. L. Japan: Under land and under water. *Building* **226:** 87–88 (1974).
92. Japan, underground movement. *The Economist* **221**/6435: 1323 (1966).
93. Japanese firm markets portable "pod houses" as disaster shelters. *Underground Space* **4**/5: 320 (1980).

94. Fujita, K. *et al.* An empirical proposal on stability of rock cavern wall during construction in Japan. In *Storage in Excavated Rock Caverns, Rockstore* **77,** Magnus Bergman (ed.). Vol. 2. Oxford, England: Pergamon Press, 1978, pp. 309–314.

Canada:

95. Legget, Robert F. 75-year old underground compressed air plant still in use in Canada. *Underground Space* **4**/1: 29–31 (1979).
96. Ponte, Vincent. Montreal's multi-level city center. *Traffic Engineering* **41**/12: 20–25ff (1971).
97. ———. Montreal high on its underground. *Kansas City Star,* December 19, 1976.

Military:

98. Blaschke, Theodore O. Norad's underground command center. *Civ. Eng.* **34**/5: 36–39 (1964).
99. Jones, Lloyd S. Non-traditional military uses of underground space. *Underground Space* **2**/3: 153–158 (1978).
100. Klass, Philip J. Norad Operations Center. *Aviat. Week Space Technol.* **82:** Parts I and II, February 1, 8, 1965, pp. 66–74 and 65–68.
101. Lally, Elaine K. The strange language of Cheyenne Mountain: The Norad underground. In *Underground Utilization,* Truman Stauffer, Sr. (ed.). Vol. 7, pp. 965–974.
102. Margison, Arthur D. Canadian underground Norad economically achieved. *Underground Space* **2**/1: 9–17 (1977).
103. Olson, J. J., *et al. Ground Vibrations from Tunnel Blasting in Granite: Cheyenne Mountain (NORAD), Colorado.* Minneapolis, Minn.: Bureau of Mines RI 7653, Twin Cities Mining Research Center, 1972.

Other Countries:

104. Bergman, Sten G. A. Underground construction in Sweden. Stockholm: Swedish Underground Construction Mission, 1976.
105. Clausen, O. Sweden goes underground. *New York Times Magazine,* May 22, 1966: 23–25.
106. Crawford, Harriet (ed.). *Subterranean Britain—Aspects of Underground Archeology.* New York: St. Martin's Press, 1980.
107. Duffaut, Pierre. Ground pressure and tunnelling from the nineteenth century to the present. *Underground Space* **1**/3: 185–200 (1977).
108. ———. Past and future of the use of underground space in France and Europe. *Underground Space* **5**/2: 86–91 (1980).
109. Fitch, James M. and Branch, D. Primitive architecture and climate. *Sci. Am.* **203**/6: 134–144 (1960).
110. Follenfant, H. G. *Reconstructing London's Underground.* London: London Transport, 1974.
111. Invisible architecture in the Paris underground; UNESCO three story underground buildings. *Fortune* **73**/1: 176 (1966).
112. Kovač, Milan. Subterranean museum for Pharaoh's funeral vessel at Giza, Egypt. *Underground Space* **3**/6: 303–308 (1979).
113. Labs, Kenneth. Terratecture: The underground design movement of the 1970s. *Landscape Arch.* **67:** 244–249 (1977).
114. ———. The use of earth covered buildings through history. In *Alternatives in Energy Conservation,* Frank Moreland (ed.), pp. 7–19.
115. LaNier, Royce. *Geotecture, Subterranean Accommodation and the Architectural Potential of Earthworks.* South Bend, Ind.: Royce LaNier, 1970.
116. Mozayeni, Manootchehr. In praise of indigenous man-made environments: Massuleh in Iran. *Ekistics* **45**/271: 304–307 (1978).
117. O'Reilly, M. P. Some examples of underground development in Europe. *Underground Space* **2**/3: 163–178 (1978).

118. Owen, Christopher R. L. Lalibala: The rugged highlands of Ethiopia hold a cache of extraordinary rock-hewn churches. *Architecture Plus* **2**/6: 44–49 (1974).
119. Prussin, Labelle. *Architecture in Northern Ghana.* Berkeley, Calif.: University of California Press, 1969.
120. Rudofsky, Bernard. *The Prodigious Builders.* New York: Harcourt Brace Jovanovich, 1977.
121. ———. Troglodytes. *Horizon* **9**/2: 39 (1967).
122. Russell, Dick. K. C. cave men. *Topeka Capital-Journal,* February 18, 1973.
123. The Secrets of Spirit Cave. *Time,* February 8, 1970.
124. Wolfe, Nancy H. *The Valley of Bamiyan, Kabul.* Kabul, Afghanistan: Afghanistan Tourist Organization, 1963.
125. Woodcock, George. Cave temples of Western India. *Arts Magazine* **36**/8–9: 74–81 (1962).

Section II: Subterranean House Design

Psychology:

126. Bechtel, Robert B. Psychological aspects of earth covered buildings. In *Earth Covered Buildings and Settlements,* Frank Moreland (ed.), pp. 71–77.
127. Collins, Belinda L. Review of the psychological reaction to windows. In *Underground Utilization,* Truman Stauffer, Sr. (ed.). Vol. 4, pp. 532–540.
128. Paulus, Paul B. On the psychology of earth covered buildings. In *Alternatives in Energy Conservation,* Frank Moreland (ed.), pp. 65–69. Also in *Underground Space* **1**/2: 127–130 (1976).
129. Seybert, Jeffrey A. Psychological factors and future expansion of underground space utilization. In *Underground Utilization,* Truman Stauffer, Sr. (ed.). Vol. 7, pp. 1018–1019.
130. Wunderlich, Elizabeth. Psychology and underground development. In *Underground Utilization,* Truman Stauffer, Sr. (ed.). Vol. 4, pp. 526–529.

Design:

131. Anderson, Bruce N. with Michael Riordan. *The Solar Home Book—Heating, Cooling and Designing with the Sun.* Harrisville, N.H.: Cheshire Books, 1976.
132. Bligh, Thomas P. Building underground. *Build. Syst. Des.,* October/November 1976.
133. Boyer, Lester L. (ed.). *Proceedings: Earth Sheltered Building Design Innovations.* Stillwater, Okla.: Oklahoma State University, 1980.
134. Duke, Buford W. Jr. Incorporating earth shelter considerations into the design process for nonresidential buildings. In *Proceedings: Earth Sheltered Building Design Innovations,* L. L. Boyer (ed.). Stillwater, Okla.: Oklahoma State University, 1980, pp. V.23–V.30.
135. Gorman, James. Underground architecture. In *Underground Utilization,* Truman Stauffer, Sr. (ed.). Vol. 3, pp. 329–331.
136. Labs, Kenneth. The architectural underground, Parts 1 and 2. *Underground Space* **1**/1 and **1**/2: 1–8 and 135–156 (1976).
137. ———. The architectural use of underground space: Issues and limitations. 147 Livingston St., New Haven, Conn. 06511: Kenneth Labs, 1975. Also Master's Thesis, Washington University, St. Louis, Mo., 1975.
138. Langley, John B. The barrel shell—structural rethinking in earth-sheltered design. *Underground Space* **5**/2: 92–101 (1980).
139. Metz, Don. The latest-not the last-word in underground house design. *Solar Age* **4**/10: 36–39 (1979).
140. Newman, Jerry and Godbey, Luther C. *Design Considerations for Below Grade-Level Houses.* St. Joseph, Michigan: American Society of Agricultural Engineers, 1978.

141. Newmark, N. M. *et al.* Analysis and Design of Flexible Underground Structures. Corps of Engineers, Contract no. DA-22-079-ENG-225, October 31, 1962.
142. Pennsylvania State University. Shelter Research and Study Program. *Planning, Analysis, and Design of Shelters.* 5 Vols. CDM-OS-60-54. State College, Pa.: Pennsylvania State University, 1961.
143. Powell, Evelyn J. Design considerations for the elderly and handicapped. In *Earth Covered Buildings and Settlements,* Frank Moreland (ed.), pp. 149–151.
144. Shih, Jason C. The optimization of earth-covered shell structures. In *Earth Covered Buildings: Technical Notes,* Moreland, Higgs and Shih (eds.), pp. 92–114.
145. ———. Simplified design procedure of underground shells. *Underground Space* **4**/1: 45–49 (1979).
146. ——— and Kumar, S. Optimization of Subsurface and Underground Shells. U.S. Army Research Office, April 1970.
147. University of Minnesota. Department of Civil and Mineral Engineering. *Preliminary Design Information for Underground Space.* Minneapolis, Minn.: University of Minnesota, Department of Civil and Mineral Engineering, August 1975.
148. Wells, Malcolm B. *Underground Designs.* Brewster, Ma: Malcolm Wells, 1977.
149. ———. Down under, down under, . . . or how not to build underground. *Progres. Arch.* **49**/5: 164–165 (1968).
150. ———. Gentle architecture catches on. *Earth Shelter Digest and Energy Report,* No. 9: 18–19 (May/June 1980).
151. ———. Underground architecture. *The CoEvolution Quarterly,* No. 11 Fall 1976.
152. ———. Why I went underground. *The Futurist:* 21–24 (February 1976).

Soil Property:

153. Algoode, J. R. Structures in soil under high loads. *Proc. Am. Soc. Civ. Eng.* **97:** 565–579 (1971).
154. Baver, Leonard D. *Soil Physics.* 4th edition. New York: John Wiley & Sons, 1972.
155. Bergman, S. Magnus. Geo-planning-the Key to successful underground construction. *Underground Space* **2**/1: 1–7 (1977).
156. Bodonyi, J. Engineering geology related to underground openings. *UNESCO International Post-Graduate Course on the Principles and Methods of Engineering Geology.* Budapest, 1975.
157. Druker, E. F. and Haines, J. T. A study of thermal environment in underground survival shelters using an electronic analog computer. *ASHRAE Trans.* **70:** 7–20 (1964).
158. Flathau, W. J. and Balsara, J. P. Soil-structure interaction—An overview. In *Earth Covered Buildings: Technical Notes,* Moreland, Higgs & Shih (eds.), pp. 13–33.
159. Jumikis, Alfreds R. *Introduction to Soil Mechanics.* Princeton, N.J.: D. Van Nostrand Co., 1967.
160. Lytton, Robert L. Soil and ground water considerations. In *Alternatives in Energy Conservation,* Frank Moreland (ed.), pp. 257–262.

Soil Temperature:

161. Ashbel, D. *et al. Soil Temperature in Different Latitudes and Different Climates.* Jerusalem, Israel: The Hebrew University, 1965.
162. Blick, Edward F. A simple method for determining heat flow through earth covered roofs. In *Proceedings Earth Sheltered Building Design Innovations,* L. L. Boyer (ed.). Stillwater, Okla.: Oklahoma State University, 1980, pp. III.17–III.23.
163. Bligh, Thomas P. Thermal energy storage in large underground systems and buildings. In *Storage in Excavated Rock Caverns, Rockstore* **77,** Magnus Bergman (ed.). Vol. 1. Oxford, England: Pergamon Press, 1978, pp. 63–71.
164. Bliss, Donald E. *et al.* Air and soil temperatures in a California date garden. *Soil Sci.* **53:** 55–64 (1942).
165. Boileau, G. G. and Latta, J. K. Calculation of basement heat losses. In *Underground Utilization,* Truman Stauffer, Sr. (ed.), Vol. 5, pp. 754–764. Also Ottawa: Technical Paper No. 292 of the Division of Building Research, National Research Council, Canada (December 1968), NRC 6477.

166. Carson, James E. Analysis of soil and air temperatures by Fourier techniques. *J. Geophys. Res.* **68**/8: 2217–2232 (1963).
167. ———. *Soil Temperature and Weather Conditions* (ANL-6470). Lemont, Ill: Argonne National Laboratory, 1961.
168. Crabb, G. A., Jr. and Smith, J. L. Soil-temperatures comparisons under varying covers. In *Soil Temperature & Ground Freezing,* Highw. Res. Board. Bull. 71. Washington, D.C.: National Academy of Sciences, National Research Council, 1953, pp. 32–80.
169. Davies, G. R. Thermal analysis of earth covered buildings. In *Proceedings of the 4th National Passive Solar Conference,* G. Franta (ed.). Newark, Del.: American Section of the International Solar Energy Society, 1979, p. 744.
170. Davis, William B. Earth temperature: Its effect on underground residences. In *Earth Covered Buildings: Technical Notes,* Moreland, Higgs & Shih (eds.), pp. 205–209.
171. de Vries, D. A. Thermal properties of soils. In *Physics of Plant Environment,* Willem R. van Wijk (ed.). New York: John Wiley & Sons, 1963, pp. 210–235.
172. Drapeau, F. J. *Mathematical Analysis of Temperature Rise in the Heat Condition Region of an Underground Shelter.* R8663. Washington, D.C.: National Bureau of Standards, U.S. Department of Commerce, April 1965.
173. Elliot, J. M. and Baker, M. Heat loss from a heated basement (Paper No. 1724). *ASHVE Trans.* **66:** 400–413 (1960).
174. Fluker, B. J. Soil temperatures. *Soil Sci.* **86:** 35–46 (1958).
175. Fuchs, Marcel and Hadas, Amos. The heat flux density in a non-homogeneous bare loessial soil. *Boundary Layer Meteorology* **3**/2: 191–200 (1972).
176. Givoni, Baruch. Modifying the ambient temperature of underground buildings. In *Earth Covered Buildings: Technical Notes,* Moreland, Higgs & Shih (eds.), pp. 123–138.
177. Gold, L. W. Influence of Snow Cover on Heat Flow from the Ground (Research Paper No. 63). Ottawa: National Research Council of Canada, Division of Building Research, 1958.
178. Jumikis, A. R. *Thermal Soil Mechanics.* New Brunswick, N.J.: Rutgers University Press, 1966.
179. Kusuda, T. *Earth Temperature beneath Five Different Surfaces.* Report 10373. Washington, D.C.: U.S. Dept. of Commerce National Bureau of Standards, February 1971.
180. ———. The effect of ground cover on earth temperature. In *Alternatives in Energy Conservation,* Frank Moreland (ed.), pp. 279–303.
181. ———. Least squares technique for the analysis of periodic temperature of the earth's surface region. *J. Res. Natl. Bur. Stand.* **71C**/1 43–50 (1967).
182. ——— and Achenbach, P. R. *Comparison of Digital Computer Simulations of Thermal Environment in Occupied Underground Protective Structures with Observed Conditions.* Report 9473. Washington, D.C.: National Bureau of Standards, December 1966.
183. ———, and Achenbach, P. R. Numerical analyses of the thermal environment of occupied underground spaces with finite cover using a digital computer. *ASHRAE Trans.* **69:** 439–452 (1963).
184. ——— and Powell, F. J. *Heat Transfer Analysis of Underground Heat Distribution Systems.* Report 10-194. Washington, D.C.: National Bureau of Standards, April 9, 1970.
185. Labs, Kenneth. The underground advantage: Climate of soils. In *Proceedings of the Fourth National Passive Solar Conference.* Newark, Del.: American Section of the International Solar Energy Society, 1979.
186. Langbein, W. B. Computing soil temperatures. *Trans. Am. Geophys. Union* **30**/4: 543–547 (1949).
187. Latta, J. K. and Boileau, G. G. *Calculation of Basement Heat Loss,* NRC 10477. Ottawa: National Research Council of Canada, Division of Building Research, December 1968.
188. Lettau, H. H. A theoretical model of thermal diffusion in non-homogeneous Conductors. *Gerlands Beitr. Geophys.* **71:** 257–271 (1962).
189. Maxwell, Robert K. Temperature measurements and the calculated heat flux in the soil. Minneapolis, Minn.: University of Minnesota, 1964. Thesis, Master of Science.
190. McBride, M. F. *et al.* Measurement of subgrade temperatures for prediction of heat loss in basements. *ASHRAE Trans.* **85,** Part 1, 1979.

191. Monitoring system defined. *Earth Shelter Digest and Energy Report* No. 9: 34–35 (May/June 1980).
192. Penrod, E. B. Variation of Soil Temperature at Lexington, Kentucky from 1952–1956. Bulletin No. 57. Lexington: University of Kentucky Engineering Experiment Station, September 1960.
193. Peters, A. *Soil Thermal Properties: An Annotated Bibliography*. Philadelphia, Pa.: The Franklin Institute, 1962.
194. Pollack, H. A. and Chapman, D. S. The flow of heat from the earth's interior. *Sci. Am.* **237**/2: 60–76 (1977).
195. Shipp, P. H. Thermal characteristics of large earth sheltered structures. Minneapolis: University of Minnesota, 1979. Ph.D.Thesis.
196. ———, Meixel, G. D., and Ramsey, J. W. Analysis and measurement of the thermal behavior of the walls and surrounding soil for a large underground building. *Underground Space* **5**/2: 121–125 (1980).
197. Singer, Irving A. and Brown, R. M. The annual variations of sub-soil temperatures about a 600-foot circle. *Trans. Am. Geophys. Union* **37**/6: 743–748 (1956).
198. Smith, A. Diurnal, average and seasonal soil temperature changes at Davis, California. *Soil Sci.* **28:** 457–468 (1929).
199. Smith, W. O. The thermal conductivity of dry soils. *Soil Sci.* **53:** 435–459 (1942).
200. Survival shelters. In *ASHRAE Guide and Data Book Applications 1971*. New York: American Society of Heating, Refrigerating and Air-Conditioning Engineers, 1971, pp. 177–202.
201. U.S. Department of Commerce, Weather Bureau. *History of Soil Temperature Stations in the U.S.* Documentation No. 1.4. Washington, D.C.: U.S. Department of Commerce, U.S. Weather Bureau, 1961.
202. Van Straaten, J. F. *Thermal Performance of Buildings*. Amsterdam: Elsevier Publishing Co., 1967.
203. Willi, W. O. Bibliography on Soil Temperature. Mandan, N. D.: U.S. Department of Agriculture, Agricultural Research, Northern Great Plains Field Station, 1964.

Structure:

204. Alterman, I. Stresses against underground structure cylinders. *Proc. Am. Soc. Civ. Eng.* **95**/SM3: 916–919 (1969).
205. Bello, Arturo A. Simplified method for stability analysis of underground openings. In *Storage in Excavated Rock Caverns, Rockstore* **77,** Magnus Bergman (ed.). Vol. 2. Oxford, England: Pergamon Press, 1978, pp. 289–294.
206. Bodonyi, Jozsef and Fekete, S. Calculation and design of large underground openings. In *Storage in Excavated Rock Caverns, Rockstore* **77,** Magnus Bergman (ed.). Vol. 2. Oxford, England: Pergamon Press, 1978, pp. 295–300.
207. Chicago Urban Transportation District. Underground construction problems, techniques & solutions. In *Proceedings of a Seminar* sponsored by the Urban Mass Transportation Administration, U.S. Department of Transportation, October 20–22, 1975, Chicago, Ill. Springfield, Va.: National Technical Information Service.
208. Darvas, Robert M. Structural problems in earth covered buildings. In *Earth Covered Buildings: Technical Notes,* Moreland, Higgs and Shih (eds.), pp. 70–73.
209. *Earth-Sheltered Construction:* A Special Issue of *Concr. Constr.* **25**/9. (September 1980).
210. Getzler, Z., Gellert, Menachem, and Eitan, Ruben. Analysis of arching pressures in ideal elastic soil. *J. Soil Mech. Found. Div. Am. Soc. Civ. Eng.* **96**/SM4: 1357–1372 (1970).
211. ——— and Lupu, L. Experimental study of buckling of buried domes. *J. Soil Mech. Found. Div. Am. Soc. Civ. Eng.* **95**/SM2: 605–624 (1969).
212. Gill, H. L. Model study of arching above buried structures. *Proc. Am. Soc. Civ. Eng.* **95**/SM4: 1120–1122 (1969).
213. ———. Static Loading of Small Buried Arches (R278, 20-11-062). Port Hueneme, Ca: US Naval Civil Engineering Laboratory, January 1965.
214. Giramonti, A. J. Preliminary Feasibility Evaluation of Compressed Air Storage Power Systems. United Technologies Research Center Report R76-952161-5, Supported by

National Science Foundation Grant AER74-00242, Technical Oversight by Energy Research & Development Administration 1976.

215. Lenzini, Peter A. Ground stabilization: Review of grouting and freezing techniques for underground openings. *Underground Space* **1**/3: 227–240 (1977).
216. Shih, Jason C. Literature review on the structural optimization for the possible application of earth-covered buildings. In *Earth Covered Buildings: Technical Notes,* Moreland, Higgs and Shih (eds.), pp. 115–122.
217. Sterling, Ray. Structural systems for earth sheltered housing. In *Earth Covered Buildings: Technical Notes,* Moreland, Higgs and Shih (eds.), pp. 60–69. Also in *Underground Space* **3**/2: 75–81 (1978).
218. Vadnais, Kathleen. Double envelope concept applied to earth-sheltering. *Earth Shelter Digest,* No. 13, January/February 1981.

Roofs:

219. Bargabus, Dave and Laubach, D. Covered vs. conventional roofs: What's best? *Earth Shelter Digest and Energy Report,* No. 10: 26–28 (July/Aug. 1980).
220. Behr, Richard; Kiesling, Ernst W.; and Boubel, Gary. Thin shell roof systems and construction techniques for earth sheltered housing. In *Proceedings: Earth Sheltered Building Design Innovations,* Lester Boyer (ed.). Stillwater, Okla.: Oklahoma State University Press, April 1980, pp. vii–17—vii–27.
221. Gordon, Ann. Rooftop plantings for earth covered buildings in temperate climates. In *Proceedings Earth Sheltered Building Design Innovations,* Lester Boyer (ed.). Stillwater, Okla.: Oklahoma State University, April 1980, pp. 111.11–111.16
222. Roy, Robert L. Earth roof thickness: The tradeoffs. *Earth Shelter Digest,* No. 13, January/February, 1981.

Drainage, Waterproofing and Insulation:

223. Bligh, Thomas P.; Shipp, Paul; and Meixel, George. Where to insulate earth protected buildings and existing basements. In *Earth Covered Buildings: Technical Notes,* Moreland, Higgs and Shih (eds.), pp. 251–272.
224. Edvardsen, Knut I. *New Method of Drainage of Basement Walls.* Ottawa: National Research Council of Canada, 1972.
225. An experimental earth-insulated house is being monitored in rural South Carolina. *Underground Space* **3**/3: 148–149 (1978).
226. Foute, Steven J. and Cargo, Douglas B. Earth covered housing: Hydrologic and pollution considerations. In *Earth Covered Buildings and Settlements,* Frank Moreland (ed.), pp. 108–124.
227. Gill, G. W. Waterproofing buildings below grade. *Civ. Eng.* **29**/1: 3–5 (1959).
228. Gratwick, Reginald T. *Dampness in Buildings.* London: Lockwood Press, 1966.
229. Hagerman, Tor H. Groundwater problems in underground construction. In *Underground Utilization,* Truman Stauffer, Sr. (ed.), Vol. 3, pp. 451–452. Also in *Large Permanent Underground Openings,* Tor Brekke and Finn Jorstad (eds.). Oslo, Norway: Scandinavian University Books, 1970, pp. 319–321.
230. Klepper, M. R. Underground space: An unappraised natural resource: Some geologic and hydrologic considerations. In *Underground Utilization,* Truman Stauffer, Sr. (ed.), Vol. 1, pp. 31–33.
231. Lane, Charles A. Waterproofing earth-sheltered houses. *Fine Home-building* No. 2: 35–37 (April/May 1981).
232. McGroarty, Bryan. Waterproofing: Design to work. *Earth Shelter Digest and Energy Report,* No. 12: 23–26 (November/December 1980).
233. Moyers, John C. *Value of Thermal Insulation in Residential Construction.* Oak Ridge, Tenn.: Oak Ridge National Laboratory, 1971.
234. Randall, Frank A. Jr. Construction recommendations for moisture control. *Concrete Construction* **25**/9: 675 (September 1980).
235. Remson, I., Hornberger, G. M., and Molz, F. J. *Numerical Methods in Subsurface Hydrology.* New York: John Wiley & Sons, Inc., 1971.

236. Robinsky, E. I. and Bespflug, K. E. Design of insulated foundations (Paper 1000–9). *J. Soil Mech. Found. Div. Am. Soc. Civ. Eng., Proc. Am. Soc. Civ. Eng.* **79**/Sm9: 649–667 (1973).

Energy and Solar:

237. Barnes, Paul R. and Shapira, Hanna. Passive solar heating and natural cooling of an earth-integrated design. In *Proceedings: Earth Sheltered Building Design Innovations,* Lester L. Boyer (ed.). Stillwater, Okla.: Oklahoma State University Press, 1980, pp. vii-29–vii-36.
238. Bedell, Berkley (Congressman). Earth sheltered homes help to achieve energy independence. *Earth Shelter Digest and Energy Report,* No. 9. 48–49 (May/June 1980).
239. Bennett, David and Bligh, Thomas P. The energy factor—A dimension of design. *Underground Space* **1**/4: 325–332 (1977).
240. Bligh, Thomas. A comparison of energy consumption in earth covered vs. non-earth covered buildings. In *Alternatives in Energy Conservation,* Frank Moreland (ed.), pp. 85–105.
241. ———. Energy conservation by building underground. *Underground Space* **1**/1: 19–33 (1976).
242. ——— and Hamburger, Richard. Conservation of energy by use of underground space. In *Legal, Economic, and Energy Considerations in the Use of Underground Space.* Washington, D.C.: National Academy of Sciences, 1974, pp. 103–119.
243. Coffee and Crier, Architects. An underground solar heated residence (plan). In *Alternatives in Energy Conservation,* Frank Moreland (ed.), p. 149.
244. Cook, Jeff. Passive design for desert homes. *Solar Age* **6**/2 (1981).
245. Cunningham, Kim. Underground architecture: An alternate approach to architecture and energy conservation. Stillwater, Okla.: School of Architecture, Oklahoma State University, 1977. Master's Thesis.
246. The down-to-earth way to beat the energy problem. *Medical Economics:* 150–151 October 22, 1978.
247. Drucker, Eugene E. and Cheng, Herbert S. Y. Analog study of heating in survival shelters. In *Symposium on Survival Shelters.* New York: American Society of Heating, Refrigerating and Air-Conditioning Engineers, 1962, pp. 35–80.
248. Earth-sheltered Colorado house is 100% solar heated. *Underground Space* **4**/6: 403–404 (1980).
249. Eckert, E. R. G., Bligh, T. P., and Pfender, E. Energy exchange between earth sheltered structures and the surrounding ground. In *Earth Covered Buildings: Technical Notes,* Moreland, Higgs and Shih (eds.), pp. 226–250.
250. Emery, A. F., Heerwagen, D. R., and Kippenhan, C. J. Earth sheltered passive solar structures—Occupant comfort and energy use. In *Earth Sheltered Building Design Innovations,* Lester L. Boyer (ed.). Stillwater, Okla.: Oklahoma State University, 1980, pp. IV-33–IV-40.
251. The energy initiative program: An earth-covered military plant. *Underground Space* **3**/5: 269–270 (1979).
252. Energy saving ideas for heating and cooling a House. *National Geographic,* March 1976, p. 393.
253. Golany, Gideon. Free-energy cooling systems for houses in the desert. In *Innovations for Future Cities,* G. Golany (ed.). New York: Praeger Publications, 1976, pp. 246–262. Also in Proceedings of the International Conference of Energy Use Management, Rocco A. Fazzolare and Craig B. Smith (eds.). New York: Pergamon Press, 1977, Vol. 2, pp. 339–346.
254. Jackewicz, Shirley A. New cave dwellers save on energy bills, keep comfortable. *Wall Street Journal,* May 24, 1979.
255. Labs, Kenneth. Earth tempering as a passive design strategy. In *Earth Sheltered Building Design Innovations,* Lester L. Boyer (ed.). Stillwater, Okla.: Oklahoma State University, April 1980, pp. 111.3–111.10.
256. ———. Underground building climate. *Solar Age* **4**/10: 44–50 (1979).
257. Lambert, Brian. Planning for energy needs: A look at three new communities. *Underground Space* **5**/6: 362–369 (1981).

258. Schutrum, L. F. and Ozisik, N. Solar heat gains through domed skylights. *ASHRAE J.* **3**/8: 51–60 (1961).

Ventilation:

259. Allen, F. C. *Ventilating and Mixing Processes in Nonuniform Shelter Environments.* Springfield, Va.: National Technical Information Service, Contract No. DAHC20-68-C-0155, SRI Project No. MU-7327, January 1970.
260. Chester, C. V. *et al.* An earth-covered residential concept for the humid continental region. In *Earth Covered Buildings: Technical Notes,* Moreland, Higgs and Shih (eds.), pp. 171–194.
261. Ducar, G. J. and Engholm, G. Natural ventilation of underground fallout shelters. *ASHRAE Trans.* **71,** Part 1: 88–100 (1965).
262. Hori, T. *Feasibility of Low Cost Ventilation Techniques* (SRI 494251). Menlo Park, Ca.: Stanford Research Institute, December 1967.
263. Hylton, Joe. Underground-solar house wind tunnel test. In *Earth Covered Buildings: Technical Notes,* Moreland, Higgs & Shih (eds.), pp. 195–204.
264. Melaragno, Michele G. *Wind in Architectural and Environmental Design.* New York: Van Nostrand Reinhold, 1981.
265. Orlowski, Henry. Thermal chimneys and natural ventilation. In *Earth Covered Buildings: Technical Notes,* Moreland, Higgs and Shih (eds.), pp. 220–225.
266. Rice, Irvin. Heating, ventilating and air conditioning of underground installations. In *An Introduction to the Design of Underground Openings for Defense,* Le Roy Goodwin, (ed.) **46**/1: 259–304. Golden, Colo.: Colorado School of Mines, 1951.
267. Thompson, Robert. Air quality maintenance in underground buildings. *Underground Space* **1**/4: 355–364 (1977).

Windows and Light:

268. Beckett, H. E. and Godfrey, J. A. *Windows: Performance, design and Installation.* New York: Van Nostrand Reinhold, 1974.
269. Boyer, Lester L. Subterranean designs need daylighting. *Earth Shelter Digest and Energy Report,* No. 4: 32–34 (July/August 1979).
270. Burts, E. and McDonald, Eva. Opinions differ on windowless classrooms. *NEA Journal* **50:** 13–14 (October 1961).
271. Chambers, J. A. A study of attitudes and feelings toward windowless classrooms. In *Dissertation Abstracts,* Vol. 24, 1963–1964, p. 4498, Ed. D. dissertation, University of Tennessee, 1963.
272. Collins, Belinda Lowenhaupt (ed.). *Windows and people: A Literature Survey, Psychological Reaction to Environments With and Without Windows.* Gaithersburg, Md.: U.S. Department of Commerce, National Bureau of Standards, June 1975.
273. Imamoglu, V. and Markus, T. A. The effect of window size, room proportion and window position on spaciousness evaluation of rooms. In *Proceedings CIE Conference on Windows & Their Function in Architectural Design,* Istanbul, 1973.
274. Karmel, L. J. Effects of windowless classroom environment on high school students. *Perceptual and Motor Skills* **20:** 277–278 (1965).
275. Lynes, J. A. *Principles of Natural Lighting.* Essex, England: Applied Science Publishers, 1968.
276. Markus, T. A. and Gray, A. Windows in low rise, high density housing—The psychological significance of sunshine, daylight, view and visual privacy. In *Proceedings CIE Conference on Windows and Their Function in Architectural Design,* Istanbul, 1973.

Natural Hazards: Earthquake and Tornado:

277. Disasters point way to earth shelters. *Earth Shelter Digest and Energy Report* No. 8: 11–14 (March/April 1980).
278. Dowding, Charles H. Seismic stability of underground openings. In *Storage in Excavated Rock Caverns, Rockstore* **77,** Vol. 2. Oxford, England: Pergamon Press, 1978, pp. 231–238.

279. Fukushima, Kunio, Hamada, M., and Hanamura, T. Earthquake resistant design of underground tanks. In *Storage in Excavated Rock Caverns, Rockstore* **77,** Vol. 2. Oxford, England: Pergamon Press, 1978, pp. 315–320.
280. Melaragno, Michele G. *Tornado Forces and Their Effects on Buildings.* Manhattan, Kans.: Kansas State Printing Service, Kansas State University, 1968.
281. Pratt, H. R., Hustrulid, W. A., and Stephenson, D. E. *Earthquake Damage to Underground Facilities* (DP-1513). Department of Energy Report Prepared by E. I. Du Pont De Nemours & Co., 1978.
282. Tamura, C. Design of underground structures by considering ground displacement during earthquakes. *Proceedings of U.S.-Japan Seminar on Earthquake Engineering Research with Emphasis on Lifeline Systems,* Tokyo, 1976.
283. Yamahara, Hiroshi; Hisatomi, Y.; and Morie, T. A study on the earthquake safety of Rock Cavern. In *Storage in Excavated Rock Caverns, Rockstore* **77,** Vol. 2. Oxford, England: Pergamon Press, 1978, pp. 377–382.

Social and Economic:

284. Aughenbaugh, N. B. and Rockaway, John D. Go underground for low cost housing. In *Proceedings of IAHS International Symposium on Housing Problems,* Atlanta, Ga., 1976, Parviz F. Rad *et al.* (eds.), Vol. 2. Coral Gables/Miami, Fla.: International Association for Housing Science, 1976, pp. 1229–1244.
285. Davidoff, Linda. Social issues in community planning for earth covered shelter. In *Earth Covered Buildings and Settlements,* Frank Moreland (ed.), pp. 17–24.
286. Dorum, Magne. Energy economy in rock stores. A study of heat requirement in air conditioned stores, freestanding and in rock caverns. In *Storage in Excavated Rock Caverns, Rockstore* **77,** Vol. 1. Oxford, England: Pergamon Press, 1978, pp. 73–77.
287. Foster, Eugene L. Social assessment—A means of evaluating the social and economic interactions between society and underground technology. *Underground Space* **1**/1: 61–63 (1976).
288. Fox, Greg. Low cost earth-sheltered housing. *Alternative Sources of Energy,* No. 48, March/April 1981.
289. Garrison, W. L. Social, economic, and planning impacts of rapid excavation and tunneling technology. In *Underground Utilization,* Truman Stauffer, Sr. (ed.). Vol. 6, pp. 854–858.
290. Hoch, Irving. Economic trends and demand for the development of underground space. In *Legal, Economic, and Energy Considerations in the Use of Underground Space.* Washington, D.C.: National Academy of Sciences, 1974, pp. 68–86.
291. Isakson, Hans R. Institutional constraints on the marketing and financing of earth covered settlements. In *Earth Covered Buildings and Settlements,* Frank Moreland (ed.), pp. 7–11.
292. Korell, Mark L. Financing earth-sheltered housing: Issues and opportunities. *Underground Space* **3**/6: 297–301 (1979).
293. McKown, Cora and Stewart, K. Kay. Consumer attitudes concerning construction features of an earth-sheltered dwelling. *Underground Space* **4**/5: 293–295 (1980).
294. McWilliams, Donald B. and Findley, Stephen M. A life cycle cost comparison between a conventional and an earth covered home. In *Earth Covered Buildings and Settlements,* Frank Moreland (ed.), pp. 94–107.
295. Muller, C. A. and Taylor, R. A. No cause for apprehension about costs of insuring earth-sheltered homes. *Underground Space* **5**/1: 28–30 (1980).
296. Newcomb, Richard T. Dynamic analyses of demands for underground construction. In *Legal, Economic, and Energy Considerations in the Use of Underground Space.* Washington, D.C.: National Academy of Sciences, 1974, pp. 87–103.
297. Parker, Albert D. *Planning and Estimating Underground Construction.* New York: McGraw-Hill, 1970.
298. Rockaway, John D. and Aughenbaugh, N. B. Go underground for low cost housing. In *Underground Utilization,* Truman Stauffer, Sr. (ed.). Vol. 4, pp. 614–618.
299. Roy, Robert L. *Underground Houses: How to Build a Low-Cost Home.* New York: Sterling Publishing, 1979.

300. Smay, V. E. Underground living in this ecology house saves energy, cuts building costs, preserves the environment. *Popular Science:* 88–89ff. (June 1974).
301. Williams, John E. The application of life cycle cost techniques for earth covered building analysis: A preliminary evaluation. In *Underground Utilization,* Truman Stauffer, Sr. (ed.), Vol. 5, pp. 765–780.
302. ———. Comparative life cycle costs. In *Alternatives in Energy Conservation,* Frank Moreland (ed.), pp. 43–64.

Legal, Policy, Zoning and Code:

303. American Society of Civil Engineers. *The Use of Underground Space to Achieve National Goals.* New York: American Society of Civil Engineers 1973.
304. Barker, Michael B. Earth-sheltered construction: Thoughts on public policy issues. *Underground Space* **4**/5: 283–288 (1980).
305. Colvin, Brenda. *Land and Landscape: Evolution, Design and Control,* 2nd ed. London: J. Murray, 1973.
306. Duffaut, Pierre. Site reservation policies for large underground openings. *Underground Space* **3**/4: 187–193 (1979).
307. Fischer, Hans C. National & International Cooperation in Underground Construction. Stockholm: Swedish Underground Construction Mission, 1976.
308. Green, Melvyn. Building codes and underground buildings. In *Earth Covered Buildings & Settlements,* Frank Moreland (ed.), pp. 25–29.
309. Guinnee, John W. and Gunderman, William G. Terraspace: In consideration of a policy. In *Underground Utilization,* Truman Stauffer, Sr. (ed.), Vol. 6, pp. 834–836.
310. Hamburger, Richard. Public policy considerations and earth covered settlements. In *Earth Covered Buidlings & Settlements,* Frank Moreland (ed.), pp. 1–6.
311. ———. Strategies for legislative change. In *Alternatives in Energy Conservation,* Frank Moreland (ed.), pp. 243–246.
312. Harza, Richard D. National commitment to better urban life assures a rapid growth in underground construction. Washington, D.C.: paper presented at the Annual Meeting of the Geological Society of America, 1971.
313. Higgs, Forrest S. Integrating earth covered housing into existing energy efficiency codes structures. In *Earth Covered Buildings and Settlements,* Frank Moreland (ed.), pp. 44–53.
314. Labs, Kenneth. Underground development, zoning and you. *Earth Shelter Digest and Energy Report,* No. 4: 4–7 (July/August 1979).
315. LaNier, Royce and Moreland, Frank L. Earth sheltered architecture and land-use policy. *Underground Space* **1**/4: iii–iv (1977).
316. *Legal, Economic, and Energy Considerations in the Use of Underground Space.* Washington, D.C.: National Academy of Sciences, 1974.

Section III: Integration of Underground Placement within Urban Design

City Planning:

317. Birkerts, Gunnar. *Subterranean Urban Systems.* Ann Arbor, Mich.: University of Michigan, Industrial Development Division, Institute of Science and Technology, 1974.
318. Jansson, Birger. City planning and the urban underground. *Underground Space* **3**/3: 99–115 (1978).
319. House 'revealed' itself on slope (case study). In *Earth Shelter Digest and Energy Report,* No. 11: 11–14 (September/October 1980).
320. Legget, Robert F. The underground of cities. In *Underground Utilization,* Truman Stauffer, Sr. (ed.), Vol. 2, pp. 184–188.
321. Sterling, Ray et al. Site considerations for earth-sheltered structures. *Concr. Constr.* **25**/9: 653 (1980).

Arid Zone Design:

322. Barber, E. M. Adequacy of evaporative cooling and shelter environmental prediction (Report 10689). Washington, D.C.: National Bureau of Standards, February 1, 1972.
323. Behr, Richard; Kiesling, Ernst; and Boubel, Gary. Earth sheltered housing potentials for West Texas. *Earth Shelter Digest and Energy Report,* No. 5: 24–25 (September/October 1979).
324. Gelder, John. Underground desert shelter. Unpublished B. Arch. thesis, University of Adelaide, Australia, 1977.
325. Golany, Gideon S. Subterranean settlements for arid zones. In *Earth Covered Buildings and Settlements,* Frank Moreland (ed.), pp. 174–202. Also in *Housing for Arid Lands,* G. Golany (ed.). London: The Architectural Press, 1980, pp. 109–122.
326. Green, Kevin W. Passive cooling: Designing natural solutions for summer cooling loads. *Research and Design* **2**/3: 4 (Fall 1979).
327. Hummell, John D., Bearint, David E., and Eibling, James A. Survival shelter cooling: Conventional and novel systems (Paper No. 1919). *ASHRAE Trans.* **71**/Part 1: 125–133 (1965).
328. Kiesling, E. W. and Boubel, Gary A. Ashford earth-covered residence, Muleshoe, Texas. In *Earth Covered Buildings and Settlements,* Frank Moreland (ed.), pp. 273–277.
329. ———, ———, and Behr, R. Earth forms Texas house. In *Earth Shelter Digest and Energy Report,* No. 4: 8–11 (July/August 1979).
330. Knauer, Virginia and Branscomb, Lewis M. *Eleven Ways to Reduce Energy Consumption & Increase Comfort in Household Cooling* (Consumer Bulletin C13.2 EN2). Washington, D.C.: National Bureau of Standards, 1971.
331. Labs, Kenneth. Terratypes: Underground housing for arid zones. In *Housing in Arid Zones: Design and Planning,* G. Golany (ed.). Architectural Press, Ltd., London, 1980, pp. 123–140.
332. Langley, John B. Sun belt subsoil studied. *Earth Shelter Digest and Energy Report,* Number 12: 20–21 (November/December 1980).
333. ——— and Gray, James L. *Sun Belt Earth Sheltered Architecture.* Winter Park, Fl: Sun Belt Earth Sheltered Research, 1980.
334. Shih, J. C. and Kumar, S. A case for subsurface and underground housing in hot-arid lands. U.S. Army Research Office, April 1976.
335. Van Der Meer, Wybe J. The possibility of subterranean housing for arid zones. In *Housing for Arid Lands: Design and Planning,* G. Golany (ed.). London: Architectural Press, 1980, pp. 141–150.

Schools:

336. Allen, Phelan G. Fremont Elementary School air conditioning. In *Underground Utilization,* Truman Stauffer, Sr. (ed.). Vol. 4, pp. 562–563.
337. Bailey, Edwin R. Alternatives to educational space needs. In *Underground Utilization,* Truman Stauffer, Sr. (ed.), Vol. 2, pp. 180. 180–183.
338. Carter, Douglas N. Community and building official reaction to earth covered buildings: A case study Terraset Elementary School, Reston, Virginia. In *Earth Covered Buildings and Settlements,* Frank Moreland (ed.), pp. 78–81.
339. ———. Terraset Elementary School, Reston, Virginia. In *Alternatives in Energy Conservation,* Frank Moreland (ed.), pp. 135–149.
340. ———. Terraset School. *Underground Space* **1**/4: 317–323 (1977).
341. Cooper, James G. and Ivey, Carl H. *A Comparative Study of the Educational Environment and the Educational Outcomes in an Underground School, A Windowless School and Conventional Schools.* Defense Civil Preparedness Agency, Department of Defense, Washington, D.C., Contract #OE-3-99-033, August 1964.
342. ——— and ———. *Final Report of the ABO Project.* Santa Fe: New Mexico Department of Education, August 1964.
343. Demos, G. D. Controlled physical classroom environments and their effect upon elementary school children (Windowless classroom study). Riverside County, Ca.: Palm Springs School District, 1965.
344. Larson, C. Theodore. *The Effect of Windowless Classrooms on Elementary School Children.* Ann Arbor: Architectural Research Laboratory, University of Michigan, Nov. 1965.

345. Lutz, Frank W. Studies of children in an underground school. In *Alternatives in Energy Conservation,* Frank Moreland (ed.), pp. 71–77. Also in *Underground Space* **1**/2: 131–134 (1976).
346. ——— and Lutz, Susan B. *The Preliminary Report on the ABO Project.* Defense Civil Preparedness Agency, Washington, D.C., 1964.
347. ———; Lynch, Patrick D.; and Lutz, Susan B. *Abo Revisited: An Evaluation of the Abo Elementary School and Fallout Shelter* (Final Report). Washington, D.C.: Defense Civil Preparedness Agency, Contract #DAHC20-72-C-0155, June 1972.
348. Platzker, J. The first 100 windowless schools in the U.S.A. Paper presented at educational workshop, *28th Annual Convention of National Association of Architectural Metal Manufacturers,* 1966.
349. Tikkanen, K. T. Window factors in schools. In Part 2, *Overhead Natural Lighting.* Helsinki: University of Technology, February 1974.
350. Three underground schools. *Progres. Arch.* **56**/10:30 (1975).

Industry and Commercial Uses:

351. Building underground: Factories and offices in a cave. *Eng. News Rec.* **20**/166: 58–59 (1961).
352. Callahan, S. J. The development of limestone mines into underground warehouses and manufacturing plants. In *Underground Utilization,* Truman Stauffer, Sr. (ed.), Vol. 1, pp. 22–26.
353. Carlier, R. Design of windowless factories in Belgium. *Institut Technique du Bâtiment et des Travaux Publics, Annales* **22**/262: 1441–1483 (1969).
354. Chryssafopoulos, Nicholas. Employee attitudes. In *Underground Utilization,* Truman Stauffer, Sr. (ed.), Vol. 4, pp. 530–531.
355. Fairhurst, Charles. U.S. develops new, commercial uses for earth-sheltering. *Underground Space* **5**/1: 31–35 (1980).
356. Keighley, E. C. Visual requirements and reduced fenestration in office buildings—A study of window shape. *Build. Sci.* **8:** 311–320 (1973).
357. Lonsdale, Richard E. Underground space as a locational consideration in industry. *Underground Space* **1**/1: 59–60 (1976). Also in *Underground Utilization,* Truman Stauffer, Sr. (ed.), Vol. 5, pp. 687–688.

Utilities, Highways and Parking:

358. Alders, Charles. A study of earth covered buildings by an electric utility company. In *Alternatives in Energy Conservation,* Frank Moreland (ed.), pp. 107–115.
359. Berg, N. Aspects on Underground Location of Urban Facilities—Power Supply, Oil Storage, and Sewage Treatment. Washington, D.C.: Association of Engineering Geologists, Symposium on Geological and Geographical Problems of Areas of High Population Density, October 1970.
360. Branyan, S. G. Operation and maintenance of underground storage. In *Symposium on Salt.* Cleveland, Ohio: Northern Ohio Geological Society, Inc., 1963, pp. 609–615.
361. Brown, Verne. Underground location of utilities as part of underground development. In *Underground Utilization,* Truman Stauffer, Sr. (ed.), Vol. 2, pp. 254–256.
362. Hoffman, George A. Urban Underground Highways and Parking Facilities (Memorandum RM-3680-RC). RAND Corporation, 1963.
363. Homes and Narver Inc. *Engineering Study of Underground Highway & Parking Garage & Blast Shelter for Manhattan Island.* Oak Ridge, Tenn.: Oak Ridge National Laboratory, March 1966.
364. Hunt, F. R. Underground power transmission. *Underground Space* **3**/1: 19–33 (1978).
365. Meyer, Kirby. Utilities for underground structures. In *Alternatives in Energy Conservation,* Frank Moreland (ed.), pp. 165–181.
366. Mitchell, Ansel. Operation and maintenance of underground facilities. In *Underground Utilization,* Truman Stauffer, Sr. (ed.), Vol. 1, pp. 74–76.
367. Paulson, Boyd C., Jr. Research and development needs for systems and management in underground transportation construction. *Underground Space* **2**/2: 81–89 (1977).

368. Rosander, A. Underground sewage treatment plant. *Underground Space* **2**/1: 39–45 (1977).
369. Smith, Wilbur and Associates. *Transportation and Parking for Tomorrow's Cities.* New Haven, Conn.: Automobile Manufacturers Association, 1966.

Environmental Issues:

370. Carleton, Joseph G. An environmentalist's views on underground construction. In *Underground Utilization,* Truman Stauffer, Sr. (ed.). Vol. 6, pp. 894–899.
371. Dasler, A. R. and Minard, D. Environmental physiology of shelter habitation. *Transactions. Proceedings of the ASHRAE Semiannual Meeting,* Chicago, Illinois, 1965. Chicago, Illinois: American Society of Heating, Refrigerating & Air-Conditioning Engineers, Inc. **71,** Part I: 115–124 (1965).
372. LaNier, Royce. Assessing environmental impact of earth covered buildings. *Underground Space* **1**/4: 309–315 (1977).
373. ———. Earth covered buildings and environmental impact. In *Alternatives in Energy Conservation: The Use of Earth Covered Buildings,* Frank Moreland (ed.), pp. 269–278.
374. Lasitter, H. A. *Acoustics Noise Reduction in Shielded Enclosures* (R549). Port Hueneme, Ca.: Naval Civil Engineering Laboratory, November 1967.
375. Wells, Malcolm. To build without destroying the earth. In *Alternatives in Energy Conservation,* Frank Moreland (ed.), pp. 211–232. Also in *Underground Utilization,* Truman Stauffer, Sr. (ed.), Vol. 4, pp. 546–552.

Index